T0318562

VARIABLE GENERATION, FLEXIBLE DEMAND

VARIABLE GENERATION, FLEXIBLE DEMAND

Edited by

FEREIDOON SIOSHANSI

ACADEMIC PRESS

An imprint of Elsevier

Academic Press is an imprint of Elsevier
125 London Wall, London EC2Y 5AS, United Kingdom
525 B Street, Suite 1650, San Diego, CA 92101, United States
50 Hampshire Street, 5th Floor, Cambridge, MA 02139, United States
The Boulevard, Langford Lane, Kidlington, Oxford OX5 1GB, United Kingdom

Notices
Knowledge and best practice in this field are constantly changing. As new research
and experience broaden our understanding, changes in research methods, professional
practices, or medical treatment may become necessary.

Practitioners and researchers must always rely on their own experience and knowledge
in evaluating and using any information, methods, compounds, or experiments
described herein. In using such information or methods they should be mindful of
their own safety and the safety of others, including parties for whom they have a
professional responsibility.

To the fullest extent of the law, neither the Publisher nor the authors, contributors, or
editors, assume any liability for any injury and/or damage to persons or property as a
matter of products liability, negligence or otherwise, or from any use or operation of
any methods, products, instructions, or ideas contained in the material herein.

Library of Congress Cataloging-in-Publication Data
A catalog record for this book is available from the Library of Congress

British Library Cataloguing-in-Publication Data
A catalogue record for this book is available from the British Library

ISBN: 978-0-12-823810-3

For information on all Academic Press publications visit our website at
https://www.elsevier.com/books-and-journals

Publisher: Joseph P. Hayton
Acquisitions Editor: Graham Nisbet
Editorial Project Manager: Hannah Makonnen
Production Project Manager: Kiruthika Govindaraju
Cover Designer: Victoria Pearson

Typeset by TNQ Technologies

Working together
to grow libraries in
developing countries

www.elsevier.com • www.bookaid.org

Contents

Part 1 Variable renewable generation

Part 2 Flexible demand

Part 3 Coupling flexible demand to variable generation

15. Demand flexibility and what it can contribute in Germany 355
Dierk Bauknecht, Christoph Heinemann, Matthias Koch and Moritz Vogel, Oeko-Institut, Freiburg, Germany

16. Industrial demand flexibility: A German case study 371
Sabine Löbbe, and André Hackbarth, Reutlingen University, Heinz Hagenlocher and Uwe Ziegler, AVAT, Tübingen, Germany

Part 4 Implementation, business models, enabling technologies, policies, regulation

17. Market design and regulation to encourage demand aggregation and participation in European energy markets 393
Juan José Alba, Carolina Vereda, Julián Barquín and Eduardo Moreda, Endesa, Spain

Author biographies

Juan Jose Alba is in charge of regulatory affairs at Endesa, the Spanish utility. He has been chairman of the Markets and Investments Committee of Eurelectric, the association of European utilities, and is a member of the Management Committee of Aelec, the Spanish electricity industry association.

He has been member of the board of directors of the European Federation of Energy Traders and co-chairman of its Working Group on financial regulation as well as of the Supervisory Boards of Powernext, the French Power Exchange, and Gielda Energii S.A., the Polish Power Exchange.

He holds a PhD in electrical engineering from Comillas Pontifical University.

Ross Baldick is Professor Emeritus in the Department of Electrical and Computer Engineering at The University of Texas at Austin. His current research includes optimization, economic theory, and statistical analysis applied to electric power systems, particularly in the context of increased renewables and transmission.

Dr. Baldick has published over 100 refereed journal articles and has research interests in several areas in electric power and is a Fellow of the IEEE and the recipient of the 2015 IEEE PES Outstanding Power Engineering Educator Award.

He received his BSc and BE degrees from the University of Sydney, Australia, and his MS and PhD from the University of California, Berkeley.

Ken Baldwin is the inaugural Director of the Australian National University's (ANU) Energy Change Institute, an inaugural ANU Public Policy Fellow, and Director of the ANU Grand Challenge: *Zero-Carbon Energy for the Asia-Pacific*. The main focus of his work is to help drive the energy transition, particularly for Australia's future export industries based on renewable energy.

He has had a number of prominent appointments including Steering Committee of the Australian Energy Technology Assessment, the CSIRO Hydrogen RD&D Committee and currently serves on the Australian Energy Transition Research Plan. Professor Baldwin is a Fellow of the

American Physical Society, the Institute of Physics (UK), the Optical Society of America, and the Australian Institute of Physics.

Prof. Baldwin holds a BSc and MSc in Physics from ANU, PhD from Imperial College, University of London, and DIC (Diploma of Imperial College).

Julian Barquin works at the Regulatory Affairs Department of Endesa, the Spanish utility. He specializes in wholesale market regulation, network and operation codes, as well as techno-economical analysis and simulation.

Previously he was a Professor at Comillas Pontifical University in Madrid, Spain. He has been Visiting Scientist at MIT and Visiting Scholar at the University of Cambridge. He is author or coauthor of several books and more than 100 papers in peer-reviewed journals and a frequent speaker at conferences.

He holds a PhD and Power Engineering degrees from Comillas Pontifical University and a Physics degree from UNED.

Dierk Bauknecht is a senior researcher with Öko-Institut's Energy and Climate Division and professor for sustainability and transformation research at the University of Freiburg. Dierk has led a broad range of national and European projects on the integration of renewables into the power system; the governance of electricity system transformation; infrastructure regulation and smart grids; flexibility options and market design; power system modeling; European renewables policy.

Before joining the Öko-Institut in 2001, he was modeling manager with a UK-based power market consultancy.

Dierk graduated in political science at the Freie Universität Berlin and holds an MSc in Science and Technology Policy and a doctorate from the University of Sussex, UK.

Simona Benedettini is Senior Manager at PwC Italy. Before joining PwC, Simona worked as energy economist for leading industry associations, consultancies, and research institutions.

Simona has published several articles on regulation and competition of the energy sector in academic journals and leading Italian newspapers.

She earned a PhD in Law and Economics from the University of Siena and has been visiting scholar at the University of Illinois.

Joseph Bowring is the President of Monitoring Analytics and, as the Independent Market Monitor for PJM, is responsible for all market monitoring activities of PJM Interconnection. He has extensive experience in applied energy and regulatory economics.

He has a PhD in Economics from the University of Massachusetts.

Kelly Burns is a Senior Research Fellow at the Victoria Energy Policy Centre, Victoria University. She has an extensive background in economics, law, and finance across the public and private sectors including Australian Bureau of Statistics, Colonial First State, Department of Premier and Cabinet (Vic), and Department of Justice (Vic).

Kelly holds a PhD in Economics from RMIT, L.L.B (hons) and B.Eco (hons) from LaTrobe University, and Graduate Certificate in Financial Planning from FINSIA. Her primary specialism is in the application of econometric modeling and forecasting in the field of energy economics.

Prabpreet Calais is a Senior Adviser in the Australian Energy Market Commission's Markets team. Mr. Calais has led and worked on several electricity market reforms and reviews including five-minute settlement, global settlements, the operation of the retail energy market, and gas and electricity financial contract markets.

Prab has previously worked at the Climate Change Authority, Commonwealth Department of the Environment, and Commonwealth Department of Climate Change and Energy Efficiency. Previous work includes an extensive assessment of several climate change policies for the Australian electricity sector and the development of carbon credit methodologies.

Prab holds a Masters of Integrated Resource Management (with Distinction) from the University of Edinburgh, and a Bachelors of Agricultural Economics (Honors) from the University of Sydney.

Laurens de Vries is an Associate Professor at the Faculty of Technology, Policy and Management of Delft University of Technology where he teaches and performs research in energy market design and regulation in relationship to the physical infrastructure. He is involved in research projects such as market design for a 100% renewable power system, the design of European balancing markets, direct-current microgrids, congestion in distribution networks, and urban energy system integration.

Laurens is the Coordinator of the European Energy Research Alliance Joint Program on Energy Systems Integration. He has an MSc in Mechanical Engineering from Delft University of Technology, an MA in Environmental Studies from The Evergreen State College, and PhD from Delft University of Technology.

Gerard Doorman is presently market design specialist at Statnett SF, the Norwegian system operator where he is mainly focused on the integration of European power markets. Previously he was a professor in electric power systems at the Norwegian University of Science and Technology (NTNU) in Trondheim working on optimization of power systems including those with large-scale hydro reservoirs, market design, and balancing.

He is active with CIGRE on market design for short-term flexibility and power system operation and previously worked at SINTEF Energy Research on optimization and simulation of power systems.

Gerard has an MSC and a PhD in electric power engineering from NTNU.

Ahmad Faruqui is a principal with The Brattle Group. He works on modernizing tariff design, enhancing energy efficiency, improving the ability to forecast demand, carrying out experiments with innovative prices, and cost—benefit analysis of new technologies such as distributed energy resources, electric vehicles, energy storage, combined heat and power, and microgrids.

He is the author, coauthor or editor of four books and more than 150 articles, papers, and reports on energy economics.

He holds BA and MA degrees from the University of Karachi and an MA in agricultural economics and a PhD in economics from The University of California at Davis.

Lynne Gallagher is the Chief Executive Officer of Energy Consumers Australia (ECA), which is a national voice for residential and small business consumers. ECA's priorities include enabling increased consumer participation in energy markets, improved affordability, individualized energy services, and optimization of the energy system.

She is a member of the Monash Energy Institute's Advisory Council and Chair of the Advisory Council for the *Digital Energy Futures* at the Emerging Technologies Research Lab at Monash University. She has over 25 years experience in the energy sector including social research and demand forecasting.

Lynne has a Bachelor in Economics degree from the Australian National University, majoring in econometrics and pure mathematics.

André Hackbarth is a researcher and lecturer at Reutlingen University, School of Engineering/Distributed Energy Systems and Energy Efficiency, Germany. His research interests include business models in the energy sector and consumer attitudes, preferences, and decision-making concerning energy-related behaviors and products.

Prior to his current position, he was research associate at the Institute for Future Energy Consumer Needs and Behavior (FCN) at RWTH Aachen University, Germany.

He is PhD candidate at RWTH Aachen University and studied Economics at Heidelberg University.

Heinz Hagenlocher is Director of Business Unit, Energy Automation Solutions at AVAT Automation GmbH, an Energy Engineering company in Tübingen, Germany. Since 2007 Heinz and his team provide comprehensive services to vertically integrated, decentralized energy systems. Typical range of services includes analysis, planning, development, implementation, and operational support for fully digitalized solutions for utilities, buildings, and smart grid/microgrid infrastructures such as energy communities. Using artificial intelligence, AVAT enables integration of renewable energies while optimizing performance.

Heinz' prior experience includes management positions in the electric power sector and in automation business.

Heinz graduated in Electronics and Information Technology at the University of Stuttgart.

Christoph Heinemann is a Senior Researcher at the Freiburg branch of Öko-Institut's Energy and Climate Division. His key activities include the topics integration of renewable energy, flexibility options, regulation and modeling of the future power system.

Before he joined the Institute, he worked in the field of smart grids and innovative power products for 2 years.

He studied geography, political economy, and business studies at the University of Freiburg.

Udi Helman is an independent consultant advising a range of clients including commercial, research, governmental organizations on emerging technologies and wholesale markets.

Before becoming a consultant, he worked for over 15 years at the Federal Energy Regulatory Commission (FERC), the California ISO, and in the utility-scale solar industry. He has advised, and created optimization models for, demand response companies. His publications include many in peer-reviewed journals as well as research studies with different organizations.

Udi has a PhD in energy economics from The Johns Hopkins University.

Ryan Hledik is a Principal in The Brattle Group's San Francisco office. His consulting practice is focused on regulatory and planning matters related to emerging energy technologies and policies. His research on the "grid edge" has been cited in federal and state regulatory decisions, as well as by *Forbes*, *The New York Times*, *Utility Dive*, *Vox*, and *The Washington Post*. He has supported clients across 35 states and 9 countries in matters related to energy storage, load flexibility, distributed generation, electrification, retail tariff design, energy efficiency, and grid modernization.

Ryan received his MS in Management Science and Engineering from Stanford University, where he concentrated in Energy Economics and Policy. He received his BS in Applied Science from the University of Pennsylvania, with minors in Economics and Mathematics.

Richard Lee Hochstetler is Director of Economic and Regulatory Issues at Instituto Acende Brasil, a think tank focused on electric power policy and regulation in Brazil and promoter of the biennial *Brazil Energy Frontiers* conference. He recently published a book on market design for the Brazilian electric power industry.

Prior to joining Acende, he worked as a Public Utility Specialist at the Federal Energy Regulatory Commission, and as Project Leader and Partner at Tendências Consultoria Integrada, a consultancy in Brazil.

Richard obtained his BA in Economics from Goshen College, and a Doctorate in Economics from the University of São Paulo, Brazil.

Matthias Koch is a senior researcher within Öko-Institut's Energy and Climate Division, which he joined in 2009. He contributes to a broad range of European and national projects, especially in energy system modelling with the energy market model PowerFlex-Grid-EU, including flexibility options like demand-side management or storage.

Before joining Öko-Institut, he worked as a research fellow and PhD student at the French-German Institute for Environmental Research at the University of Karlsruhe.

His doctoral thesis at the faculty of Economics and Business Engineering, University of Karlsruhe, focused on a warehouse location problem for biogas plants under economic and ecological constraints.

Tony Lee was a Consultant in The Brattle Group's San Francisco office until recently. Currently, he is pursuing a Masters degree in Technology and Policy at the Massachusetts Institute of Technology (MIT). He specializes in wholesale market design, environmental policy analysis, resource planning, and economic analysis of generation, transmission, and demand-side resources. He has extensive experience developing and operating economic models of electricity systems.

Mr. Lee holds a BA in Economics and a BS in Engineering, both from Swarthmore College.

Sabine Löbbe is a professor at Reutlingen University, School of Engineering/Distributed Energy Systems and Energy Efficiency, Germany, and lectures in masters programs at the universities of Chur and St. Gallen, Switzerland. Her consulting company advises utilities in strategy and business development and in organizational issues.

Prior to her current position, she was Director for Strategy and Business Development at swb AG Bremen, project manager at Arthur D. Little Inc., and at VSE AG, Saarbrücken.

She holds a doctorate in business administration from the University Saarbrücken and studied business administration in Trier, Saarbrücken, and EM Lyon/France.

Luca Lo Schiavo is the Deputy Director of the Infrastructure Regulation Department at the Italian Regulatory Authority for Energy (ARERA), where he has worked since 1997.

He is an expert of quality of service and innovation in the power system, a member of the ACER Infrastructures Task Force, of CEER Distribution

Systems Working Group, and coauthor of a book on service quality regulation (2007) and numerous articles on smart metering regulatory policies with focus on quality of service and innovative incentives.

Luca earned an MSc in Industrial Engineering from the Technical University of Milano and a specialization qualification in Public Policy Analysis at IDHEAP in Lausanne.

Reinhard Madlener is full professor of energy economics and management at the School of Business and Economics, RWTH Aachen University, Germany, and adjunct professor at the Department of Industrial Economics and Technology Management, Norwegian University of Science and Technology (NTNU).

His main research interests are in energy economics and policy, sustainable energy transition, technological diffusion, and investment under uncertainty. He serves as Senior Editor of *Energy Policy*, is on editorial boards of scientific journals, and is the director of the Institute for Future Energy Consumer Needs and Behavior at RWTH's E.ON Energy Research Center.

He holds two masters degrees and a PhD in the social sciences and economics from the Vienna University of Economics and Business, and a postdoctoral diploma in economics from the Institute for Advanced Studies and Scientific Research, Vienna, Austria.

Eduardo Moreda works in Endesa's Regulatory Affairs department as the Subdirector in charge of Generation, Natural Gas, and Wholesale Markets. His responsibilities include regulation of generation and gas wholesale markets, system operation procedures, and regulation for the energy transition toward a fully decarbonized system.

Formerly, he was at Compañía Sevillana de Electricidad working on transmission and generation planning, distribution operations and planning, and R&D generation and transmission projects.

Eduardo holds an Engineering degree from the University of Seville.

Bruce Mountain is the Director of the Victoria Energy Policy Centre at Victoria University.

He has many years of experience as an advisor and researcher on a wide range of issues in the economics of energy and regulation in Australia, Britain, South Africa, and other countries.

He has a PhD in Economics from Victoria University, a Bachelor's and a Master's degree in Electrical Engineering from the University of Cape Town, and qualified as a Chartered Management Accountant in England.

Tim Nelson is Executive General Manager, Energy Markets at Infigen Energy and an Associate Professor at Griffith University.

Tim joined Infigen Energy from his previous position as Executive General Manager, Strategy and Economic Analysis at the Australian Energy Market Commission. Prior to that, he was Chief Economist at AGL, where he managed the company's public policy advocacy as well as sustainability and strategy. He is a member of Westpac's Stakeholder Advisory Council and the Centre for Policy Development's Research Committee. He previously held roles with NSW Government and the Reserve Bank of Australia.

Tim holds a PhD in Economics from the University of New England for which he earned a Chancellor's doctoral research medal.

Alan Rai is a Director at Baringa Partners, where he leads their investment advisory, regulatory, and policy practices. Previously, he was a Senior Economist at the Australian Energy Market Commission, where he led projects related to grid integration of variable renewables, emissions reduction policies, and reforms to retail and wholesale markets.

Alan is also a Senior Fellow at Macquarie University, an Industry Fellow at the University of Technology Sydney, and a Fellow at the Australian Institute of Energy. Prior roles included work at CSIRO, at Macquarie University as Assistant Professor, and at the Reserve Bank of Australia.

Alan holds a PhD in Economics from the University of New South Wales and a Bachelor of Commerce and Economics (first Class Honors).

David Robinson is a Senior Research Fellow at the Oxford Institute for Energy Studies (OIES). He works primarily on decarbonization with a focus on public policy and corporate strategy in the energy sector.

Before joining the OIES, he led the European operations of NERA and The Brattle Group. He is still an academic advisor to the Brattle Group.

David has a DPhil in economics from the University of Oxford, a Master's Degree in Economic Policy and Planning from the Institute of Social Studies, Erasmus University, and a Joint Honors BA in Economics and Political Science from McGill University.

Elisabeth Ross is an independent consulting economist and energy policy expert. She works with businesses, governments, regulators, and advisory groups on regulatory and policy matters affecting infrastructure utilities focusing on energy markets.

Elisabeth was previously a Director at the Australian Energy Market Commission, with responsibility for a number of strategic policy reforms. Elisabeth has also previously held consulting roles with KPMG and NERA Economic Consulting.

Elisabeth holds a Masters of Economics (with honors) from the University of Sydney and a Bachelor of Commerce in Economics from the University of Canterbury.

Oliver Ruhnau is a PhD Candidate and Research Associate at the Hertie School, Berlin. His research focuses on the economics of flexible electricity demand in the context of variable renewable energy sources.

Previously, he was Data Scientist at the energy service company Digital Energy Solutions.

Oliver studied Engineering and Economics at the RWTH Aachen University and the KTH Royal Institute of Technology, Stockholm.

Fereidoon Sioshansi is the founder and president of Menlo Energy Economics, a consulting firm advising clients on strategic issues in the electricity sector. He is the editor and publisher of *EEnergy Informer*, a monthly newsletter with international circulation.

His prior work experience includes working at So. Calif. Edison Co., EPRI, NERA, and Global Energy Decisions, acquired by ABB. This is his 13th edited volume since 2006 on different subjects including Innovation and disruption at the grid's edge, Consumer, prosumer, prosumagers and, most recently, Behind and beyond the meter.

He has a BS and MS in civil and structural engineering, an MS and PhD in economics from Purdue University.

Carlo Stagnaro is Director of the Observatory on the Digital Economy at Istituto Bruno Leoni, a Milan-based think tank, where he manages the Index of Liberalization, a benchmarking report on market openness in the European Union. Prior to that, he was chief of the Minister's technical staff at the Italian Ministry of Economic Development.

He is a member of the editorial board of the journal Energia and a regular columnist to Il Foglio. He is also a member of the Institute of

Economic Affairs' Academic Advisory Council and a fellow of the Italian Observatory on Energy Poverty at the Padua University's Levi-Cases Center. His publications include *Power Cut? How the EU Is Pulling the Plug on Electricity Markets*, 2015. With Alberto Saravalle he wrote *Contro il sovranismo economico*, 2020.

Carlo earned an MSc in Environmental Engineering from the University of Genoa and a PhD in Economics from IMT Alti Studi Lucca. He was awarded the Innovation Award at Eurelectric PowerSummit 2019.

Mike Swanston heads The Customer Advocate, a consultancy advising regulators and governments on the imperative of a customer focus in meeting the needs of change in the energy industry. A strong supporter of the remarkable shift to customer-owned renewable energy resources, Mike's focus is that customers remain a top priority as networks and markets make the transition to our energy future.

Before forming The Customer Advocate, Mike held executive positions in electricity distribution authorities, including accountability for customer and stakeholder engagement.

Mike holds an honors degree in Electrical Engineering, postgraduate qualifications in business management, and is a fellow of the Institution of Engineers Australia.

Giacomo Terenzi is a Grid and Market Analyst working in the System Strategies Division of Terna Spa, the Italian transmission system operator (TSO).

He currently works on future scenarios at national and international level focusing on the gradual electrification of the energy system and the evolution of generation mix. He is also engaged in network studies for the long-term developments in Mediterranean TSO framework. Before joining Terna, he worked as a market analyst at Gestore dei Mercati Energatici SpA, the Italian Power Exchange, in Day Ahead and Intraday market design and operations.

Giacomo graduated in Economics at Sapienza University of Rome and holds an MSc in Finance from the Strathclyde Business School, Glasgow, UK.

Carolina Vereda works in Endesa's regulatory affairs department supporting interface with the DSOs and retail market businesses as well as topics such as flexibility markets, aggregation models, DSO retribution, network development plans, and customers.

In 2018, Carolina was elected Chair of the WG Business Models and Networks Customers at Eurelectric. Before that and for 3 years, she worked for Eurelectric as Policy Adviser. Her main activities were to build relations with stakeholders and implementing Eurelectric's advocacy before the European institutions.

She holds an industrial engineering degree from Universidad Alfonso X el Sabio, Madrid.

Moritz Vogel is a research associate at the Freiburg branch of Öko-Institut's Energy and Climate Division. His key activities are located in the areas electricity market design, decentralized energy systems, integration of renewable energies, as well as flexibility mechanisms.

He studied environmental sciences at the University of Lüneburg and finished his studies with a Bachelor of Science degree. Afterwards Moritz Vogel began his Master studies "Sustainability Economics and Management" at the University of Oldenburg, which he finished with a master's thesis at the German Institute for Economic Research in Berlin.

Kate Wild is an Acting Director in the Australian Energy Market Commission's Markets team. Ms. Wild has led the AEMC's annual review of retail energy competition, developed rules to strengthen protections for customers in hardship, and led the AEMC's recent work on digitalization and the potential to move to a two-sided national electricity market.

Prior to the AEMC, Kate spent several years at Sydney Water providing advice on regulatory matters, including competition and access arrangements, recycled water policy, and funding frameworks for urban development.

Kate holds a Masters of Environmental Planning and a Bachelor of Arts/ Bachelor of Laws from Macquarie University.

Greg Williams is a Senior Economist at the Australian Energy Market Commission who has over 20 years' experience working on electricity market design in both the Australian NEM and New Zealand electricity markets.

Greg's experience in market design is both deep and broad, covering network cost allocation and rate design, wholesale market design, ancillary service markets, and resource adequacy.

Greg holds a Bachelor of Commerce and Administration, majoring in Economics, from Victoria University in Wellington, New Zealand.

Uwe Ziegler is team leader of the Energy Network team at AVAT Automation GmbH in Tübingen, Germany, responsible for the development of optimization methods for utilizing flexible demand in the electricity market. He has substantial experiences as a project leader of smart energy solutions, virtual power plants, and power management systems. He is also lecturer in the master degree program on Distributed Energy Systems and Energy Efficiency at Reutlingen University.

His prior experience includes design of SCADA systems and PLC devices for automation solutions.

Uwe studied Electrical engineering (Dipl.Ing degree) at the University of Stuttgart.

Foreword

The world is undergoing an unprecedented transformation of its economic and social systems as we shift rapidly toward a decarbonized world over the next few decades. Leading this shift is the increasing adoption of zero-carbon electricity sources, principally solar and wind, to replace fossil fuel—based generation, within an economic and infrastructure setting that is geared to traditional paradigms of centralized, large-scale generation to satisfy unfettered demand.

Given the intermittent nature of disseminated renewable generation, a paradigm shift is needed to create a smooth, efficient, and low-cost transition to this new world order. This can be achieved either by adjusting demand to match the supply or by storing the renewable energy to meet demand. A flexible and adaptive system will enable the best outcome through a combination of both, which requires the creation of options on both sides to optimize the result.

This book examines the range of options available to provide flexible demand to match variable supply—and to couple the two via the most appropriate policy and market mechanisms. What is appropriate depends on the economic, physical, and jurisdictional context, which is why the book surveys the prospects in a range of countries which have different requirements for their future electricity systems.

Nowhere is this choice more stark than in Australia—perhaps a limiting case in the world context. An island of continental scale, Australia cannot depend upon neighbors to meet shortfalls in demand within its domestic economy. It also has the longest and thinnest electricity supply network in the world, stretching over 5000 kms from end to end. Nevertheless, the tyranny of distance can be turned to geographic advantage by expanding the availability of renewable generators in regions experiencing partially correlated energy resources, potentially enabling reduction in variability.

Of even more significance is that Australia is transitioning from one of the highest levels of fossil generation in the world (84% in 2017[1]) at the fastest rate of any country. Through a combination of utility-scale wind, plus utility-scale and domestic rooftop solar, Australia leads the world as the

---

[1] Australian Energy Update 2018, Department of the Environment and Energy, https://www.energy.gov.au/publications/australian-energy-update-2018.

fastest per capita installer of renewables at over 200W per person per year. This is four to five times faster than the United States, the European Union, Japan, or China, and around 10 times faster than the world average.[2,3]

In addition, Australia has the highest level of domestic rooftop solar penetration in the world with a quarter of dwellings possessing solar installations.[4] In states like South Australia, renewable generation has reached half the electricity supply, and in the Australian Capital Territory surrounding the Australian capital of Canberra, the local jurisdiction has purchased all of its electricity needs from renewable generators using a pioneering reverse auction system coupled with contracts-for-difference. At this rate, the nation has the potential to reach an electricity sector dominated by renewables in the second half of the decade.

To address the issues created by high levels of renewable penetration being achieved at the world's fastest renewable intensity rate, a Future Electricity Market Summit was held in Sydney in November 2019, cohosted by the Australian Energy Security Board, the Australian National University Energy Change Institute, in collaboration with the Energy Research Institutes Council for Australia—ERICA, and the International Energy Agency.

The Summit canvassed a number of issues covered by this volume, and to which the book's editor and a number of contributors were key participants. In subsequent developments, the Australian Energy Security Board has examined options for two-sided energy markets[5] and ahead markets,[6] along with trialing a range of other policy options such as virtual power plants to enable scheduling for demand-side participants to enter the market.

The empowerment of the consumer, combined with their new roles as producers and storers of energy, will require these and many other novel

[2] *Powering ahead: Australia leading the world in renewable energy build rates*, Matthew Stocks, Ken Baldwin and Andrew Blakers, https://energy.anu.edu.au/files/Renewable%20energy%20target%20report%20September%202019_1_0.pdf.

[3] *Australia is the runaway global leader in building new renewable energy*, Matthew Stocks, Andrew Blakers and Ken Baldwin, https://theconversation.com/australia-is-the-runaway-global-leader-in-building-new-renewable-energy-123694.

[4] Australian Photovoltaic Institute (2019). *Mapping Australian Photovoltaic installations*, https://pv-map.apvi.org.au/historical#4/-26.67/134.12.

[5] Energy Security Board. *Moving to a Two-sided Market*, http://www.coagenergycouncil.gov.au/publications/two-sided-markets.

[6] Energy Security Board. *System Services and Ahead Markets*, http://www.coagenergycouncil.gov.au/post-2025/system-service-and-ahead-markets.

frameworks to make the transition to the carbon-free energy systems of the future.

This book, with its focus on demand flexibility, presents a key part of that journey.

Professor Ken Baldwin
Director, The Energy Change Institute
Australian National University

Preface[1]

As suggested by the title, this book is focused on the topic of demand flexibility in the current framework of the power systems transitioning toward decarbonization and decentralization. Regulation is an extremely relevant part of this transition, although most of the drivers of this transition—for instance, RES penetration targets and incentives—are not set by independent regulators but at the national or higher levels. This is certainly the case for the European Union (EU) and its Member States where challenging targets are set at the EU level in Brussels, and each member state must find the most appropriate means to meet the targets.

In the case of Italy, as further explained by Terenzi in Chapter 3, the RES-related targets have resulted in wind and solar generation capacity to increase from less than 4 GW at the beginning of this decade to more than 30 in 2019. During the month of April 2020, due to pandemic-related loss of consumption and thanks to a very sunny spring, the amount of RES-produced electricity reached 52% of the total electricity consumption—this fraction typically fluctuates around 35% on a yearly basis in recent years.

The result of having so much distributed generation is illustrated in the accompanying figure for Italy's sunny south, where there is excess generation not only on Sundays and Holidays during the middle of the day (as already happens since 2015) but also on weekdays—a visual reason for the urgency of developing more flexible demand.

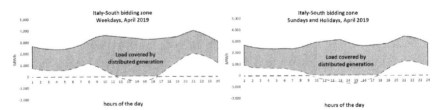

Source: ARERA, Yearly report on renewables at www.arera.it/it//docs/rapporti.htm

---

[1] **Disclaimer**: The opinions expressed in this Preface are the personal opinions of the author and don't necessarily represent the official position of the Authority nor commit the Authority to any course of action in the future.

The graphs show the dramatic increase of distributed as well as utility-scale solar generation resulting in an Italian version of the *duck curve* and the *residual load*, with challenging ramps in late afternoon, as further explained in Chapter 1 by Sioshansi.

In such a "perfect storm" impacting power systems all over the world, it's clear that we need a *change in regulation to regulating the change.*[2] In short, regulators cannot cope with the rapid pace of change and innovation taking place with the regulatory tools conceived decades ago, when power systems were a technologically stable and mature sector, when demand was deemed almost completely rigid and foreseeable, both for its level and pattern, when networks were designed with contemporariness coefficients related to traditional electro-appliances, and when the focus of regulation was essentially on maintaining static efficiency on a static system.

Since all these assumptions are impacted by the dramatic transformation of the power sector, it is a true blessing that now Fereidoon Sioshansi has assembled some of the best worldwide experience to investigate in detail the issue of demand flexibility in a changing—or in the case of Italy an already changed—context of decarbonization, decentralization, and digitalization. The challenge, in this context, is one of dynamic regulation.[3]

As a regulator, I believe that we must abide by simple yet robust principles when demand is on the dissection table, to avoid false and inefficient solutions. Moreover, I hope that the regulatory experience in the EU with full ownership unbundling of TSOs and full competition in the retail market can serve as an advanced laboratory of system transformation—especially in States, like Italy, with huge penetration of renewables and can be helpful for the readers of this book.

With this as my guiding principle, let me share a few ideas.

First, the reason why demand hasn't been flexible enough up to now may be attributed to many factors including the following three, which I believe are important:

- Historical energy efficiency culture;
- Capillarity of natural gas networks; and
- Fuel mix.

---

[2] L. Lo Schiavo, Delfanti M., Fumagalli E., Olivieri V., *Changing the regulation for regulating the change. Innovation-driven regulation in Italy*", Energy Policy, vol. 57, 506—517, 2013.

[3] Council of European Energy Regulators, *CEER's 3D Strategy to foster European energy markets and empower consumers: Digitalisation, Decarbonisation, Dynamic regulation*, Ref. C18-BM-124-04, 9 January 2019, www.ceer.eu/1740.

Looking at statistics we can observe that a simple indicator as consumption per household varies a lot among developed countries, mainly due to these reasons that are all largely dependent on national historical paths of energy development. In Italy, for instance, we did have, for many decades since the 1970s, a policy of limiting capacity at private homes as well as small business, through a breaker on the meter that trips the customer off if the load is higher than contractually rated capacity. This concept, which happens to be covered in the chapter by De Vries and Doorman, has led, over the decades, to a very low level of unitary electricity consumption—nowadays around 2.1 MWh/year as an average for residential households—combined with a very widespread distribution network of natural gas and a fuel mix in which nuclear, for environmentalist reasons, was blocked in its very first years.

In France, on the other hand, the situation is rather different. As a result of the aggressive policy to develop nuclear power, household consumption is higher and electricity is largely used for residential space and water heating as described in Chapter 7. In Norway and Sweden the average household consumption is even higher, 5 to 10 times more than in Italy, due to the abundance of low-cost hydro. As these examples illustrate, demand flexibility can be expected to be very different among countries: a single-size-fits-all solution does not exist and regulators, as well as policymakers, have to be conscious of the different starting points, the natural fuel endowments, differences in the climate, and other factors.

Second, demand flexibility requires advanced metering, but as explained in chapter by Stagnaro and Benedettini, this is clearly not enough. Here the Italian "lab experience" can be helpful. Italy is the nation where the largest project of smart metering in the planet has been implemented with almost 35 million smart meters installed and operating since 2006. Since 2009, all household customers that do not choose an electricity supply offer in the free market are mandatorily billed, as for energy, with time-of-use (ToU) rates. Nonetheless, the effect of the largest case of ToU in the planet—more than 20 million households impacted—on electricity demand pattern has been rather low, probably due to "static timebands" used so far,[4] but there may be other reasons for such limited customer response, a topic explored by Burns and Mountain.

---

[4] M. Benini, M. Gallanti, W. Grattieri, S. Maggiore *Impact of a mandatory time-of-use tariff on the Italian residential customers,* 12th IAEE European Energy Conference, September 9–12, 2012, Venice, Italy.

Recent surveys from other European countries demonstrate that an excess of faith has been put on smart metering per se as a trigger of demand flexibility.[5] Although a strong push toward dynamic energy price has been emphasized by new European legislation,[6] success of flexible demand cannot be based only on the customer's response to electricity price variations over time, but must be accompanied by profound energy management initiatives within the houses, the buildings and the working spaces if we want flexible demand to really take off. However, smart metering systems remain an essential infrastructure that enables not only retail market competition but also new services for demand response via intelligent appliances as well as demand aggregation for market participation of demand-side resources—among the topics examined in Chapter 7 and elsewhere in this volume.

Third, market design is essential, and regulators have to be "technology neutral" when market rules are set and enforced. Even more than day-ahead market, in a context of huge RES penetration and participation of distributed energy resources, the balancing market can be the key to unlocking demand flexibility. Any discrimination among different resources, however, shall be avoided, whether on the supply or demand side. In the context of overgeneration, RES participation for downwards services, for instance, has to be allowed as for any other available resource. Participation of demand for upward services is also to be allowed, although it can be less likely to actually be available in case of very efficient usage of electricity—as it's happening in Italy, where demand-side resources often offer rather high priced bids in the balancing market and therefore are hardly called in real time.[7]

Fourth, technological and business innovation is the key for take-off of totally new forms of demand flexibility. Here I'm totally in agreement with Fereidoon and other contributors on the fact that e-mobility is the killer application for demand flexibility. This new segment of electricity demand is inherently flexible, not only thanks to batteries onboard EVs, but

---

[5] D. Fredericks, Z. Fan, S. Woolley, E. de Quincey, M. Streeton, *A Decade On, How Has the Visibility of Energy Changed? Energy Feedback Perceptions from UK Focus Groups*", Energies, 2020, 13, 2566.

[6] Council of European Energy Regulators, *CEER Recommendations on Dynamic Price Implementation*, Ref: C19-IRM-020-03-14, 03 March 2020, www.ceer.eu/1932.

[7] ARERA, Stato di utilizzo e di integrazione degli impianti di produzione alimentati dalle fonti rinnovabili e di generazione distribuita, anno 2018, Relazione 291/2019/R/efr, https://www.arera.it/allegati/docs/19/291-19.pdf.

especially due to "circadian rhythms" of EVs' usage. V2G technologies, once developed and implemented, can unleash a totally new kind of demand-side participation to system equilibrium, especially in large parking lots where commuters' cars stay parked for many hours, the same for EVs parked at home overnight. The advent of "mobile electricity consumer," as further explained in Chapter 8 may bring about new forms of demand flexibility to the past failures.

Fifth, the trend towards increasing self-consumption shouldn't be the occasion to distort network tariffs. Benefits related to self-consumption and electricity sharing among communities can—and must—be reflected in the final expenditure for electricity withdrawal from the network of prosumers and "prosumagers" (as Fereidoon named prosumers that operate their own storage in his previous book), but this shall not jeopardize the key principle of "cost-reflectiveness" that regulators must follow when setting network tariffs—topics examined in chapter by Robinson. In Europe, recently this principle has been elevated at EU level, with 943/2019, article 18, and is therefore an obligation for all national regulatory authorities of EU member states to comply with.

In some countries, like Italy, distribution costs are now collected through a fully capacity-based component—euro/kW per year—in the network tariff, while transmission costs are collected as a volumetric component—euro/kWh—plus a fixed component—euro per year—for metering costs. Such a three-part tariff structure gives appropriate incentives to consumers for being conscious of the costs they impose on the network with their consumption pattern, especially for consumers with rated capacity over 15 kW, for which the monthly maximum level of capacity actually used (in quarter of an hour, thanks to full development of smart meters) is considered for billing every month.

In a future-proof perspective, distorting network tariff or, even worse, considering the net balance between withdrawal and consumption over a long period (net metering) is not a wise option for regulators. And this is even truer for collective self-consumption, for which the public distribution network is still used for "electricity sharing" among members of the so-called "energy communities."

Last, innovation related to system transformation is going to challenge the traditional boundaries between regulated and competitive activities, even in a fully unbundled regime as the one currently in practice in Europe.

A new holistic approach, called "whole system approach,[8]" is to be adopted by regulators, especially when long-term decisions have to be taken. System integration among different sectors,[9] not only electricity and gas but also district heating/cooling and, in a wider perspective, even hydrogen, waste management, water and transportation, is a new challenge that, *inter alia,* requires cooperation among regulatory authorities that are historically organized along a traditionally "vertical" approach to the single sector or a couple of sectors that are under their jurisdiction. A lot of issues have to be addressed, at least in Europe, but this is the way to move forward.

This book is a plunge into the future. I wish that everybody could emerge, after a deep dive in its chapters, with new ideas for coping with the challenges of energy decarbonization and power system decentralization.

Luca Lo Schiavo
Deputy Director for Energy Infrastructure Regulation
Italian Regulatory Authority for Energy, Networks and
Environment (ARERA)

---

[8] Council of European Energy Regulators, *CEER paper on Whole system approach,* Ref: C19-DS-58-03, 30 June 2020, www.ceer.eu/1913.

[9] C. Cambini, R. Congiu, T. Jamasb, M. Llorca, G. Soroush, *"Energy Systems Integration: Implications for public policy",* Energy Policy 143 (2020) 111609.

Introduction

Traditionally the power sector predicted demand and dispatched genera-tion to meet it. "Utilities," most of which were vertically integrated and regulated monopolies, maintained a portfolio of plants—base-load, inter-mediate, and peaking—to meet demand as it varied from hour to hour, day to day, and across the seasons. Historically, demand was a "given," namely the sum of the load from all electricity-using devices connected to the network. There was little or no attempt to manage demand, either its quantity or when or where it was consumed—topics covered in Chapter 5. Finally, all consumers bought all the kWhs they consumed from the network and paid a regulated bundled tariff.[1]

In the future, this paradigm is likely to change or has already changed—at least on some systems—with the market operator predicting variable renewable generation and scheduling demand, or attempting, to match it. This is likely to happen in more places as the percentage of re-newables reaches higher levels, already exceeding 50% and higher in many networks, as further explained in Part One. As the pressure to go net zero carbon increases, many countries are aiming for 100% renewable electricity generation by 2045-50.

Already, many networks experience **overgeneration**—with prices going negative. When this happens, the excess generation must be exported, stored or—as a last resort—*curtailed*. During such episodes, in places like California, renewable generation must be curtailed because it exceeds load, because there is congestion on the transmission network, and/or due to lack of sufficient storage. Fig. 1 illustrates places where negative prices were experienced in the United States in 2017.[2]

The situation is likely to get worse over time as more solar and wind—both of which are non-dispatchable—are added. A 2019 report from the Lawrence Berkeley National Laboratory (LBL) projects continued installation of solar capacity in the 16—18 GW range per annum in the United States for the next few years[3] (Fig. 2).

---

[1] For further discussion of the traditional utility business paradigm refer to Behind and beyond the meter, F. Sioshansi (Ed.), Academic Press, 2020.
[2] Refer to Utility-scale Solar, Lawrence Berkeley Lab, 2019 at https://emp.lbl.gov/sites/default/files/lbnl_utility_scale_solar_2019_edition_final.pdf.
[3] The 2020 global pandemic is likely to impact this and other projections.

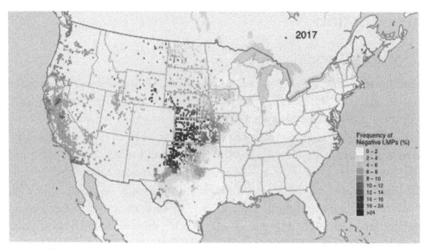

Figure 1 Frequency of negative prices in the United States in 2017.

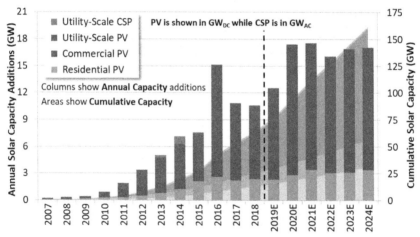

Figure 2 Projections of additional solar capacity installations in the United States. *Source: Utility-scale Solar, Lawrence Berkeley Lab, 2019.*

More impressive is the fact that solar additions compromised 40% of all new capacity additions in the United States in 2019[4]—a trend that is likely to be replicated in more places over time.

The distribution of solar installations in the United States, however, has been highly skewed, mostly concentrated in a handful of states. At the end

---

[4] US Solar Market Insight 2019Year-in-Review, Solar Energy Industries Association (SEIA) and Wood Mackenzie, Mar 2020.

Table 1 Top five solar states in the United States, 2018 data.

State	% from solar
California	19%
Nevada	12.7%
Hawaii	11.2%
Vermont	11.0%
Massachusetts	10.7%

Source: Utility-scale Solar, LBL, 2019.

of 2018, the top five states were already getting more than 10% of their generation from the sun (Table 1) while the average for the country as a whole was 2.3%.

A similar picture applies to wind, which is rising in many parts of the world, and which is also nondispatchable. Not unlike solar energy, wind installations tend to be highly concentrated in certain areas where wind is plentiful and steady. In the United States, a number of states now get more than 20% of their generation from wind. Moreover, nearly 30 states now have renewable portfolio standards (RPS) as illustrated in Fig. 3. In places where the targets are ambitious and mandatory such as in Hawaii and California, this has led to massive investments in renewable generation.

Similar trends are, of course, happening in other parts of the world— South Australia, Denmark, Germany, the North Sea, Ireland, Portugal, Spain, the United Kingdom, and many others—where wind and/or solar generation overwhelm the network during many hours of the day, days of the week, or months of the year. In China, wind generation is routinely curtailed for lack of adequate transmission capacity.

It is *not* uncommon to reach 100% levels during extremely windy and/ or sunny periods especially on weekends or when demand is low, in which case the wholesale prices go negative—that is not free electricity but electricity you are paid to use.

But the opposite also happens with regularity. If there is lot of demand when the wind does not blow and/or the sun does not shine, prices spike, signaling scarcity. During such periods, consuming electricity can be rather expensive. The fact that the contribution of renewables can vary from very high levels to virtually zero is a serious cause of concern.[5]

---

[5] In places like California and Germany, the grid operator has to consider extensive periods when there may be little or no wind and little or no sun for a few days in a row.

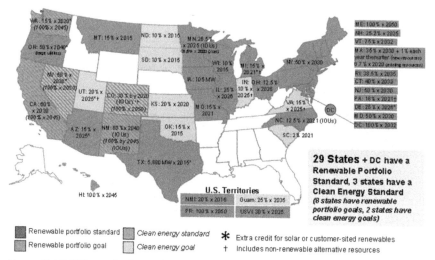

Figure 3 US RPS standards as of June 2019. *Source: https://www.dsireusa.org/resources/detailed-summary-maps/.*

Flexible demand, the topic of this book, can be applied to this frequently occurring feast or famine situation nearly everywhere resulting from the abundance or disappearance of variable renewable generation. But what makes flexible demand timely and topical today?

As described in Chapter 5, up to now, grid operators have done their best to use the traditional tools at their disposal, namely turning flexible thermal plants—typically gas-fired peaking units—up or down to accommodate the variable renewable generation. This works up to a point but is approaching limits that are either too expensive and/or operationally challenging for the grid operators who must maintain the system's reliability. Moreover, peaking plants are expensive to maintain and operate, are polluting, and are not necessarily profitable because they are only used for short durations and only when there is a shortfall in renewable generation. In most instances, flexible demand can serve the same purpose at lower cost, with zero emissions, and often at much faster response rate.

A well-known example of the feast-and-famine phenomenon is the famous California "duck curve (Fig. 4) where 13 GW of ramping is required to meet the evening's peaks on many sunny days.[6] Unsurprisingly, wholesale prices fluctuate from very low—or negative—levels to very high ones corresponding to the peaks and valleys of the duck curve.

---

[6] More on this topic may be found in Chapter 1.

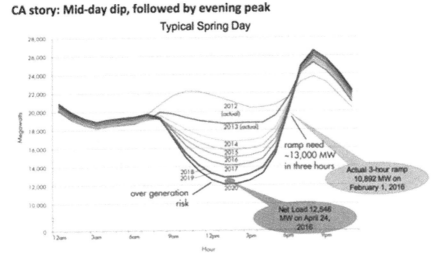

Figure 4 California duck curve. *Source: CAISO.*

During the midday sunny hours of the year, especially in the spring and the fall when the temperatures are mild and there is no air-conditioning load, wholesale prices on CAISO routinely plunge and occasionally go negative. The value of solar—and wind—whether from utility-scale or rooftop solar panels during these period is essentially zero or worse.

The opposite, of course, happens during the "neck" of the duck curve after the sun sets and peak demand is typically experienced. It is not rocket science to see that saving some of the excess solar generation in midday for use after sunset makes sense. And this explains the growing interest to pair solar—and wind—with storage. It also explains why more *prosumers* are investing in distributed storage, thus becoming *prosumagers*.[7]

While interest in storage is definitely on the rise, harnessing flexible demand makes an even more compelling business sense. Most customer loads have some inherent flexibility, and it is simply a question of shifting some of this flexible consumption to low price, and away from the high cost, periods—topics further explored in the following chapters of this book.

It must be noted that many versions of the California duck are now routinely experienced in other parts of the world with similar implications.

---

[7] For further discussion refer to Consumer, prosumer, prosumager, F. Sioshansi (Ed.), Academic Press, 2019.

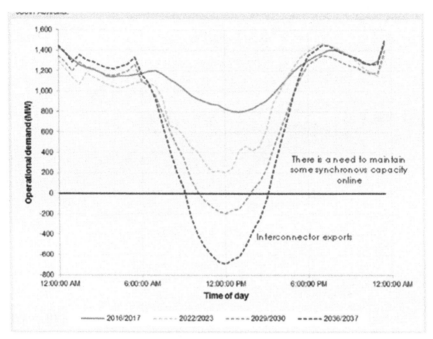

Figure 5 South Australian duck curve. *Source: Jenny Riesz, AEMO, presented at Future El. Mkt. Summit, Sydney, 18 Nov 2019.*

For example, **South Australia's** duck curve (Fig. 5) is already difficult to manage and is likely to get worse.

Terna, the grid operator in Italy, is expected to face a similar fate as the percentage of renewables increases in Italy — more fully described in chapter by Terenzi. Unsurprisingly, Italy's duck curve, illustrated in Fig. 16 in Chapter 3 looks familiar to those of California and So. Australia.

In some places, the traditional method of ramping thermal plants up and down is approaching the limits of its usefulness. Exporting the excess energy and/or storing it are gaining traction but can only go so far. As countries adopt ever more ambitious renewable targets, new ways must be found to balance supply and demand—but what?

One obvious answer is flexible demand. As time goes on new methods and functionalities must be found to explore how flexible demand can be aggregated and managed to meet variable generation. Concepts such as demand response (DR) as well as schemes to aggregate and manage flexible demand are being developed. These ideas are not new, but were historically considered "nice-to-do." They are now becoming "must-do."

As it happens, three developments are enabling their implementation on scale that was not imaginable in the past:

- Smart meters are becoming the norm in many places[8];
- Smart tariffs provide incentives to respond to price signals; and
- Smart connected devices are easier to monitor and manage.

The first two are essential, even if slow in the making. The third, sometimes called *digitalization*, is expected to make it easier for individual devices within the customers' premises to be remotely monitored and managed. Moreover, enterprises are emerging who can aggregate large portfolios of behind-the-meter devices and using sophisticated algorithms—artificial intelligence (AI) and machine learning (ML) among them—to optimize how, when, and how much energy is used.

The key to harnessing flexible demand—especially in the commercial and residential sectors where it has been barely explored to date—is *scale*. Small customers have flexible demand, but it comes in small increments, which only become worth the bother if they can be aggregated across many customers.

For example, take electric water heaters, prevalent in many countries. While managing the inherent flexibility of a single unit is hardly interesting, managing a million such devices could add up to a sizeable chunk of flexible load that can be worth a lot, especially if it can reliably respond to price signals from the grid operator.

Similarly, a large number of electric vehicles (EVs) can soak up some of the excess generation during sunny hours of the "belly" of the duck curve. Perhaps, some of the stored energy in the batteries can be discharged back into the network during the peak demand hours after the sun has set, a functionality called vehicle-to-grid (V2G). California, for example, is expected to have 5 million EVs by 2030. That represents an enormous form of flexible *generation* and *load* if it is properly managed, further discussed in Chapter 8.

Similar things can be done with a multitude of other devices with flexible demand including precooling or preheating buildings, making ice and storing it for later use, pumping or desalinating water, shifting electricity-intensive processes to low price hours of the day, and many more.

---

[8] For further details refer to Stagnaro & Benedettini in Chapter 12 of Behind and beyond the meter, F. Sioshansi (Ed.), Academic Press, 2020 as well as Chapter 6 in this volume.

With a few exceptions, virtually all customer demand has some *flexibility*. The industry traditionally ignored this valuable resource either because it was difficult to manage—it still is—or because it was easier to adjust generation. In the future, demand has to learn to tango with variable generation simply because the alternatives are becoming impractical, difficult, or too expensive.

Organization of this book

This book, which comprises of 22 chapters from contributors from different disciplines and from different parts of the world, is entirely focused on the topic of demand flexibility and why it must be harnessed to balance variable renewable generation.

In the book's **Foreword, Ken Baldwin** notes that given the variable nature of renewable generation, a paradigm shift is needed to create a smooth, efficient, and low-cost transition to a new world order and says that this can be achieved either by adjusting demand to match the supply, or by storing the renewable energy to meet demand.

A flexible and adaptive system will enable the best outcome through a combination of both, topics examined in this book, which surveys the prospects in a range of countries which have different requirements for their future electricity systems.

Baldwin notes that nowhere is this choice more stark than in Australia which is transitioning at the fastest rate of any country from one of the highest levels of fossil generation in the world, 84% in 2017. In this context, the Australian Energy Security Board is examining a number of options including two-sided energy markets to empower consumers with their new roles as producers and storers of energy to make the transition to the carbon-free energy systems of the future.

In the book's **Preface, Luca Lo Schiavo** points out the important role of regulation in the transition of the power systems towards decarbonization and decentralization and makes a case for the urgency of developing more flexible demand.

He explains that power systems all over the world need a *change in regulation to regulating the change*—and says that regulators cannot cope with the rapid pace of change and innovation taking place with the regulatory tools conceived decades ago, when power systems were technologically stable and mature, when demand and network flows were predictable, and when the focus of regulation was essentially to maintain efficiency on a static system.

Lo Schiavo believes that regulators must abide by simple yet robust principles in considering how best to make demand more flexible while avoiding inefficient solutions. He shares ideas on how Italy's experience can be helpful for the readers of this book.

The balance of the book is organized into four parts as outlined below.

Part one: Variable renewable generation

This part expands on how the rising variability of renewable generation in many parts of the world is leading to challenging balancing requirements for the grid operator as thermal plants are currently ramped up and down to compensate for fluctuations in generation—rather than flexible demand—which is covered in part 2.

In **Chapter 1, Challenges of California's variable generation, Fereidoon Sioshansi** describes how the rising variability of generation, primarily from solar resources, is exacerbating the already difficult "duck curve" problem, where ramps of 13—16 GW are required in the morning and afternoon to compensate for the daily rise or fall of solar generation. These daily evening ramp ups are projected to grow to 25 GW by 2030 as California marches ahead to meet its 2045 carbon neutrality goals.

The chapter also outlines the current thinking on how schemes such as time-of-use pricing, storage, growing penetration of electric vehicles, and increased emphasis on demand response may address some of these challenges.

The chapter's main contribution is to illustrate that without additional demand flexibility, the challenges of maintaining grid reliability and security will continue to increase.

In **Chapter 2, Variability of generation in ERCOT and the role of flexible demand, Ross Baldick** describes some of the challenges associated with the rising penetration of renewables in the Electric Reliability Council of Texas (ERCOT) and the increasing role of flexible demand particularly focusing on the coupling of flexible demand to variability of renewables to enable high levels of renewable integration.

The chapter describes several types of demand response including low and high effort adaptations. The former includes precooling of residences in air-conditioning intensive climates to better match photovoltaic production; the latter includes storage of end-use products in chemical production.

The chapter's main conclusion is that incorporating flexible demand is likely to be a cost-effective part of integrating high levels of renewables as compared to battery storage or carbon capture and sequestration.

In **Chapter 3, Rising variability of generation in Italy: The grid operator's perspective, Giacomo Terenzi** focuses on the current trends of renewable penetration in the Italian electrical system and the challenges that it imposes on Terna, the Italian Transmission System Operator (TSO).

The chapter examines projections of the growth of renewables over time and their impact on how the Italian grid will have to be operated and managed. Ambitious renewable targets, the expected retirements of thermal plants, plus the impact of climate change are likely to give rise to an Italian version of the "California Duck" by 2030 requiring challenging daily ramping and concerns about the reliability and security of the grid.

The chapter's main contribution is not only to examine the changes in the shape and patterns of the "net load" on the Italian network but, more important, discuss strategies and attempts by Terna to cope with these challenges over time.

In **Chapter 4, Integrating the rising variable renewable generation: A Spanish perspective, Juan Jose Alba, Julian Barquin, Carolina Vereda,** and **Eduardo Moreda** address the challenges caused by massive addition of variable renewable generation on the power system in Spain, and how new demand and storage assets can alleviate some of the problems.

The authors examine the latest Spanish decarbonization plan, which requires massive investments in renewables consistent with the EU climate and energy policy goals which aim to make the entire continent net zero carbon by 2050. The Spanish system is simulated along the planned trajectory with special attention to carbon emissions, energy spillage, and flexibility needs.

The chapter's main conclusion is that while managing the daily cycles are difficult the inter- and intraseasonal fluctuations in generation are far more daunting requiring longer-term storage technologies, such as green gases.

Part two: Flexible demand

This part explores what is flexible demand, where it can be found, how can it be aggregated, how much there is, and why hasn't it been exploited to date, among other topics.

In **Chapter 5, What is flexible demand: What demand is flexible? Fereidoon Sioshansi** examines the two key questions in the chapter's title including why the topic has not received the attention it deserves to date—hence the motivation for this volume.

Examining the literature and the empirical evidence, the author concludes that virtually *all* customer demand has *some* flexibility in the sense that not every electricity using device needs to be on at all times, and certainly not at the time of the peak demand on the network. Historically, the industry's mindset was to invest in sufficient infrastructure to meet customers' aggregate demand at all times, regardless of the costs. But as explained in other chapters in this volume, this is not going to be practical or economically justified in the future.

The chapter's main insight is to offer a new business and operating paradigm in which customers can choose if they wish to be on at any time they wish or would they rather lower their individual bill—and society's collective cost—by shifting some demand to times when supplies are plentiful and inexpensive.

In **Chapter 6, Who are the customers with flexible demand, and how to find them? Carlo Stagnaro** and **Simona Benedettini** examine the potential contribution that active customers may provide to flexible demand, either individually or through aggregators, while examining the reasons for low levels of customer engagement in competitive retail markets.

The authors ask whether electricity customers behave as rationally as they do when confronted with other economic choices, in particular the extent to which short-term variations of electricity prices result in behavioral responses and whether retail competition can contribute to making the demand more price-responsive.

The chapter's main contribution is policy suggestions with the aim of promoting customer engagement in electricity markets.

In **Chapter 7, How can flexible demand be aggregated and delivered to scale? Fereidoon Sioshansi** describes how a number of companies are emerging to develop innovative approaches to engage customers with flexible demand to participate in demand response programs. Among the many challenges is to convince reluctant and skeptical customers that participating in such programs is win-win-win; for them, for the aggregator, and for the society at large.

The author presents a number of anecdotal examples and case studies that describe the typical engagement process, which frequently involves convincing customers with flexible demand to offer their demand flexibility in exchange for lower electricity bills—universally important to customers everywhere.

The chapter's main insight is to suggest the scope and scale of practical and cost-effective demand flexibility that typically remains untapped among large commercial and industrial customers as well as residential customers, whose flexibility potential remains largely untapped.

In **Chapter 8, Electric vehicles: The ultimate flexible demand, Fereidoon Sioshansi** points out that the expected penetration of EVs in key global markets over the next couple of decades not only promises to eliminate a major source of carbon emissions but also offers a tremendous source of demand flexibility for storing—and potentially discharging—electricity.

The author examines California's efforts to manage how, when, and where the EVs are charged—a critical issue given that at least 5 million are expected to be on the roads by 2030; the critical charging infrastructure is grid-to-vehicle integration.

The chapter's main contribution is to point out that EVs can be a blessing in balancing the daily cycles of supply and demand because of the collective sheer capacity of their batteries to store—and potentially discharge—vast amounts of energy. No other electricity using device comes close to offering this level of demand flexibility.

In **Chapter 9, Load flexibility: Market potential and opportunities in the US, Ryan Hledik** and **Tony Lee** present the results of a major study by The Battle Group which estimates the total amount of cost-effective load flexibility potential in the United States.

Using Brattle's LoadFlex model and a national database of market data, the authors examine how much of the aggregate US demand may indeed be flexible—including the costs and benefits of this increased flexibility, the programs and customer segments that could provide the greatest load flexibility opportunities, and factors influencing the achievement of this potential.

The chapter's main contribution is to reinforce the notion that load flexibility has the potential to play an important role in facilitating the transition to a decarbonized power grid with large renewable penetration, given proper regulations, incentives, and market designs.

In **Chapter 10, Demand response in the US wholesale markets: Recent trends, new models, and forecasts, Udi Helman** provides a comparative perspective on the recent evolution of wholesale DR and other types of responsive wholesale demand, and reviews the nationally aggregated and region-specific data on market performance and the entities who provide these services.

The author reviews current wholesale DR programs and market participation, offers explanation of the factors, which have facilitated DR's success while identifying some of the barriers that continue to constrain further applications. Of particular interest are the details on how economic and reliability demand response has been adapted to the different primary market products providing capacity, ancillary services, and energy.

The chapter's main contribution is to show that DR participation has fluctuated across various US markets, sometimes by a fair amount in recent years, as a function of continued changes in market design, market prices, and competition from other resources.

In **Chapter 11, What's limiting flexible demand from playing a bigger role in the US organized markets: the PJM experience, Joseph Bowring** examines efforts by the PJM Interconnection LLC to deliver more flexible demand to participate in one of the biggest wholesale electricity markets in the world.

The author examines past and present demand response schemes as well as proposals for the future that may be incorporated in the PJM market rules, allowing more active participation of flexible demand in the market.

The chapter's main insight is that introducing proper incentives to generate more active demand response in a market like PJM is not easy but given the scale of savings, it is definitely worth pursuing.

Part three: Coupling flexible demand to variable generation

This part explores how flexible demand and variable generation can be matched to produce a lower cost, more manageable, reliable, and ultimately resilient outcome in balancing supply and demand.

In **Chapter 12, Valuing consumer flexibility in electricity market design, Laurens de Vries** and **Gerard Doorman** describe how a different electricity market design based on capacity subscription places the right value on consumer flexibility.

The authors examine how a capacity subscription market design can ensure the adequacy of a future electricity system that is dominated by variable renewable generation. Such a largely weather-driven system is characterized by high price volatility and significant differences between yearly average electricity prices. The authors argue that capacity subscription induces an economically optimal mix of flexible generation, storage, and consumer flexibility by making the demand for flexibility explicit and creating a consumer-driven market for reliable resources.

The chapter's main contribution is to illustrate how such a capacity subscription uses a market mechanism to find an optimal trade-off between consumer-side and market-provided investments.

In **Chapter 13, Variable renewables and demand flexibility: Day-ahead versus intraday valuation, Reinhard Madlener** and **Oliver Ruhnau** examine the impact of intraday markets and accurate short-term forecasting of electricity generation from variable renewable resources on their market value and on the value of demand response.

The authors point out that the uncertainty of renewable electricity supply drives the volatility of electricity prices, especially in the intraday market, and leads to balancing cost for renewables, which can be optimized through forecasting and trading. The increased intraday price volatility can be exploited and moderated by demand response, which thereby helps to balance the uncertainty of renewables.

The chapter's main contribution is an overview of the relevant literature, a theoretical framework for the analysis of day-ahead and intraday trading, and illustrative quantitative examples.

In **Chapter 14, The value of flexibility in Australia's national electricity market, Alan Rai, Prabpreet Calais, Kate Wild, Greg Williams,** and **Tim Nelson** discuss the growing need for supply- and demand-side resources that are flexible and dispatchable in Australia's national electricity market (NEM) given the historical and likely future penetration of small- and utility-scale variable renewable energy (VRE).

The chapter provides estimates of the value of being dispatchable and flexible in South Australia, the region with the highest VRE penetration worldwide. It shows the "flexibility and dispatchability premium" has grown in tandem with the rise of VRE. It notes that both flexible supply- and demand-side resources can earn this premium, with automation, digitalization, and the Internet of Things acting as critical enablers for the latter.

The chapter provides estimates of the flexibility premium, outlines the key enablers of dynamic demand side, and identifies promising business models that can bring about a truly two-sided market.

In **Chapter 15, Demand flexibility and what it can contribute in Germany, Dierk Bauknecht, Christoph Heinemann, Matthias Koch,** and **Moritz Vogel** focus on demand flexibility in Germany and present key insights from a range of research projects.

The chapter looks at how the demand for flexibility increases with an increasing share of renewables and discusses the role that demand flexibility can play. What are the pros and cons of demand flexibility compared to other flexibility options? What are the trade-offs between demand flexibility and efficiency? What role can new consumers play? And why hasn't it happened yet?

The chapter shows that demand flexibility can make an important contribution to provide flexibility, but there also limitations and pitfalls, such as the trade-off between flexibility and efficiency.

In **Chapter 16, Industrial demand flexibility: A German case study, Sabine Löbbe, André Hackbarth, Heinz Hagenlocher,** and **Uwe Ziegler** describe business opportunities for smaller utilities enabling flexible industrial demand provided by small and medium-sized enterprises (SMEs) within a virtual power plant (VPP).

The authors refer to results of a cooperative project with industrial companies, utilities, and researchers, resulting in a platform to connect SMEs to a VPP. Unlike other VPPs, the focus is on participation, data control, and sovereignty for the SMEs. An application of the concept for a cement mill demonstrates positive profit margins, suggesting opportunities for further applications.

The chapter's main contribution is to exemplify viable business models and processes in value chain ecosystems for small and medium-sized utilities. Digitalization and cooperative approaches are identified as game changers to integrate flexible industrial demand.

Part four: Implementation, business models, enabling technologies, policies, regulation

This part examines practical issues in implementation of flexible demand including new technologies, new business models, as well as emerging regulatory and policy directives that can encourage or hamper further development and implementation of flexible demand on large scale.

In **Chapter 17, Market design and regulation to encourage demand aggregation and participation in European balancing markets, Juan José Alba, Carolina Vereda, Julián Barquín,** and **Eduardo Moreda** examine different regulatory and policy models to encourage independent aggregators with different approaches to participate in wholesale markets, providing balancing services and flexible demand.

The chapter provides a comprehensive overview of the role of aggregators in the European Union and identifies the benefits they could bring to the electricity system if properly implemented.

The chapter's main contribution is an assessment of the useful role that independent aggregators can play in offering balancing services in different markets such as day-ahead, intraday, balancing and reserves, and the financial impact of this on customers, suppliers, and other stakeholders.

In **Chapter 18, Do time-of-use tariffs make residential demand more flexible? Evidence from Victoria, Australia, Kelly Burns,** and **Bruce Mountain** examine whether households respond to Time-Of-Use (TOU) tariffs based on evidence from around 7000 residential electricity bills in Victoria, Australia.

The study finds residential customers respond weakly to the difference in peak and off-peak prices; households with rooftop photovoltaics (PVs) do not respond any differently; and households in the lowest socioeconomic regions do not respond to TOU tariffs.

Although demand that responds to price signals is increasingly valuable as the penetration of renewables and Electric Vehicles (EVs) rises, existing TOU tariffs are not effective and new approaches, perhaps with shorter peak periods could be more effective.

In **Chapter 19, Empowering consumers to deliver flexible demand, Lynne Gallagher** and **Elisabeth Ross** explain why Australian consumers are well placed to provide flexible demand and describe the barriers to this happening.

The chapter examines Australia's experience with flexible demand and why it is increasingly important as the industry decarbonizes. The authors identify some of the barriers to market participants providing flexibility services and conclude that frameworks need to support the entry of new business models and third-party intermediaries to extract the value of flexible demand and achieve demand flexibility at scale.

The chapter's main contribution is to consider the critical role of enablers and opportunities for empowering consumers to deliver flexible demand in the future energy system.

In **Chapter 20, Markets for flexibility: product definition, market design, and regulation, Richard Hochstetler** discusses the different strategies that can be employed to harness the various types of demand response services.

The chapter provides a general conceptual framework that is helpful to determine how markets can incentivize consumers to provide demand response. As the share of noncontrollable variable generation increases, flexibility will become increasingly scarce. Demand response has great potential to meet this need. A wide range of mechanisms can be utilized to meet this challenge. Practical examples are presented to illustrate different market solutions.

To fully engage consumers, markets will need to evolve. Markets will need to provide more precise signaling, adapt so as to better accommodate consumer participation and provide better remuneration for consumers' response.

In **Chapter 21, Energy communities and flexible demand, David Robinson** examines the potential for energy communities to accelerate decarbonization and improve overall social welfare.

The chapter draws on recent European legislation and experience to identify policy reforms needed to support *efficient* demand-side flexibility both within and between energy communities and the rest of the network. It encourages the use of regulatory pilots and sandboxes to encourage business model innovation, understand consumer preferences, and stimulate the growth of products and services supporting demand flexibility.

The chapter's main thrust is that consumers should be encouraged to drive the process rather than rely on central decision-makers. Energy communities need a suitable policy framework to ensure their incentives are efficient and aligned with those of the system. This will reduce the risk that they may separate themselves from the network and markets, restricting competition and raising the costs of the energy transition.

In **Chapter 22, Flexible demand: What's in it for the customer? Mike Swanston** examines the pros and cons of participating in flexible demand schemes offered by retail companies, DSOs, or innovative aggregators from the perspective of large and small customers. The key question, as in the case of investing in rooftop solar PV panels or storage batteries, is what's in it for individual customers, what motivates them to engage in such schemes.

The author examines the available empirical evidence and data from a number of ongoing trials and pilot projects as well as historical experience with schemes such as hot water heating systems to draw conclusions on what types of incentives and marketing messages resonate with "typical" customers, enticing them to volunteer or engage in demand response programs.

The chapter's main insight is that customers *do* respond to trusted players with a compelling easy-to-understand message plus the right incentives and appropriate tools that does not inconvenience them while offering savings to them and/or the society at large.

In the book's **Epilogue**, **Ahmad Faruqui** notes that over time, the power supply will become increasingly variable as renewable resources dominate the generation mix with electricity prices fluctuating in the wholesale markets based on the circumstances.

Simultaneously, an increasing number of buildings will generate some of their energy needs with solar panels with many pairing the solar generation with storage and/or electric vehicles. Additionally, just about every major appliance is likely to become smart and connected.

Faruqui predicts that instantaneous load flexibility will emerge as the perfect complement to inherent variability of renewable resources—with prices seamlessly flowing to appliances turning all electricity using devices into flexible or controllable loads. This will lead to a new emerging paradigm of a demand-side merit order and agrees with Luca Lo Schiavo's statement in the Preface, that "This book is a plunge into the future," and says the book offers new ideas for the challenges that lie ahead.

PART 1

Variable renewable generation

CHAPTER 1

The evolution of California's variable renewable generation

Fereidoon Sioshansi

Menlo Energy Economics, Walnut Creek, CA, United States

1. Introduction

California's "duck curve" is well known the world over. Moreover, versions of the same phenomenon are now appearing with regularity in other markets where the penetration of renewables is on the rise, as further described in the following chapters.

When and how did the "duck curve" become recognized, how has it evolved since 2012 when it was first revealed, and what is likely to happen in the years ahead as California marches toward even higher renewable penetration levels are among the vexing questions facing policymakers.

This chapter provides the context for California's duck curve, its implications, and the challenges it poses for the state's grid operator. More important, it provides the context to examine how flexible demand can contribute to balance supply and demand, the book's main topic.

The chapter is organized as follows:

- Section 2 provides the genesis of the California duck curve, its evolution to date, and its prospects for the future;
- Section 3 describes the challenges associated with the duck curve as the grid operator confronts ever-rising ramping rates, more frequent episodes of negative prices, and higher renewable curtailment; and
- Section 4 argues that the traditional means of balancing supply and demand—namely relying on more flexible generation—are likely to go only so far and suggests that fundamentally different solutions, including increased reliance on flexible demand and storage will be needed to keep supply and demand in balance; followed by the chapter's conclusions.

Variable Generation, Flexible Demand
ISBN 978-0-12-823810-3
https://doi.org/10.1016/B978-0-12-823810-3.00018-2

3

2. The genesis of the "California duck curve"

The California Independent System Operator (CAISO) covers roughly 80% of California's load (Fig. 1.1). It is responsible for operating the grid while maintaining reliability of supply. Over the years, the percentage of renewable generation in California's electricity mix has continued to rise in response to the state's mandatory renewable portfolio standard (RPS).[1]

Figure 1.1 *CAISO's service area map. (Source: CAISO.)*

[1] The main driver of the changing electricity mix in CA is Senate Bill 100 or SB 100, with further details found at https://leginfo.legislature.ca.gov/faces/billNavClient.xhtml?bill_id=201720180SB100.

In 2012, CAISO conducted a study to determine the likely evolution of the trends and their impact on the network over time.[2] Among a number of scenarios examined was one, which assumed a 40% renewable penetration in the electricity mix by 2020—consistent with the RPS requirements at the time.

The most notable feature of the study was the stunning prediction of the "net load,"[3] suggesting that it would progressively dip during the sunny hours of the day—the *belly* of the duck curve—requiring ever rising ramping rates in the morning, as the sun rises, and even steeper ones in the late afternoon, when the sun sets—the neck of the duck curve (Fig. 1.2).

Three features of the curve were particularly alarming:
- The growing morning and evening ramps;
- The sharp peak demand, which was gradually moving into later evening hours when no solar generation is available; and
- The emergence of "overgeneration" and "minimum load" in midday on sunny days resulting in negative prices and more frequent episodes of renewable curtailment.

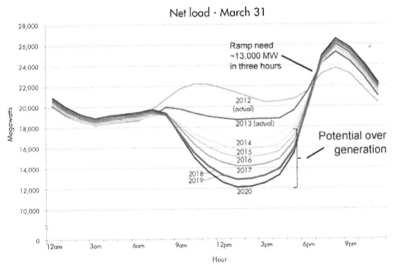

Figure 1.2 *The original 2012 California "duck curve" projections for 2020. (Source: CAISO.)*

[2] Refer to https://www.caiso.com/Documents/FlexibleResourcesHelpRenewables_FastFacts.pdf.

[3] Definition of net load at https://www.lawinsider.com/dictionary/net-load.

While everyone is familiar with the shape of the duck curve today, it was a revelation when first publicized in 2012. As illustrated in Fig. 1.2, the curve was projected to end up with a bulging belly and a steep neck by 2020 with a couple of noteworthy—and troubling—features:

- First, the *minimum load*, which refers to the fact that the belly of the curve was projected to dip to levels that would require virtually all thermal plants with any flexibility or ramping capability to be shut down during the sunny midhours of the day;
- Second, and more troubling was the *daily ramping requirements*, which projected that CAISO might need as much as 13 GW of ramping capacity during a 3 hr window around sunset to make up for the disappearance of the solar generation.

Naturally, CAISO's management was alarmed by these results. In subsequent years, it turned out that CAISO had in fact been *conservative* in projecting the 13 GW evening ramp by 2020. The dreaded ramps arrived as early as 2016, 4 years ahead of schedule, as illustrated in Fig. 1.3, along with a worsening minimum load.

As reported in the press, the CAISO management wasted no time in alerting the governor, the regulator, California Public Utilities Commission (CPUC), the California Energy Commission (CEC), the utilities,

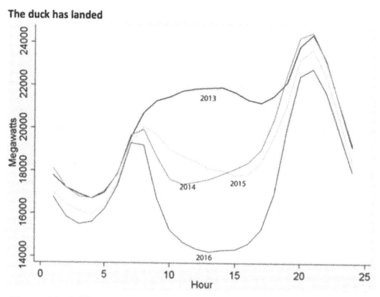

Figure 1.3 *California duck arrives earlier than predicted. (Source: CAISO.)*

generators, and other stakeholders about the implications of the evolving duck curve[4] and the operational challenges it imposed on keeping the network balanced and reliable.

This led to the California energy storage mandate[5] and other regulatory decisions to accelerate the introduction of variable tariffs and make adjustments in the state's net energy metering law.

Unable to expand its *geographical* footprint physically,[6] for political reasons, CAISO redoubled its efforts to expand *virtually* by introducing the *energy imbalance market* or EIM,[7] which allows utilities in neighboring states to participate in the real-time energy imbalance market. At the time of this writing, the voluntary EIM market has expanded far beyond California's borders as illustrated in Fig. 1.4.

CAISO is keen to gradually expand into a much larger *western* market, eventually covering much of the Western Electric Coordinating Council[8] or WECC, which covers the 14 states and the 2 Canadian provinces west of

Active and planned participants in CAISO's energy imbalance market

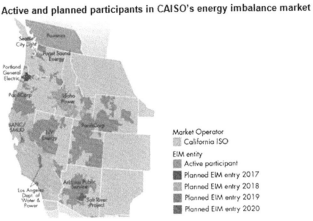

Market Operator
California ISO
EIM entity
Active participant
Planned EIM entry 2017
Planned EIM entry 2018
Planned EIM entry 2019
Planned EIM entry 2020

Figure 1.4 *CAISO's energy imbalance market. (Source: CAISO.)*

---

[4] Duck curve soon found a citation on Wikipedia at https://en.wikipedia.org/wiki/Duck_curve.

[5] In 2013, the CPUC issued Decision D.13-10-040 which set an energy storage procurement target of 1325 MW by 2020, for details refer to https://www.cpuc.ca.gov/General.aspx?id=3462.

[6] Some of the neighboring states have concerns about joining CAISO for jurisdictional issues—mostly due to the *California centricity* of CAISO's board and its primary mission.

[7] Further details on the first 20 years of CIASO history may be found at http://www.caiso.com/about/Pages/20-year-anniversary.aspx.

[8] Refer to https://www.wecc.org/Pages/home.aspx.

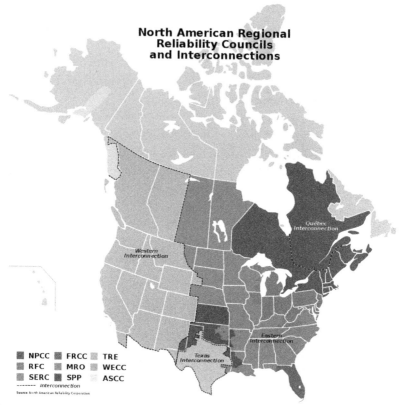

Figure 1.5 *Western Electric Coordinating Council (WECC) covers 14 western states plus 2 Canadian provinces on the left of the map. (Source: WECC.)*

the Rockies, the area highlighted in the left of the map in Fig. 1.5. Moreover, CAISO is offering a range of balancing and scheduling services to anyone who needs it—both within and beyond the state's borders.[9]

Clearly, having more participants from a broader geographical area such as the WECC would help smooth the variations in both supply and demand while taking advantage of different load patterns, different time zones, and other factors that may smooth out the duck curve to some extent.[10] Achieving this vision, however, is a daunting political and

---

[9] For list of products and services refer to http://www.caiso.com/participate/Pages/MarketProducts/Default.aspx.

[10] For details on the cost savings resulting from the EIM refer to https://www.utilitydive.com/news/caiso-western-eim-benefits-surpass-100m-since-2014/429459/.

jurisdictional endeavor for a number of reasons—most notably a require-ment to change the *governance* and the board of CAISO to make it a *regional* grid operator as opposed to a California-centric one—a topic that goes beyond the scope of this chapter.[11]

3. The challenges associated with the duck curve

The initial CAISO study that resulted in the famous duck curve is now history. Since 2012, numerous subsequent studies by CAISO have iden-tified additional challenges associated with the rising percentage of variable—i.e., nondispatchable—renewables over time.

In the mean time, the California duck curve has continued to evolve much faster than originally projected. The evening ramp rate was already approaching 16 GW during a 3 hr window as early as 2019 (Fig. 1.6).

Among the serious challenges facing CAISO is *overgeneration*, which occurs when generation exceeds demand on the network. As illustrated in the simulation for 2024 in Fig. 1.7, the overgeneration is most pronounced

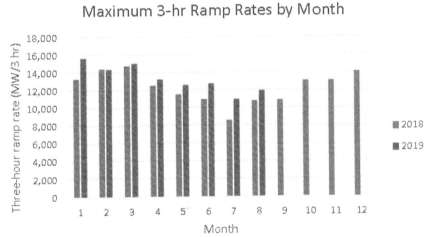

Figure 1.6 *CAISO 3-hr ramp rates, 2018–19. (Source: CAISO.)*

[11] For obvious reasons, neighboring states want more representation while some are con-cerned about California's sheer market size, which dominates other less populous states. Additionally, there is friction among and between states within WECC on a number of priorities such as those who are moving toward high renewable mix versus those who wish to maintain coal and other fossil fuel–fired generation. California, by contrast, is moving toward a virtual fossil fuel–free future with restrictions on imports of carbon-loaded electricity from out of state.

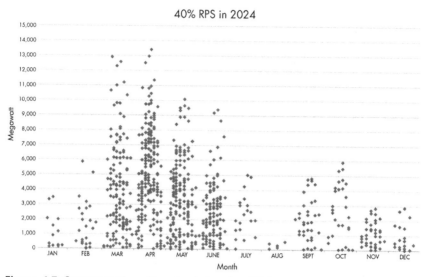

Figure 1.7 *Overgeneration.* The simulation for 2024 assuming 40% RPS, predicts significant overgeneration in the spring months of March, April, and May when renewable generation peaks but there is little air conditioning load. *(Source: CAISO.)*

during the spring and the fall when renewable generation can be exceptionally high because all renewables—hydro, solar, as well as wind—tend to be plentiful while demand tends to be modest because of moderate temperatures and little or no air conditioning load. This combination results in massive excess generation during the spring, and to a lesser extent in the fall. During the hot summer months, however, there is little overgeneration because usually most solar and other renewable generation is used and useful to serve the heavy air conditioning load.

As difficult as it is to manage the daily load cycles, addressing the seasonal imbalances in supply and demand is far more daunting. For example, while a variety of solutions currently exist for storing the excess generation during the belly of the duck curve and release it during the duck's neck period—say by using battery storage or charging electric vehicles (EVs)—few technologies currently exist that can store vast amounts of energy, say in the spring, for use 3 months later in the summer. Pumped hydro is among a few known and proven technologies that can do this but California currently has limited pumped hydro capacity.

It must be noted that managing the daily and seasonal cycles is a universal concern as more investment is made in renewable generation. Box 1.1 describes a study by Eurelectric looking at similar issues in Europe.

Box 1.1 The challenge of intraseasonal imbalances in supply and demand in Europe[12]

Bowing to the increased pressure to address climate change, the European Commission (EC) in Dec 2019 set a target to achieve carbon neutrality for the continent's economy by 2050.[13] What is not clear is how quickly will the member states deliver on what is now official EC policy. With its legendary slow bureaucracy, decisions made in Brussels move at *glacial speed* before they are implemented where it matters. However, now that the glaciers are actually melting rather fast, the pressure is on to deliver tangible results more quickly. A number of countries including the United Kingdom—no longer an EU member—as well as Germany have already adopted a similar target for 2050 with other countries likely to follow.

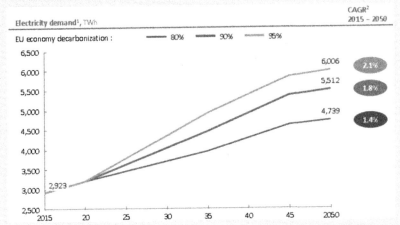

Deep decarbonization of the EU economy must be accelerated. (Source: The electric power investment challenge, Juan Jose Alba Rios at Eurelectric's E-Invest event, Brussels, 26 Sept 2019.)

The political pressure is felt by lobbying groups such as Eurelectric, which represents the European electricity sector. As with all such associations, when the political pressure can no longer be ignored or postponed, the industry must finally act.

[12] This box is based on an article titled "Eurelectric: $100 billion per year and counting", which appeared in the Nov 2019 issue of *EEnergy Informer*, pp. 8–10, available at www.eenergyinformer. com. The article is based on a presentation by Endesa's Juan Jose Alba Rios at a Eurelectric event in Brussels on 26 Sept 2019. The topic is also covered in Chapter 4.

[13] Refer to https://ec.europa.eu/info/sites/info/files/european-green-deal-communication_en.pdf.

Continued

Box 1.1 The challenge of intraseasonal imbalances in supply and demand in Europe—cont'd

Presenting the results of a study of the future of the European electric power sector at E-Invest in Brussels on 26 Sept 2019, Juan Jose Alba Rios, vice president, regulatory affairs at Endesa and the chair of Eurelectric's Markets and Investment Committee, said that the European power sector must invest some €100 billion ($110) per year to make the transition to a carbon-neutral future.

Alba's message was straight out of the deep decarbonization playbook such as those by the International Renewable Energy Agency (IRENA) and others. To reach economy-wide carbon neutrality, most energy must be converted to electric, which in turn must be generated from low carbon resources, and fast.

As illustrated in the accompanying visual, the recent pace of EU investment in renewables must be accelerated by a factor 3 or 4 during the decade of 2020s to somewhere between 50 and 70 GW of additional wind and solar capacity.

Adding so much variable renewable generation to the generation mix will result in wild variations in supply, particularly with the solar contribution spiking on sunny days and disappearing between sunset and the next day's sunrise and during extended periods of cloud cover. According to the Eurelectric's study, the intraday variations could be as much as 345 GWhs for a mid-sized country in the EU as illustrated in Fig. 4.4 of Chapter 4.

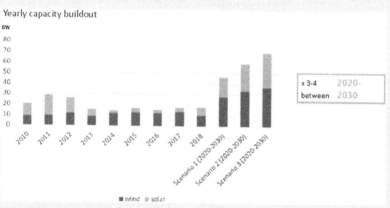

Renewable investment must increase 3—4 times between 2020 and 30. (Source: The electric power investment challenge, Juan Jose Alba Rios at Eurelectric's E-Invest event, Brussels, 26 Sept 2019.)

This well-known daily phenomenon along with predictable patterns for wind generation could be handled through proven technologies including storage and increased demand response. More transmission capacity will allow more diversity in supply, dampen the price swings, and enhance the reliability of the network.

While this is not a trivial challenge, it is a relatively simple one compared to the variations in *seasonal generation*—a far more demanding issue to address. The European interconnected network is likely to experience increasing supply—demand

Continued

Box 1.1 The challenge of intraseasonal imbalances in supply and demand in Europe—cont'd

imbalances over time, requiring massive amounts of long-term storage capacity by 2050.

As illustrated in Fig. 4.5 in Chapter 4, the excess renewable generation from solar, wind, and hydro during the spring and summer must somehow be stored to fill the deficit during the winter months. Until now most of this interseasonal balancing has been provided by varying thermal generation. But this capacity is expected to diminish as more thermal plants are retired. Aside from large hydro pumping systems, there are currently no known cost-effective technologies to store vast amounts of energy for, say, 3—6 months. Moreover, the size of this challenge is orders of magnitude greater than the daily cycles—well beyond what the current storage technologies can deliver. A key challenge of decarbonizing the EU's energy system will be to develop firm and flexible carbon-neutral capacity to complement renewables. Options such as green hydrogen, also called *power-to-gas*, are under consideration.

In his presentation, Alba offered a number of solutions such as the need for more transmission capacity to move the excess and deficit generation around the European continent or use the excess renewable electricity to produce green gas.

Other challenges include getting the European heating, transport, and the industrial sectors to electrify and do it in a way that does not exacerbate the imbalances in supply and demand—such as large numbers of EVs charging at the wrong time. This will also require significant investments in extending and modernizing the grid.

Eurelectric's recommendations include the "usual suspects" such as better wholesale price signals, cost-reflective tariffs, better network regulations and incentives, and smart network policies that encourage efficient integration of variable renewable generation.

Additionally, the report pointed out that the current retail electricity prices across the EU essentially *penalizes* the use of electricity through excessive taxes and levies compared to fossil fuels, which are taxed less, not taxed at all, or in some cases actually *subsidized*.

According to Alba, "For the deep de-carbonization to work, electricity generated from renewable resources must ultimately be priced at its actual value, instead of being penalized by taxes and levies, which make it artificially more expensive than dirtier fossil fuels." The biggest challenge for the decarbonization of the electricity system, he says, is to replace the current carbon-emitting thermal power plants with new carbon-neutral ones generating firm and flexible capacity.

In the case of California, the seasonal imbalance is expected to get worse over time by climate change:

- First, California's summers are getting longer, drier, and hotter while the winter snow packs are shrinking, reducing available hydro generation during the summer months. Moreover, what little snow accumulates in the mountains tends to melt faster and earlier, leaving little for the critical high-demand months of August–October when the peak demand is typically experienced.

- At the same time, the historical pattern of interseasonal exchange between California and hydro-rich Pacific Northwest and British Columbia is diminished since the hydro surplus in the former regions tends to coincide with overgeneration in California. Likewise, California's excess thermal and nuclear generation in the winter is diminishing when the Pacific Northwest's demand tends to be high.
- Making matters worse, when CAISO usually experiences its peak demand, many of the neighboring states have their own demand peaks while during the periods when CAISO has excess generation, they may not have any need to import power.

These issues, plus the limited transmission capacity, transmission losses given the long distances, and the fact that many states within WECC rely on inflexible thermal generation—coal or nuclear—puts additional constraints on what can be done with both the daily and interseasonal imbalances across the region.

Another serious and growing problem is the rising frequency of hours when overgeneration leads to *negative prices* and renewable *curtailment*. Curtailment happens when the excess generation cannot be exported, stored, or given away.

While nothing is fundamentally wrong with either negative prices or curtailment, both may be considered anomalies resulting from lack of adequate storage, flexible demand, and/or transmission capacity to export.

If renewable curtailment becomes more frequent, it poses another dilemma for the state's policymakers. After all, the main reason for increasing renewables in the electricity mix is to displace fossil fuel generation to meet the state's ambitious climate law. If, having made the investment in renewable capacity, it turns out that they cannot be utilized, it would look like a wasted investment.

Another vexing issue is how realistic is the state's 100% renewable target. The state's regulators at the CPUC and policymakers at the CEC envision a future where virtually all thermal generation including flexible natural gas—fired peakers are eventually phased out, replaced by renewables, storage, and demand response.[14]

---

[14] California's two remaining nuclear plants are expected to be shut down around 2025 partly because they are not flexible and partly because they are expensive to maintain.

The debate about the feasibility and the wisdom of phasing out *all* natural gas plants is controversial. Those in favor of keeping them on point out that these plants currently offer critical ramping generation that would be far more expensive to get from storage or from current demand response programs as further explained in Box 1.2.

Box 1.2 Calpine says phase out of gas-fired peakers will be expensive[15]

California's carbon neutrality target for 2045 is not limited to its electricity sector but extends to the entire economy. To achieve this a lot must happen in the next 25 years as schematically illustrated below.

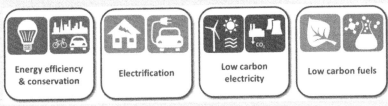

De-carbonization pathway: Use less, electrify & go low carbon. (Source: Long-run resource adequacy under deep de-carbonization pathways for California, Energy & Environmental Economics, June 2019.)

Moving toward high renewable targets is not particularly onerous to start with given the rapidly dropping costs and the abundance of renewables. Initially the grid operator can take on a bit more renewable generation in the mix so long as it does not overwhelm the network.

But as the percentage rises above a certain threshold, the inherent variability of renewables, notably wind and solar, begins to become noticeable. At times, there is more generation than load—say during sunny, breezy weekends.

[15] This box is based on an article titled "100% renewable: Nice but is it affordable?" which appeared in July 2019 issue of *EEnergy Informer*, pp. 8–11, available at www.eenergyinformer.com. The Calpine study, titled Long-Run Resource Adequacy under Deep Decarbonization Pathways for California, was released in June 2019 may be found at https://www.ethree.com/wp-content/uploads/2019/06/E3_Long_Run_Resource_Adequacy_CA_Deep-Decarbonization_Final.pdf.

Continued

Box 1.2 Calpine says phase out of gas-fired peakers will be expensive—cont'd

At other times, say cloudy but windless days or nights, there may be little renewable generation relative to demand. In the former case, wholesale prices drop or go negative, and/or the excess generation must be *curtailed*.

According to the CAISO, solar curtailment rate in 2018 was less than 2%, rising to 3%–4% in 2019–20, considered tolerable. But once it reaches double digits, it becomes harder to ignore or justify.[16]

By mid-2019, curtailment was around 530,000 MWhs with roughly 7100 MW of connected solar installations on the 3 investor-owned utilities. The opposite happens when renewable generation is insufficient to meet demand, when thermal plants must be brought on line to fill the gap.

In response CAISO, like grid operators around the world, is looking at a number of old as well as new options to deal with the imbalances including:

- More transmission capacity to import/export;
- More storage capacity;
- More flexible demand; and
- An expanded geographical footprint, which includes more transmission.

It is generally agreed that at very high renewable generation levels, the operating costs begin to escalate. Beyond some point adding more variable renewable becomes *counterproductive*, making the network hard to balance and/or expensive to operate.

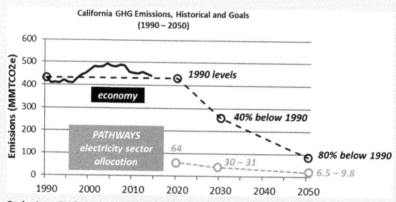

Reducing GHG emissions gets progressively more difficult. (Source: Long-run resource adequacy under deep de-carbonization pathways for California, Energy & Environmental Economics, June 2019.)

[16] Explaining the phenomenon, CAISO's Mark Rothleder noted that once curtailment "starts inching up toward 10%, and greater than 10%, you have to start looking at it and asking what else you can be doing. I don't think at that point just building more solar is the right thing to do."

Continued

Box 1.2 Calpine says phase out of gas-fired peakers will be expensive—cont'd

The question naturally boils down to how high a renewable mix is high enough. Or alternatively, can we afford a system without *any* thermal generation? Calpine Corporation, a major thermal generator in the California market, tasked Energy and Environmental Economics Inc. (E3), a consulting firm, to provide some answers.

Calpine, who currently has around 6 GW of natural gas—fired capacity in the state, wanted to know what might happen to its thermal assets in a future when the state deeply decarbonizes its entire economy to meet ambitious carbon targets as illustrated in the accompanying visual.

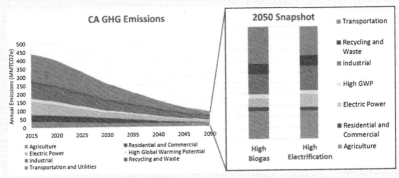

Focus on the biggest GHG emitter: Transportation. (Source: Long-run resource adequacy under deep de-carbonization pathways for California, Energy & Environmental Economics, June 2019.)

Specifically, Calpine wanted to know if California could afford to essentially phase out *all thermal* plants. California's Senate Bill 100 (SB100) is the main driver of the state's greenhouse gas reduction targets.

E3's modeling concluded that all is more or less manageable up to a point. After that the going gets progressively tougher as balancing the variable load and demand becomes more challenging *and* more expensive.

The study also concluded that the goal of SB 100, which is 80% reduction in statewide greenhouse gas emissions from 1990 level by 2050 is achievable and not horrendously expensive so long as *some* thermal plants remain in the generation portfolio. Squeezing out the last remaining thermal plants is, however, expensive.

Continued

Box 1.2 Calpine says phase out of gas-fired peakers will be expensive—cont'd

That is among the issues policymakers in California need to address. At what point the total—both capital and operating—cost of operating the grid with little or no thermal generation becomes exorbitant?

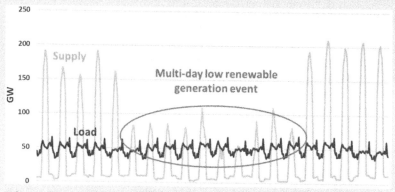

What to do when there is no sun or wind for an extended period? (*Source: Long-run resource adequacy under deep de-carbonization pathways for California, Energy & Environmental Economics, June 2019.*)

Calpine, whose combined cycle plants currently provide much needed *ramping flexibility* in California, argues that maintaining certain amount of gas-fired capacity will be the wise thing to do.

But how much? E3 examined two scenarios where between 17 and 35 GW of gas-fired capacity is maintained by 2050. The differences in costs—both capital and operations—are significant. In the High Electrification Scenario, going from a 25 to 10 GW gas-fired fleet would add $28 billion per year to costs; phasing out the entire fleet would add $65 billion as illustrated.

E3 looked at additional storage but found it to be expensive—at least with today's technology. Without more storage, there would be more renewable curtailment, wasting as much as 50% of the solar generation under some scenarios.

Referring to the E3 study, Matthew Barmack, Director, Market and Regulatory Analysis at Calpine said,

> The study illustrates that some continued reliance on gas generating capacity is cost-effective and consistent with the state's environmental goals.

One, of course, can argue that there are *other* options including better utilization of the inherent flexibility in demand, storing the excess renewable generation as hot water in tanks or ice, or precooling or preheating buildings and so on. Moreover, the intelligent charging of as many as 5 million EVs expected on

Continued

Box 1.2 Calpine says phase out of gas-fired peakers will be expensive—cont'd

California roads by 2030,[17] can absorb a lot of the excess energy during the belly of the duck curve. Even more can be done by charging the EVs with the excess solar energy in midday and discharging them during the evening peak demand.[18] Arid California will need a lot of desalination to meet its water demand—which can run when there is excess and cheap solar energy. Cheap and abundant renewables can also be used to extract carbon out of the atmosphere and store it underground, helping to meet the state's carbon neutrality goal. The list of options is long, including making renewable hydrogen or methane.

Then there is the promising potential of demand response—scarcely utilized to date—the topic of this book. Why not get demand to balance the variable generation rather than the other way around as further explained in Chapter 5?

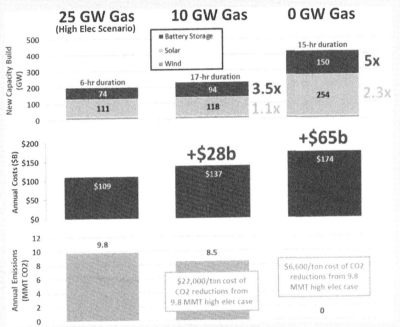

Main message: No gas, higher costs. (Source: Long-run resource adequacy, June 2019.)

[17] According to an article in *Micro-grid Knowledge*, by 2025 there may be as many as 3 million EVs in the United States that represent the equivalent storage capacity of 26 GWs. Couldn't California, currently home to roughly half of the national fleet of EVs or roughly 13 GW of virtual EV battery storage capacity, make better use of this underutilized resource?

[18] EV storage is covered in Chapter 8.

Continued

> **Box 1.2 Calpine says phase out of gas-fired peakers will be expensive—cont'd**
>
> E3's key findings were:
>
> *Natural gas generation capacity is currently the most economic source of firm capacity. The least-cost electricity portfolio to meet the 2050 economy-wide GHG goals for California includes 17–35 GW of natural gas generation capacity for reliability. This firm capacity is needed even while adding very large quantities of solar and electric energy storage.*
>
> One thing that everyone can agree on is that states like California will need more flexibility in both supply *and* demand in a future where virtually everything is electrified and virtually all generation is renewable.

More recent analysis by CAISO has identified the following additional challenges:[19]

- Meeting summer evening peaks;
- The increased ramping needs; and
- Covering potential renewable generation shortfalls during multiday weather events.

The challenge of meeting the rapid ramp rates is already acute and is projected to get worse. Fig. 1.8 shows the daily load profile for January 2019 when a 27 GW peak was experienced around 6:30 p.m. What is noteworthy, however, is not the peak—which is quite modest in January—but what it takes to meet it.

As illustrated, CAISO has to begin the evening ramp starting at 2:25 p.m.—this being winter in California. During a 3 hr period, it must secure 15,600 MW of replacement capacity to substitute for the setting sun, which on this particular day goes from providing 11% of the capacity to zero, which was replaced by:

- A 9% increase in imports—from 28% to 37%; and
- A 5% increase in gas-fired generation—from 30% to 35%.

---

[19] For example, refer to 19 Dec 2019 Briefing on post 2020 grid operational outlook, at http://www.caiso.com/Documents/BriefingonPost2020GridOperationalOutlook-Presentation-Dec2019.pdf.

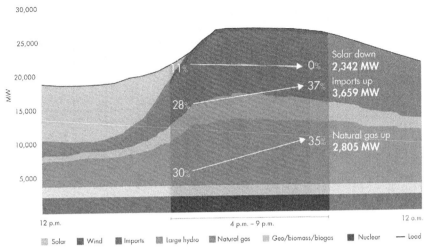

Figure 1.8 *CAISO load profile for Jan 1, 2019.*

CAISO notes two problems associated with the ramping:

- Relying on such a large contribution from natural gas—fired plants is counter to the state's target of reaching a low carbon electricity mix by 2045 as required by the SB100; and
- The state's continued reliance on imports/exports may not be sustainable since the neighboring states may not have excess capacity to offer when CAISO needs it, and vice versa.

The latest CAISO projections suggest that managing the daily cycles will continue to become more challenging by 2030 and beyond. As illustrated in Fig. 1.9, the 3 hr evening ramp may reach 26 GW by 2030, and who knows what beyond that date.

Adding to the list of concerns is what to do about the *extended* periods when the wind may not blow and the sun may not shine. What happens when little or no renewable generation is available for several days in a row? Such *multiday weather anomalies* are statistically rare but not improbable. Short duration storage will be woefully inadequate in such cases.

As shown in Fig. 1.10 solar production as a percentage of installed capacity varies greatly from day to day based on cloud cover from a low of 16% to a high of 71% for a typical week in January—a winter month with frequent overcast and/or rainy days. What happens in a week of little or no sunshine across much of California? How would CAISO fill the void?

Currently, CAISO manages solar—and to a less extent—wind's inherent variability by relying on natural gas plants and imports. As

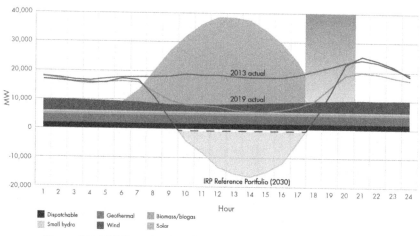

Figure 1.9 *CAISO projections for 2030 and beyond. (Source: CAISO.)*

illustrated in Fig. 1.11 this reliance varies from a normal week (on the left) to a multiple-day low solar week (on the right).[20]

To summarize, California's march toward a low carbon future has been relatively easy to date. But as the state runs out of the low-hanging fruit, it will get progressively more difficult and expensive.

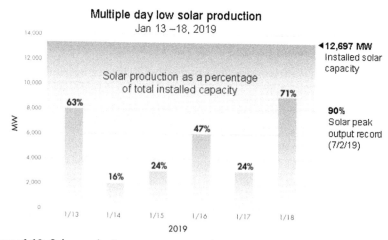

Figure 1.10 *Solar production as percentage of installed capacity.* Typical winter week, 13–18 Jan 2019. *(Source: CAISO.)*

---

[20] During multiday weather events storage batteries will not be able to recharge, further exacerbating the grid operator's ability to make it through.

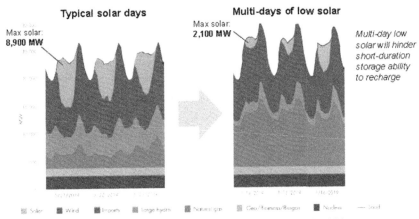

Figure 1.11 *Multiday weather events. (Source: CAISO.)*

4. Future role of flexible demand

As already explained, CAISO, not unlike other grid operators featured in subsequent chapters, is facing growing challenges for which the traditional tools and methods are operationally less practical, less effective, more expensive, or a combination of the above.

Ramping ever more peaking plants to compensate for the growing evening period after the sun goes down is expensive and heavily polluting. Nor are they compatible with a cleaner or greener future.

Storage is an option, and as the technologies improve and the costs fall, it will gain increased traction. But there are limits to how much storage can economically contribute. Moreover, while storage may work well for shifting limited generation over relatively short periods of time—say 2—5 hrs—it is not well suited to shift huge amounts of generation for very long periods, say from spring to summer as explained in Box 1.1.

Pairing renewables, particularly solar, with storage is gaining popularity for a number of obvious reasons:

- The value of solar generation in many networks during the sunny hours is steadily falling as excess generation floods the market resulting in low or negative prices. There is diminishing value in adding more solar capacity to a network such as California unless some of the generation can be stored for delivery in the evening hours when prices tend to be high, making solar + storage profitable;
- Adding storage with flexible capabilities generates additional revenue streams such as allowing the combined entity to participate in ancillary services market, providing frequency control and regulation; and

• Storage, especially batteries and flywheels, can respond to command signals much faster and more accurately than gas peakers or anything else. This makes them ideal for managing frequency and stability and offering balancing services that are increasingly critical to reliable operation of grids.

When a grid such as the CAISO faces frequent imbalances in load and generation, exporting/importing excess/deficit generation to neighboring regions is an obvious solution. But even this strategy can only go so far due to the limitations in the capacity of the transmission lines as well as line losses when transmitting power across long distances. Building new transmission lines takes a long time and is expensive.

Curtailing renewable generation when in excess of load can be used as a last resort but is considered a waste of green electricity.

This leaves flexible demand as a potentially cost-effective and largely unexplored option. As further explained in other chapters of this book, there is so much ingenuity and effort going into finding ways to aggregate, manage, and deliver flexible demand using technology and software that was previously not available, practical, or profitable. This is not to say that flexible demand is a panacea for addressing the many challenges of variable generation, but it is surely an attractive opportunity worthy of further exploration. As further explained in Chapter 5, flexible demand can play a critical role in two ways:

• By soaking up some of the excess generation when there is surplus; and
• By shifting some load away from peak demand periods.

Neither option has been widely explored to date.

5. Conclusions

The well-known and widely publicized challenges facing CAISO are being replicated to varying degrees in other parts of the world. As the percentage of renewables rises above a certain threshold in the generation mix—we can debate what that threshold may be—the variability, and to a lesser degree the predictability, becomes problematic for the grid operator. While the severity of the challenge varies from one system to another, and while the tools and options available also vary, there are significant similarities across many systems, topics covered in the following chapters.

The challenges, broadly speaking, include managing daily and seasonal cycles of feast and famine characteristic of solar and, to a lesser degree, wind generation which frequently result in overgeneration and deficits, which must be compensated by other types of flexible generation or *demand*.

CHAPTER 2

Variability of generation in ERCOT and the role of flexible demand

Ross Baldick
Department of Electrical and Computer Engineering, University of Texas at Austin, Austin, TX, United States

1. Introduction

Increase in the use of renewables is an important contributor to decarbonizing the electricity system. However, the temporal endowment of solar and wind resources, meaning the distribution over time of the resource availability, must be considered. Just as countries adapted to their fuel endowment, there is a need to adapt to the temporal endowment by:

1. developing those renewable resources that best match patterns of load, while
2. exploring ways to increase the flexibility of the timing of consumption of electricity; that is, changing the patterns of load to match the variation and intermittency of renewable resources.

At high enough penetrations, however, integration of wind and solar poses various challenges for every electricity system. For example, uncontrolled rooftop solar photovoltaics are already a challenge in California, as further explained in Chapter 1, and are likely to become so in Australia. Coping with reduced system inertia due to decommitment of thermal generation under high levels of renewables is a particular challenge for smaller interconnections such as that of Australia, Ireland, and the Electric Reliability Council of Texas (ERCOT), each of which are virtual electricity islands.

As another example of a challenge, electricity markets with relatively long dispatch intervals must procure additional ancillary services to cope with increased intrainterval variability due to renewables, resulting in both increased costs of ancillary services procurement and typically worsened dispatch efficiency. The inefficient treatment of transmission in zonal dispatch models such as the European Union tends to result in more curtailment of renewables, higher thermal dispatch costs, or the need for

Variable Generation, Flexible Demand
ISBN 978-0-12-823810-3
https://doi.org/10.1016/B978-0-12-823810-3.00010-8

25

more transmission capacity than in a nodal system. The difficulty with building new transmission in most regions limits access to remote renewables.

Any one of these or other issues could challenge deep decarbonization of a particular region's electricity system. However, the most fundamental and endemic challenge to all systems with high levels of renewables is the variability and intermittency of renewable production. Average patterns of the variation in production of renewable production over time do not perfectly match current temporal variation of load, while sample paths of renewable production exhibit randomness due to weather. Both issues must be faced: the deviation of average patterns of variability of renewable production from patterns of load over time, and the intermittent fluctuations of renewables. For brevity in this chapter, the word "variability" will be used primarily to refer to the first issue relating to average patterns of renewable production, while "intermittency" will be used to refer to the second issue of random fluctuations of renewable production from the average patterns of production. Moreover, "patterns" of load will refer to the temporal character of load generally.

Demand response and flexible demand will be defined in detail below but should be construed as adaptations on the demand side to help match supply and demand. This chapter will consider the role of demand response and flexible demand in accommodating variability and intermittency, arguing that increased flexible demand, including changing the patterns of load, will be a major part of cost-effective integration of very high penetration levels of renewables. Most electricity systems have historically utilized at least some forms of demand response, whether to adapt typical patterns of load to economical supply, as in off-peak water heating, or to respond to contingencies or emergencies; however, the requirements for renewable integration for deep decarbonization are unprecedented and will correspondingly require much greater levels of flexible demand for renewable integration. For example, with many tens of GW of installed renewable capacity and given the poor prospects currently for carbon sequestration (Cembalist, 2020), there would be a need for tens of GW of nonthermal resource, whether storage or demand response, to compensate for variability and intermittency of those renewables across various time-scales. In brief, the chapter will argue that since we cannot change the variability and intermittency of renewables, we should consider the options for changing the patterns of load by making demand more flexible and active.

This chapter will explore the possibilities for flexible demand to compensate for renewable variability and intermittency, primarily at medium timescales, with ERCOT as the focus of discussion because of the absence of existing storage options such as pumped storage hydroelectricity (PSH) and ERCOT's relatively low current wholesale electricity prices. That is, ERCOT poses a significant challenge because of the lack of cost-effective storage options and the relatively low wholesale prices, meaning that demand response options compete in a low-cost environment and typically with few subsidies.

The remaining sections are as follows:

- Section 2 provides background on renewables, variability and intermittency, storage, and demand response, particularly focusing on ERCOT.
- Section 3 presents a brief description of the ERCOT market, including relevant history, and trends relating to renewable integration.
- Section 4 describes fuel endowments and introduces the concept of temporal endowments, meaning the distribution over time of the renewable resource, to describe the challenge of high penetrations of renewables.
- Section 5 details various types of flexible demand that can facilitate renewable integration through adapting to the temporal endowment, followed by the chapter's conclusions.

2. Background on renewables, variability and intermittency, storage, and demand response

The ERCOT region, shown in Fig. 2.1, and its characteristics will be used as a concrete example to illustrate issues with renewables, storage, and demand response, with reference to other jurisdictions as appropriate. The ERCOT region consists of about 75% of the land area of Texas and represents 90% of Texas electric customers. ERCOT is the smallest of the three AC interconnections in the United States, with the other interconnections being the Western and Eastern.

ERCOT has a small amount of DC transmission capacity on ties that allows interchange of power with the Eastern Interconnection and, together with a variable frequency transformer, with Mexico. The total capacity of these ties to other interconnections is around 1.3 GW (Mele, 2018), which is a small fraction of ERCOT peak load of around 75 GW (ERCOT, 2020). Proposals to build significant additional capacity between the interconnections have not come to fruition. That is, in the absence of

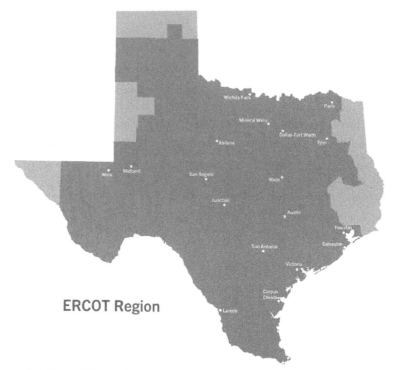

Figure 2.1 Map of Texas showing ERCOT region. *(Source: ERCOT, 2020. Fact Sheet, Available from: http://www.ercot.com/content/wcm/lists/197391/ERCOT_Fact_Sheet_5. 11.20.pdf. (Accessed 22 May 2020).)*

significant build-out of transmission capacity to other interconnections, ERCOT must balance its demand and supply, including supply of renewables, almost completely without assistance from other regions and without PSH.

Currently, most renewable resources in ERCOT are wind, with most wind located in West Texas. Fig. 2.2 shows, as of 2018, the distribution of peak load, generation capacity, and wind generation capacity in each of four load zones, West, North, South, and Houston. Over the coming years, significant growth in solar is expected, greatly expanding the current total capacity of rooftop and grid-scale solar in ERCOT of around 2 GW.

As of early 2020, ERCOT has multiple GW of various types of demand response and flexible demand (Potomac, 2019). Several GW of flexible demand in ERCOT is activated by excursions of wholesale prices, which occasionally rise to several thousand dollars per megawatt-hour

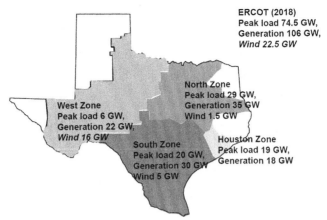

ERCOT (2018)
Peak load 74.5 GW,
Generation 106 GW,
Wind 22.5 GW

North Zone
Peak load 29 GW,
Generation 35 GW
Wind 1.5 GW

West Zone
Peak load 6 GW,
Generation 22 GW,
Wind 16 GW

South Zone
Peak load 20 GW,
Generation 30 GW
Wind 5 GW

Houston Zone
Peak load 19 GW,
Generation 18 GW

Figure 2.2 ERCOT zones and distribution in 2018 of generation, wind, and peak load. *(Source: Lee, D., Ross B., 2020. Wind Variability and Impact on Markets, Available from: http://users.ece.utexas.edu/~baldick/papers/GDFM.pdf. (Accessed 4 June 2020), based on data from Potomac Economics, 2019. 2018 State of the Market Report for the ERCOT Wholesale Electricity Markets, Available from: www.potomaceconomics.com, (Accessed 18 September 2019).)*

(Silverstein, 2020; Baldick et al., 2020). In the "old world" of primarily thermal generation, this level of flexible demand would have been sufficient to compensate for variations in supply due to the lumpiness of thermal generation investment, unit commitment, and typical outages. In principle, such a level of demand response allows for a relatively "pure" version of an energy (and reserves)-only market without regulatory interventions, as is currently practiced in ERCOT, while avoiding involuntary curtailment except in extraordinary conditions.

In particular, when generation supply is inadequate to fully serve all requirements for energy and reserves then the price in an energy-only market should "clear on the demand side," reflecting marginal willingness-to-pay of consumers, rather than marginal generation costs, under scarcity of supply.[1] Efficient levels of such prices are generally considerably higher than marginal generating costs, and these high prices might occur for a few tens of hours in an average year, resulting in rents during these periods above generator production costs that remunerate generator capital investment.

---

[1] It should be understood that the lack of explicit price formation through demand-side bidding in many current markets prevents direct observation of the marginal willingness to pay.

The availability of several GW of demand response reduces the likelihood of involuntary demand curtailment without the need for a large reserve margin and despite mismatches over time between load and the hypothetical perfectly adapted level of generation capacity. However, it should be recognized that the ERCOT market design, to be discussed in more detail in Section 3, has only relatively recently incorporated features such as an "operating reserve demand curve" (ORDC) that seek to approximate efficient pricing under scarcity (Hogan, 2012). The ORDC supplanted the previous ad hoc approach that tolerated exercise of market power by smaller market participants (Baldick et al., 2020). In recent years, ERCOT has managed tight supply conditions through demand response and without involuntary curtailment (Potomac, 2020).

The advent of significantly greater variability and intermittency under very high renewable penetration challenges the energy-only paradigm because the mismatch over time between demand and supply will tend to be much larger due to the variability and intermittency. As observed in the Introduction, this mismatch might be tens of GW, which is far more than could be accommodated by existing demand response in ERCOT. This chapter argues that an energy-only market such as ERCOT will require much more significant flexible demand and/or storage compared to historical levels in order to compensate for the variability and intermittency under deep decarbonization. The variability and intermittency has multiple timescales, which will be roughly divided into:

- Short, which will be shorthand in this chapter for subsecond to minutes,
- Medium, which will be used to mean from minutes to several days, and
- Long, which will be used to mean longer than several days.

Occasionally, "very long" will be used to refer to the seasonal to multiyear timescale.

Demand response and various types of storage have cost structures and features that make them most suited to particular timescales and applications. For example, chemical battery storage is currently competitive with existing thermal generation in the provision of services that compensate for variability at short timescales. This includes the provision of frequency regulation services and may also be the case in smoothing short-term fluctuations in renewable production. Chemical battery storage may also be a cheaper alternative to some transmission and distribution system build-out (Chang et al., 2014; Sue et al., 2014).

In contrast, chemical batteries are relatively expensive at current and near future costs for compensating renewable fluctuations at medium and long timescales in, for example, energy arbitrage. A simple calculation shows the reason given the following assumptions:

- Forecast 2023 lithium-ion battery costs of around $100 per kWh of energy capacity (Henze, 2019),
- An assumed lifetime equivalent to about 15 years of daily round-trip charge and discharge cycles (Cole and Frazier, 2019) to compensate for medium timescale renewable variation and intermittency, and
- Ignoring finance costs.

With these optimistic assumptions, the average cost of charging and discharging a lithium-ion battery in terms of use of its lifetime is around $20 per MWh of energy stored for medium timescale storage. This does not include the cost of the energy to charge the battery and just reflects the capital cost of the battery. If chemical batteries were instead used to compensate for long timescale variation and intermittency, then far fewer charge-discharge cycles would be possible over the life of the battery. Hledik et al., 2020, highlights the case for 4 days of storage; however, as an extreme, if chemical batteries were used as seasonal storage (for example, charging and discharging four times per year with a 25 year lifetime) then there might be as few as 100 charge—discharge cycles over the lifetime, resulting in a cost per charge—discharge cycle of around $1000 per MWh of energy stored.

This range of costs from $20 to $1000 per MWh of energy stored compares to a typical wholesale electricity cost in ERCOT of around $35/MWh, which, given the ERCOT energy-only market to be discussed in more detail in Section 3, is a rough estimate of the average total operating and annualized capital costs of generating electricity in ERCOT. This implies that storing a large fraction of energy produced by renewables in lithium-ion batteries to help match production to load at medium and long timescales would result in a significant increase in the wholesale cost of resources to provide electricity in ERCOT. Conversely, steep reductions in battery costs would be necessary to make storage viable for medium and long timescales (Ziegler et al., 2019). Lower cost alternative battery chemistries to lithium-ion are unlikely to change this basic observation in the absence of significant reductions in manufacturing costs and materials requirements.

While it is true that wholesale energy costs may only form less than half of the total for residential retail bills, storing a significant fraction of the electricity produced in ERCOT would significantly increase (unsubsidized) delivered prices in the commercial and industrial sector given current and foreseeable battery costs. It is, however, acknowledged that the level of residential retail prices in other states, particularly California and Hawaii, together with Federal and State subsidies, may make storage and other innovative products, such as the Tesla Virtual Power Plant, competitive with retail purchased electricity in those jurisdictions for compensating medium timescale variability and intermittency (Tesla, 2020). Furthermore, in regions such as California, Hawaii, Australia, and parts of Europe, where wholesale prices are significantly higher than in ERCOT, the cost of chemical battery storage for medium timescales may involve a relatively smaller increase in delivered prices and may be tolerable. See, for example, Hledik et al. (2020), for an analysis of California, which highlights that California has plans to increase battery capacity to several GW, in a system with peak load around 50 GW.

Even in relatively lower wholesale cost jurisdictions such as ERCOT, some developers are including storage with renewable projects, reflecting in part continued Federal subsidies (Sixel, 2020); however, similarly to California the total anticipated battery capacity in the coming years in ERCOT is a small fraction of total peak load. Nevertheless, as will be discussed in Section 3, there may be a role for more chemical battery storage in ERCOT to provide for occasional renewable "droughts," while chemical batteries also have a role in backup power for such applications as lighting.

Eventually, it may also be possible to repurpose chemical batteries for stationary storage that have been used in electric vehicle applications. That is, batteries that have reached their useful lifetime in electric vehicle applications may be available at costs that are much lower than new batteries and still be usable for stationary energy storage, albeit with degraded capacity or performance compared to a new battery. It should, however, be understood that this resource will not be available for many years. For example, total ERCOT electrical energy in 2019 was 384 TWh (ERCOT, 2020), for a daily average energy of around 1 TWh. The total number of electric vehicles in the United States in late 2019 was approximately 1.4 million (Wikipedia, 2020). Assuming that each vehicle has a battery pack on the order of 75 kWh and that essentially all of this energy capacity will still be available at end of vehicle life, then the total US fleet would provide about 0.1 TWh of storage at end of life. That is, repurposing the batteries from the *whole* of the current US electric vehicle fleet would still only provide for storing about 10% of the daily energy in ERCOT. Recognizing that the majority of existing electric vehicles are under 5 years old then, even with significantly increased adoption of electric vehicles, it will take many years before a large quantity of used electric vehicle batteries are available for stationary storage applications.

To summarize, there may be significant potential for battery storage in ERCOT and other jurisdictions to provide ancillary services, short time-scale services, some combined renewable and storage projects, and for other particular roles including backup power. Moreover, so-called "stacking" of services may allow for more cost-effective arbitrage at medium timescales, particularly if the storage is used to avoid transmission and distribution expansion costs (Chang et al., 2014). However, the cost of chemical battery storage is currently too high to realistically be used to fully smooth renewable fluctuations over medium and long timescales. More economical alternatives such as batteries repurposed from electric vehicles are not currently available in sufficient numbers. It will take many years, perhaps decades, before chemical battery storage is cost-effective for fully smoothing renewable fluctuations at medium and long timescales.

On the other hand, another type of storage, namely PSH, is more cost-effective for the medium and long timescale in those regions with suitable topography and water resources. A key property of PSH, which is also shared by some types of demand response and flexible demand, is that the capital cost of energy storage capacity, or its equivalent, is relatively low, while the number of charge–discharge cycles is effectively unlimited. This makes medium and long timescale PSH storage much cheaper than chemical batteries. Indeed, jurisdictions with hydro will have a significant resource to facilitate renewable integration and, for example, Australia is intending to significantly expand its existing stock of PSH (Snowy Hydro, 2020), although Elliston et al. (2012) suggests that additional storage associated with concentrating solar thermal generation would also be valuable for deep decarbonization in Australia. The Eastern and Western Interconnections of the United States, the European Union (see Chapter 4 for the case of Spain), Brazil, and several other countries also have PSH resources that can, in principle, compensate for renewable variability and intermittency at medium and long timescales. Countries with large amounts of PSH can use it to balance wind and solar variability and intermittency.

However, ERCOT has no existing PSH and, because of topography, no reasonable option for new PSH. Absent order-of-magnitude reductions in the cost of chemical storage, or development of other storage technologies such as compressed-air energy storage, concentrating solar thermal, or transmission access to regions with PSH, ERCOT will need much deeper demand response or face much greater costs (that is, including the costs of subsidies) to move toward deep decarbonization. Indeed, regions such as ERCOT without PSH or other long timescale storage options may need to maintain a significant amount of, for example, natural gas generation capacity in order to cost-effectively compensate for occasional extended wind and solar "droughts" or may need to accept the cost of chemical battery storage to achieve deep decarbonization. Elliston et al. (2012) finds that even in a region such as Australia with PSH, it may be necessary to use some gas during such "droughts." Moreover, in the United States as a whole, only about 1% of current electricity generation is stored, far less than, for example, energy generated currently from peaking plant (Cembalest, 2020). The implication is that even in the Eastern and Western Interconnections of the United States there may not be enough long timescale storage options such as PSH to compensate without thermal generation for wind and solar "droughts."

Even for those regions with access to PSH and even with continued decreases in the cost of chemical battery storage, this chapter will argue that demand response options merit consideration as part of the solution to the challenges of deep decarbonization, particularly in compensating for wind and solar variability and intermittency at medium timescales. Such demand response would need to go far beyond existing levels in ERCOT and other similar jurisdictions and far beyond what was sufficient for an energy-only market in the "old world" of primarily thermal generation. The next section will describe the ERCOT market in detail to illustrate the challenges it faces with high renewable integration.[2]

3. The ERCOT market

The ERCOT market is operated by the ERCOT Independent System Operator (ISO). The ERCOT ISO is a nonprofit corporation that manages wholesale electric power for approximately 26 million Texas electric customers in the ERCOT region.

Restructuring of the Texas electricity industry began in the late 1990s, pursuant to Texas state legislation, as implemented by the Public Utility Commission of Texas (PUCT), the regulatory body overseeing the electricity market. Open access wholesale electricity markets have existed in the ERCOT region for over two decades, and the ERCOT ISO began operating a day-ahead wholesale scheduling process and "balancing

---

[2] This section is drawn from ERCOT, 2020; Silverstein, 2020; Baldick et al., 2020.

market" in 2001 to facilitate wholesale competition. Retail competition (except for municipal utilities and rural cooperatives) was opened in 2002 (Baldick et al., 2020). Today, the ERCOT wholesale and retail markets in ERCOT are relatively robust and competitive, with transmission constraints represented explicitly in wholesale prices that vary geographically and temporally (Potomac, 2019; Baldick et al., 2020).

In contrast, the initial implementation of the ERCOT wholesale electricity market had no representation of the transmission network in its "commercial network model," so that, at any given time, there was only a single price throughout ERCOT in the balancing market that ERCOT operated. ERCOT used side payments, not reflected in balancing market prices, to compensate generators for redispatching compared to their schedules and their balancing market positions in order to prevent violations of transmission constraints. The costs of those payments were then charged to loads (Baldick and Niu, 2005). That is, the cost of resolving the transmission constraints resulted in a payment from load to generation that was not represented in wholesale prices.

Due to excessive side payments that were on the order of hundreds of millions of dollars per year in a market size of about $2 billion per year (Baldick and Niu, 2005), the commercial network model was changed in 2002 to consist of four (and sometimes five) pricing zones with a flow-based transmission constraint model for exchange of power between zones. The typical configuration consisted of four zones, West, South, North, and Houston, corresponding to geographical regions of Texas. The zonal model provided broad price signals reflecting transmission limitations between the zones (Baldick et al., 2020). Fig. 2.2 shows the zones that were in place in 2010.[3]

In the zonal market, ERCOT used side payments to pay generators to redispatch to keep intrazonal transmission flows within limits. This was analogous to the arrangement for the whole of ERCOT in the original single price design, but in the zonal market the redispatch was primarily required only for intrazonal transmission flows. However, due to various problematic issues with the zonal market, including still excessive side payments (see, for example, (Baldick, 2003), among many critiques), a decision was taken in 2004 by the PUCT to adopt a nodal representation for the commercial network model (Baldick and Niu, 2005).

---

[3] Aside from Municipal Utilities such as Austin Energy, these zones also form the basis of the current "load zones" in ERCOT, which will be discussed further below.

Implementation of the nodal market was an extended process and it was not operational until December 2010.

Today, similar to other US ISOs, ERCOT operates a day-ahead locational marginal pricing (LMP) market, which is a combination of a financial short-term forward market, together with some "physical" characteristics, particularly for commitment of generators. It also operates a real-time LMP market, with a 5 min dispatch interval that sets dispatch targets for generators to ramp toward. Deviations in real-time from day-ahead forward energy positions are charged at the real-time prices. There is also a market for transmission hedging products, called "congestion revenue rights" (CRRs). Unlike other US restructured markets, however, there is no longer-term capacity market nor any obligations on retailers to procure minimum levels of capacity. The implications of this lack of a capacity market and retailer obligations on reserve margin will be discussed later in this section.

As mentioned above, the ERCOT market is often evaluated as being highly competitive, both at the wholesale and the retail level (Potomac, 2019; Baldick et al., 2020). Wholesale prices, particularly, have been consistently low on average over multiple years, despite occasional high prices and an offer/price cap of $9000/MWh, which is the highest by far in the United States (Potomac, 2019). The high offer cap is, in principle, high enough for remuneration of capital costs under scarcity conditions from the proceeds of payments for energy and ancillary services alone.

Turning to capacity, there is over 100 GW of transmission-connected generation capacity in ERCOT, consisting, as illustrated in Fig. 2.3, primarily of natural gas (around 53% of capacity), wind (around 23%), coal (around 15%), nuclear (5%), and solar (2%) resources. By energy, the production in 2019 was: natural gas (47%), wind (20%), coal (20%), nuclear (11%), and other, including solar (2%). The contribution of coal has decreased in recent years, while the contribution of wind has grown significantly. For example, both coal and wind contributed 20% in 2019.

Figure 2.3 Generating capacity and energy in ERCOT. *(Source: ERCOT, 2020. Fact Sheet, Available from: http://www.ercot.com/content/wcm/lists/197391/ERCOT_Fact_Sheet_5. 11.20.pdf. (Accessed 22 May 2020).)*

The mix of energy will be further discussed in the context of fuel endowment in Section 4.

Of the over 100 GW of installed capacity in ERCOT, about 82 GW is expected to be available for summer peak conditions (ERCOT, 2020). The difference between installed and expected available for summer is mostly because inland wind capacity is close to its minimum availability during summer afternoons when ERCOT load peaks. As mentioned in Section 2 and illustrated in Fig. 2.2, most current ERCOT wind capacity is inland and located in the West zone.

Using expected available capacity during summer as the measure of generation capacity, the reserve margin of generation capacity above peak load less interruptible and curtailable load in ERCOT is around 10%. Under this measure, ERCOT has the tightest reserve margin of any of the US ISOs, reflecting, to a large degree, the lack of capacity market or other capacity obligations that typically enforce much higher amounts of installed thermal generation capacity relative to peak load in other US jurisdictions. (Regions with significant hydroelectric capacity also tend to have high reserve margins.) As mentioned in the Introduction, however, the significant amount of price responsive demand response in ERCOT has served to modulate demand during times of tight supply—demand balance in ERCOT without recourse to involuntary curtailment except during, for example, rare extreme weather conditions associated with multiple correlated generator failures (Silverstein, 2020).

As mentioned in the Introduction and in contrast to the other US interconnections, ERCOT has very little hydroelectric resource and no PSH that might facilitate renewable integration. It is somewhat remarkable then that ERCOT has a far higher penetration of wind power than either of the other US interconnections, with around 20% of ERCOT energy provided by wind (ERCOT, 2020). This is considerably higher than the overall wind production in the United States, at around 7.2% (AWEA, 2020), and higher than the overall wind production in the EU, at around 11% (ENTSO-E, 2019).[4]

---

[4] It is acknowledged that some countries within the EU, such as Germany, have higher fractions of their energy produced from wind than ERCOT (Appunn et al., 2020). It should be understood, however, that each individual country such as Germany forms only a small part of the synchronous interconnection of Western Europe. Similarly, West zone wind produces over 100% of the load energy in West Texas and forms only a part of the ERCOT interconnection.

As shown in Fig. 2.2, most of the current wind capacity in ERCOT is located in the West zone, which historically had less thermal generation capacity than its peak load and therefore imported electricity on transmission lines from other regions of ERCOT (Littlechild and Baldick, 2020). The advent of significant wind meant that imports tended to be decreased due to local wind production. By 2007 wind had increased so much that there were typically significant binding limitations on *exports* from the West zone. The Texas legislature ordered an expansion of transmission capacity and the resulting "Competitive Renewable Energy Zone" (CREZ) transmission expansion is perhaps the largest single expansion of transmission capacity completed anywhere in the world (Littlechild and Baldick, 2020). Fig. 2.4 shows a map of the CREZ transmission lines. Comparing to Fig. 2.2, the geographical region

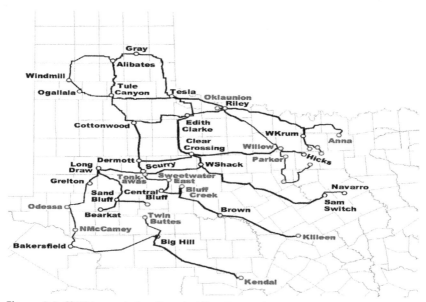

Figure 2.4 CREZ transmission lines in West Texas. *(Source: Bill Blevins, 2012. Voltage Support Service (VSS) for CREZ Region and Baseline Power Angle Ranges Post CREZ, ERCOT, Available from: http://www.ercot.com/services/training/archives. (Accessed 4 June 2020).)*

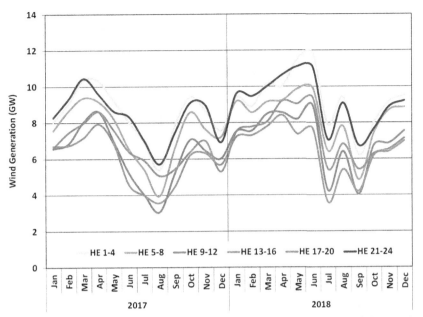

Figure 2.5 Distribution of ERCOT wind generation by month and hour of the day for 2017 and 2018. *(Source: Potomac Economics, 2019. 2018 State of the Market Report for the ERCOT Wholesale Electricity Markets, Available from: www.potomaceconomics.com. (Accessed 18 September 2019).)*

illustrated in Fig. 2.4 roughly overlaps the West Zone (shown yellow) and the Panhandle (shown in white to the North of the West Zone) in Fig. 2.2.[5]

As illustrated in Fig. 2.5, the wind resource in West Texas, however, tends to blow strongest at night and during the Autumn, Winter, and particularly the Spring.[6] That is, as mentioned above in connection with capacity to meet the peak, West Texas wind production has variability that is not correlated with the patterns of load (Lee and Baldick, 2020; Silverstein, 2020). To understand the implications, consider the concept of net load, the difference between load and wind, solar (and run-of-river hydro) production, and discussed in Chapter 1. Net load must be provided by

[5] Although the CREZ transmission extends outside of the area of ERCOT shown in Fig. 2.2, it is not synchronously connected with the existing Eastern Interconnection lines in that region.

[6] Fig. 2.5, taken from Potomac (2019), shows total wind in ERCOT, but as illustrated in Fig. 2.2, most wind farm capacity is currently located in the West zone, so the distribution shown in Fig. 2.5 primarily represents the variability of West Texas wind.

other resources; that is, by thermal generation or storage. With lack of correlation between West Texas wind and load, increasing penetration of West Texas wind tends to change the statistics of net load by decreasing its minimum significantly, while not reducing its maximum by much (Baldick, 2012; Potomac, 2019). The implication is that the capacity of thermal plus storage to meet the peak of net load will be relatively independent of West Texas wind capacity; however, the energy in the net load will tend to decrease with increasing wind capacity.

The change in the characteristics of net load has both technical and financial implications. First putting aside the implications of intermittency, the technical implications of wind variability include the need on a daily or weekly basis to decommit generation during low net load periods, while maintaining the ability to ramp up when net load increases. As an example of this in the context of solar production, and as discussed in Chapter 1, California coined the shape of its daily net load the "duck curve" and faces difficult minimum net load and net load ramping conditions, particularly in Spring.

Currently, the effect of West Texas wind on ERCOT daily net load variation has not been as problematic as solar in California. There are several reasons for this, including that solar production varies with solar irradiance, which tends to change in a highly correlated fashion across multiple solar installations, whereas wind farms tend to be somewhat less correlated, resulting in changes in total solar production often ramping more rapidly from hour to hour than typical changes in total wind production. Additionally, ERCOT has considerably more recently built and fairly flexible combined cycle gas turbines to cope with ramps in net load.

Nevertheless, significantly increased solar capacity is expected in ERCOT in coming years. While this suggests that there may eventually be similar issues to California, there are several crucial differences between ERCOT and California solar. Perhaps first and foremost, much of the solar capacity in California currently is uncontrolled rooftop capacity. This solar generation has been built pursuant to a combination of explicit state policy together with very high retail rates that encourage self-generation. The significant solar generation during the middle of the day drives wholesale prices very low or even negative, without direct financial implications for the owners of rooftop solar. This decoupling of effective prices for solar generation (whether due to net metering or explicit feed-in tariffs) from its wholesale market value is a driving force of California's difficulties with solar, and similar effects are evident in Australia and in Europe under feed-in tariffs.[7] In contrast, in ERCOT outside of Austin and San Antonio, direct policy incentives for rooftop solar are limited and, moreover, retail prices are relatively low implying that small-scale rooftop solar is unlikely to be installed in preference to large-scale solar developments.

A second difference is that because much of the new solar capacity in ERCOT will be large-scale, it will be connected at transmission level, exposed to wholesale prices, and therefore will not choose to generate if it exposed to unfavorable wholesale prices. Even if renewable resources are financially hedged against unfavorable wholesale prices, all such transmission-connected generation in ERCOT is required to offer into the market and be dispatchable. In the case of renewables, this means that it is required to be dispatchable down from the generation level that would be possible at a given time due to the renewable resource. The implication is that if steep ramps become problematic for ERCOT, then it is likely that ERCOT look-ahead dispatch, when implemented, will be able to reduce the slope of the ramps up and down through explicit dispatch instructions. This dispatchability is not possible for uncontrolled rooftop solar such as prevails in California, although this could be remedied for new solar rooftop installations through policy changes. The third difference is that, as mentioned above, ERCOT has relatively more flexible thermal capacity.

In addition to issues at the medium timescale including daily variation, as discussed above for wind and solar, renewable variation can also be problematic at long timescales. For example, West Texas wind poses challenges on a seasonal basis, since minimum wind production in Summer corresponds to maximum electrical load, and maximum wind production in Autumn, Winter, and particularly Spring corresponds to lower levels of load. Analogously, in the Western United States the Spring snow melt can result in excess hydro production that is not well correlated with load. These long timescale mismatches of variability of renewable supply to load further complicate the short and medium timescale mismatches. In regions, such as in the Western and Eastern Interconnections, Australia, and Europe, that have existing PSH, some of the available capacity can potentially be utilized for seasonal storage, expanding on current practices. As discussed in

---

[7] High feed-in tariffs may be argued to be a proxy for carbon pricing, but in the absence of industry-wide carbon prices they nevertheless distort remuneration compared to wholesale prices.

Section 2, this is not an option for ERCOT, except (possibly) through expanded transmission interconnection to the Western or Eastern Interconnections.

In the absence of existing medium and long timescale storage, there are therefore limits to decarbonization through expansion of West Texas wind. However, not all ERCOT wind is negatively correlated with load. In particular, wind farms that are near the Texas Gulf Coast in the South zone have much better correlation of production with load both on a daily and seasonal basis. Moreover, solar production in ERCOT is better correlated with load and is complementary to West Texas wind production (Silverstein, 2020).

There is currently far more West Texas wind capacity than either South Zone wind capacity or solar capacity. However, as will be discussed in Section 4, the advent of increasing amounts of solar capacity are likely to be beneficial in better matching variability of renewable production to the patterns of load at medium and long timescales (Silverstein, 2020). Adding to the three differences listed above between ERCOT and California, a fourth difference is that solar production tends to be better, albeit imperfectly, aligned with air-conditioning load, which is a large part of electricity consumption in ERCOT in Summer, meaning that solar production in ERCOT is inherently a better match to load than in California.

With increasing renewable production, the less energy is needed from thermal production. Together with persistent reductions in the market prices of natural gas due to the widespread development of "unconventional" gas resources, this has had the effect of reducing the operation of and, consequently, the operating profits of coal plants. Several coal generators have exited the ERCOT market in recent years because they were no longer financially viable (Silverstein, 2020). Although a similar pressure applies to the nuclear assets in ERCOT, they have apparently been able to maintain financial viability, meaning that operating profits are at least positive, providing low carbon production of electricity at a high capacity factor.

Besides the variability of renewable production, intermittency of renewable production also presents problems at various timescales. At short timescales, the implications include the potential need for increased frequency regulation ancillary services. Interestingly, a combination of the shift from 15 to 5 min dispatch intervals concomitant with advent of the nodal market in 2010, together with many other detailed market design changes over the years, means that short-term fluctuations due to wind intermittency have not proved problematic for ERCOT despite the major increase in wind resources (Andrade et al., 2018). Other US markets have already or are transitioning to 5 min dispatch intervals, and this change would be advantageous for renewable integration generically. The other market changes described in Andrade et al. (2018) relate to particular features of the ERCOT market, but analogous changes may also be helpful in other markets to compensate for increased intermittency at short timescales.

Variability and intermittency at medium and long timescales have nevertheless combined on occasions to result in the inability to utilize all of the available West Texas wind, underlining the observation made above that there are limits to decarbonization through integration of West Texas wind. As mentioned above, in ERCOT, all wind is effectively offered into the wholesale market and offer prices are typically low, zero, or negative, with negative offer prices reflecting the effective negative marginal cost due to Federal Production Tax Credit (PTC) subsidies (Baldick, 2012). Consequently, curtailment of wind below its potential production is concomitant with zero or negative prices in what we might call "economic curtailment," but simply reflects the outcome of a competitive market. If the reason for such economic curtailment is due to limits on export from the West zone then the zero and negative prices are generally localized to buses in that zone. However, when the issue is due to low overall load in ERCOT at a time of significant wind production then zero or negative prices can prevail throughout ERCOT (Baldick, 2012).

These two situations can be illustrated in terms of a price-duration curve, which shows the observed prices ordered from highest to lowest over a sample period, typically a year. Fig. 2.6 shows a truncated version of the price-duration curve for the four main ERCOT load zones for hours in 2018. Only the highest priced and lowest priced hours are shown. In approximately 30 of the hours of 2018, prices were negative throughout ERCOT, implying that load was low enough in ERCOT for this small number of hours so that potential wind production during those times plus minimum generation of nuclear and other units would have exceeded the load. On the other hand, for around 250 h the price was negative in the West and nonnegative elsewhere, implying that transmission constraints limited export of West Texas wind to below its potential during those times.

Zero and negative prices occurred quite regularly in ERCOT until 2013, but with the completion of the CREZ transmission in late 2013 the incidence of economic curtailment was significantly reduced, to less than

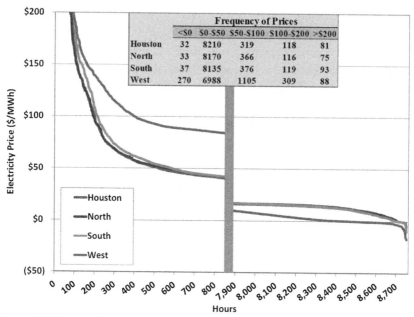

The following table appears within the figure:

	Frequency of Prices				
	<$0	$0-$50	$50-$100	$100-$200	>$200
Houston	32	8210	319	118	81
North	33	8170	366	116	75
South	37	8135	376	119	93
West	270	6988	1105	309	88

Figure 2.6 Truncated ERCOT price-duration curve for hours in 2018 in each of the load zones. (Source: Potomac Economics, 2019. 2018 State of the Market Report for the ERCOT Wholesale Electricity Markets, Available from: www.potomaceconomics.com. (Accessed 18 September 2019).)

1% of the year. However, the incidence of economic curtailment is beginning to increase again as wind capacity is further increased in the West zone, so that as illustrated in Fig. 2.6, West Texas wind economic curtailment occurred during more than 3% of the year 2018 (Potomac, 2019).

Wind also varies at very long timescales, with significant variation from year to year, and wind "droughts" potentially persisting for months or years. For example, average wind speeds in 2015 through 2017 were about 1 km/h below the 10 year average from 2009 from 2018 (Potomac, 2019). Although some countries, such as Brazil, have multiyear PSH storage, there is no such multiyear resource in ERCOT. In the absence of PSH with multiyear storage, renewable "droughts" presumably imply either occasional involuntary curtailment of load or the need to maintain some combination of:
- Other storage devices, including chemical batteries, and
- Some thermal generation used at very low capacity factors.

A modest amount of such resources could be maintained that compromised between their cost and the cost of involuntary curtailment. Maintaining some thermal generation would, however, prevent complete decarbonization of the electricity system.

To summarize, ERCOT has accommodated a significant amount of West zone wind, despite the variability of production not matching patterns of load. West zone wind tends to increase in the evening and peak at night, so that the time in the day of the peak of net load has shifted to somewhat earlier in the day as West zone capacity has increased. Future increases in solar production and coastal South zone wind in ERCOT will likely shift the peak of net load back to later in the day, and will tend to better match the variation of total renewable resources to the patterns of load. Net load ramps due to solar production may be less problematic than in California because of the better correspondence of solar variation with patterns of load, particularly in Summer, the dispatchability of large-scale solar, and the greater flexibility of the thermal fleet, particularly gas plants.

The next section will discuss the variation in use of fuels across electricity systems internationally. This will help to explain the reasons underlying the use of particular fuels in particular jurisdictions, including ERCOT, and understand directions for further development of renewable resources to most cost-effectively support decarbonization.

4. Fuel endowments and temporal fuel endowments

Internationally, there is great variation in the fuels used to generate electricity. The clearest historical driver of this diversity is what might be called the "fuel endowment" of a region, namely the available fuels, with another, and typically more recent, driver due to the availability of imported fuels. For example, as discussed in Section 3, ERCOT historically has used coal and gas to generate electricity, reflecting the availability of these two local fuel resources, with a growing amount of wind production since 2000. Prior to 2006, a liquefied natural gas import facility was being developed in Texas to augment local gas supply with imports. However, the large-scale use of hydraulic fracturing to access unconventional gas in Texas has significantly increased gas supply and decreased gas prices (and therefore increased the contribution of gas to electricity production) in ERCOT since around 2006.

In contrast to its endowment of coal and natural gas, the terrain of Texas is relatively flat, so water resources and suitable topography are scarce and there is consequently little hydroelectric production and no PSH. Texas is, however, very windy, and so, as discussed in Section 3, in the modern era of wind turbines it has developed a significant amount of wind resources, albeit with significant first investments pursuant to Federal subsidies and subsequently also benefitting from the socialization of the CREZ transmission investment costs. Texas also has considerable solar resources and, as discussed in Section 3, the amount of solar capacity is expected to significantly increase in the coming years.

Turning to Australia, natural gas has always been relatively expensive in the east coast states of Victoria, New South Wales, and Queensland, hydraulic fracturing has attracted controversy, while there are also huge coal resources. Consequently, the east coast of Australia has historically predominantly used coal for electricity production with some hydroelectric generation depending on the specific topography and hydrology in the Snowy Mountains and the island of Tasmania.

In partial contrast, South America has topography and hydrology that support significant hydroelectric resources and has therefore relied more on hydroelectricity than fossil fuels, reflecting its endowment. For example, Brazil, with huge water resources, has predominantly used hydroelectricity, developing large reservoirs with multiyear capacity.

In Europe, France, with few fossil resources, developed its nuclear industry (with imported nuclear fuels) together with hydroelectric resources. In contrast, Germany, which shares a border with France, has coal resources and therefore also developed coal-fired generation (as well as, now closing, nuclear facilities). It also has biomass resources and some hydroelectricity, reflecting endowments of those resources, and also uses imported natural gas together with some locally produced biomethane. In the last several decades, Germany has added considerable wind energy and solar production. However, because the solar resource is poor, its solar production is relatively low compared to the installed capacity, below 10% of electricity production, whereas wind now contributes over 20% of electricity production (Aupunn et al., 2020).

Several Asian countries, such as Japan and South Korea, have relatively little fossil fuel resources so that, as with France, they have developed nuclear facilities, and also imported coal and natural gas for electricity production. China and India have local sources of coal, and so their electricity has historically been significantly derived from coal, although with growth outstripping their local endowment, they now both import significant fractions of their total coal usage.

The above discussion exemplifies that the fuel endowment of various countries, and regions within countries, including ERCOT, has plainly had a significant effect on the historical development of their electricity industries. Utilization of renewable resources has also, unsurprisingly, depended on the endowment of natural resources such as solar, wind, and water. However, these renewable resources have an additional dimension due the variability and intermittency of these resources. That is, there is a temporal character to renewable endowments, since the energy is not stored in stable chemical or nuclear bonds, but rather is delivered as a time-varying flux of power.

In the case of hydroelectricity, the gravitational potential energy in the water can be stored in a suitably constructed reservoir without changing the form of the energy. In contrast, solar and wind power cannot be stored directly, and all storage of solar and wind generated electricity requires transformation to another type of energy, incurring additional capital costs and conversion losses that are over and above that required for electricity production itself.

To summarize, the fuel endowment and import options have driven the choices of generation worldwide. Moreover, the stored energy in fossil and nuclear fuels and reservoir hydroelectric resources allowed relatively unfettered temporal distribution of consumption, given enough generation capacity. Because of:

- the typically high inherent value of the products and services derived from electricity consumption compared to electricity production costs,[8]
- the inability to store electricity without converting the energy to another form, and
- the inflexibility in timing of many of the uses of electricity,

most developed countries have a pattern of consumption of electricity that has significant peaks and valleys on a daily basis. This is particularly the case with climates that use significant air-conditioning, such as ERCOT, which results in both significant diurnal variation of consumption, together with large seasonal variation. To avoid demand curtailment, generation capacity must be sufficient and that capacity must be dispatched to match the time-varying patterns of load.

As mentioned in the Introduction, increase in the use of renewables is an important contributor to decarbonizing the electricity system. However, the temporal endowment of solar and wind resources must be considered.

[8] See Chapter 20 by Hochstetler for further discussion of this issue.

Just as countries historically adapted to their fuel endowment, there is a need to adapt to the temporal endowment by:

1. developing those renewable resources that best match patterns of load, while
2. exploring ways to increase the flexibility of the timing of consumption of electricity to best match patterns of renewable production.

As an example of first type of adaptation, consider the cases in ERCOT of solar generation and near coastal South zone wind generation. These solar and wind resources are a better match to current patterns of load than is West zone wind. As suggested in Section 3, it would be prudent to, where possible, increase the relative shares of these resources (Silverstein, 2020). Unfortunately, generally speaking, the coastal wind resources are more difficult to develop due to the greater population density in this region, the consequent greater value of land, and environmental considerations. Solar, on the other hand, is poised to grow significantly in ERCOT in the coming years.

Even if solar and coastal South zone wind could be developed to best match the variability of production to average patterns of load, the match will not be perfect since the average seasonal and diurnal patterns of load are different to the space spanned by the variation of the renewable resources. That is, no combination of solar and wind will result in an overall variation of renewables that exactly matches the patterns of load. Moreover, intermittency additionally complicates the matching of supply and demand. Concentrating solar thermal with storage would facilitate matching at such medium timescales but this technology is currently considerably more costly than solar photovoltaic without storage. That is, using concentrating solar thermal with storage has analogous cost implications to solar photovoltaic with battery storage.

In the absence of PSH or other long timescale storage in ERCOT or available through increased transmission capacity to the Eastern and Western Interconnections, there will be a need to bridge the gap between supply of renewables and patterns of load. It is conceivable that ERCOT will need to maintain some thermal capacity, albeit operated at much lower capacity factors than is common today, build concentrating solar thermal with storage, or build some additional storage capacity operated for occasional renewable "droughts." (It may also be necessary to maintain some synchronous machinery connected to the system to provide inertia and other grid services even if that equipment is not generating electricity.) The next section considers the possibilities for flexible demand to match patterns of load to variability and intermittency of renewables in order to minimize the amount of storage, thermal capacity, and fossil fuel consumption that are needed in a future ERCOT with significantly higher levels of renewables.

5. Flexible demand to facilitate deep decarbonization

As defined in (Potomac, 2019), demand response means actions by end users to reduce consumption. There are multiple types of demand response in ERCOT, including participation:

1. as the ERCOT "responsive reserve" ancillary service, by curtailing load automatically in response to an electrical frequency drop resulting from a generator contingency, amounting to up to around 1.5 GW of load, and paid through a capacity payment,

2. as emergency response service (ERS), which is load that agrees to be curtailed during emergency conditions, prior to curtailment of firm load, amounting to about 1 GW of load, in return for a capacity payment,

3. as curtailable loads such as air conditioners, under control of ERCOT but administered by transmission providers, amounting to around 0.2 GW,

4. as price-responsive load either responding directly to market conditions, such as high day-ahead or real-time prices, or through agreements with retailers, amounting to several GW of load.

Historically, all of these mechanisms have been utilized rarely, reflecting the observation made in Section 4 that electricity consumption typically has high inherent value compared to production costs. Indeed, by design, the first three are intended to only be used in emergency, close to emergency, or under contingency conditions. These three mechanisms can be thought of as having inherently rare utilization and have typically involved industrial loads. Moreover, a significant increase in their utilization would likely not be tolerable to the consumers concerned.

In contrast, some participants in the fourth mechanism respond more often, and to wholesale prices that are only on the order of a few hundred dollars per MWh. To emphasize this distinction, we will refer to the fourth mechanism as an example of "flexible demand" to distinguish it from "traditional" demand response. It was this price-based flexible demand that was referred to in Section 2 as being sufficient to compensate for variations in supply due to the lumpiness of thermal generation investment, unit commitment, and typical outages. In ERCOT currently it includes not only industrial but also commercial loads. This section will argue that such price-based flexible demand has scope for significant expansion both in terms of total capacity of response and in terms of the "capacity factor" of utilization, thereby materially adapting the patterns of demand to the variability and intermittency of renewable production. Moreover, it could be extended to include, for example, aggregation of "smart" residential consumer appliances as discussed in Silva et al. (2009).

Deep penetration of renewable resources is likely to require tens of GW of price-based flexible demand. In this context, the changes in load are not necessarily only reduction, but could include shifting of load from times of high prices to times of low prices or other approaches to modulating consumption that better align consumption with times that are more favorable for renewable production (Seel et al., 2020). This can improve efficiency of electricity production, by displacing high marginal cost production with low marginal cost production, and provide favorable payoffs to consumers. It is to be emphasized that much of this flexibility may actually be arranged through agreements between consumers and retail customers, rather than retail customers reacting to wholesale prices directly.

Before discussing this fourth mechanism in detail in the rest of this section, a fifth mechanism will be mentioned for completeness: the explicit participation of load through bidding into the ERCOT real-time "security-constrained economic dispatch" (SCED), which can also be considered flexible demand. Participation of load in SCED would provide flexible demand that was explicitly dispatched by ERCOT as opposed to responding to prices. In practice, the conditions for participating in real-time SCED have been too onerous for ERCOT market participants and no load has participated to date (Potomac, 2019; Silverstein, 2020). Although participation of load in SCED is possible for some types of demand response, the discussion in the rest of this section will primarily focus on price-based flexible demand to wholesale prices, as opposed to explicitly bidding into the real-time market, given the lack of load in SCED today.

In considering price-based flexible demand, it is important to recognize that some consumers may not be willing to participate, as will be discussed in more detail in the chapters in Parts Two and Four, and particularly Chapter 18 by Burns and Mountain. Flexible demand is, however, attractive for certain groups of consumers. For example, for some industrial consumers electricity costs are a significant fraction of their total variable costs. This includes chemical processes such as air separation into oxygen, nitrogen, and other products (Kelley et al., 2018). Since real-time wholesale prices can regularly rise to be several times typical values, industrial consumers exposed to wholesale prices can significantly control the purchase price of their electricity if they have relative flexibility in their time of consumption. Other consumers, such as electric vehicle charging applications to be discussed in Chapter 8 by Sioshansi, may have significant inherent flexibility in the time of consumption. Several other concrete examples will be detailed later in this section and in more detail in Part Two, but it should be understood that only some consumers will be willing to participate.

To enable price-based flexible demand, a key issue is having either or both of:

- retail customers exposed to prices that reflect more of the variability of wholesale prices directly to them or
- retailers mediating demand management through agreements that can utilize customer flexibility in return for favorable prices.

The former mechanism relies on the consumers themselves being charged on a basis that is closer to real-time wholesale prices, and that they respond to price excursions. This is possible, for example, for most commercial and industrial customers in the retail competition areas of ERCOT and in some other regions with retail competition such as Australia. In both ERCOT and Australia there are retail offerings that effectively pass through real-time wholesale prices. Moreover, as discussed in Chapter 14 by Rai et al., time-of-use tariffs are being introduced in South Australia that partially reflect average wholesale prices.

The latter mechanism typically involves a retailer, aggregator, or intermediary controlling a large collection of customer loads such as electric vehicle chargers or air conditioners as an indirect response to high wholesale prices. The retailer is directly exposed to the wholesale prices and mitigates its risk of having to buy at a high wholesale price for its consumers by being able to modulate their demand. Part Two will discuss examples of this type of demand response in more detail.

Variation in wholesale prices must be sufficiently large to make such modulation of demand worthwhile. In most of the US markets, the existence of capacity markets or capacity obligations tend to enforce reserve margins that are so large that wholesale prices are rarely high enough to encourage price-based flexible demand. Unsurprisingly, there is very little price-based flexible demand in most US markets and very little prospects for growth of such flexible demand. However, this very lack of flexible demand then justifies the need for imposed capacity market or obligations, resulting in a vicious circle that will effectively never allow for significant price-based demand response.[9]

ERCOT's energy-only market described in Section 3 is a fundamental departure from the US norm. Fig. 2.6 illustrates the variation in prices in ERCOT from negative to many hundreds of dollars per MWh. Prices have, on occasion, reached the cap of $9000/MWh. The ERCOT market is particularly suitable for price-based flexible demand since it allows, even embraces, occasional high prices enabled by offer/price caps that are well above those in other US markets. The resulting demand response might range from relatively inflexible consumers that adjusted consumption only on an occasional basis to others that adjust, postpone, or bring forward consumption every day. As mentioned above, moving consumption from a time with high marginal costs to one with low marginal costs represents a reduction in dispatch costs.

A relatively inflexible consumer might not be prepared to adjust consumption except during the occasional times that wholesale price was at many hundred or several thousands of dollars per MWh. On the other hand, more flexible consumers may respond at much more modest prices. Aggregators can facilitate this response with product offerings that automate the behavior of household loads such as air conditioners and electric vehicles. That is, as discussed further in Parts Two, Three, and Four, the aggregators can enable the underlying flexibility of such loads by relieving the owner of the burden of responding directly to prices.

There are some market participants in ERCOT currently that will respond at a price level of around $100/MWh, albeit involving activation of behind the meter distributed generation. For example, the retailer Enchanted Rock has installed generators in supermarkets to provide backup power during outages. These generators can also operate in nonoutage conditions when wholesale prices are high and provide effective price-based flexible demand (Silverstein, 2020).

Unfortunately, although the example of supermarket backup illustrates that flexible demand is possible at quite low wholesale prices, the utilization

---

[9] US capacity markets do allow for "demand response providers" to bid to curtail demand; however, as well as the inherent logical inconsistencies in typical "baseline" approaches (Ruff, 2002), this construct is very different from the price-based demand response as described in this chapter, which is simply taken to mean that when the price is sufficiently high some consumers will choose to not purchase.

of fossil fueled generation as part of the response may not provide a net contribution to decarbonization. Moreover, a significant fraction of the current demand response in ERCOT is due to the incentives provided by the "four coincident peak" mechanism for charging for transmission and distribution costs (Potomac, 2019) and is not directly related to supply—demand conditions. Adaptation to the temporal endowment of renewables to achieve deep decarbonization will require demand response that does not solely rely on distributed generation and which is responding to wholesale prices reflecting supply—demand balance. Such demand response will typically involve end-use storage of a product or service that can be activated daily or weekly.

With a heterogeneous portfolio of price responsive consumers, it is possible to envisage a range of prices across which diverse loads will modulate their consumption. Notionally aggregating such response results in an inverse demand curve that implies relatively smooth adjustment of total consumption in response to prices. That is, the collective response of such consumers to varying prices would be a relatively smooth variation in consumption.

There are a number of such flexible demand applications that require relatively low effort. For example, off-peak water heating has been used in many jurisdictions to adapt electric heating to supply. As mentioned above, electric vehicles can be charged at night in regions with inland wind, and during the day for regions with high solar, with charge profiles designed to match renewable production (see, for example, Kefayati and Baldick, 2012).

As another application, using air-conditioning to precool buildings can better align consumption with solar production, albeit with overall increased consumption of energy (Baldick, 2019). Fig. 2.7 shows the situation schematically for precooling, using data from Pecan Street, Inc., for houses in Austin Texas. The blue trace in the graph shows the average household net load (in this case, consumption minus rooftop solar production) of several homes in the Austin area that have air-conditioning and rooftop solar. It shows a similar pattern to the duck curve described in Chapter 1. The red trace is intended to suggest the type of change that can be expected with precooling. Instead of detailed modeling of precooling as in (German and Hoeschele, 2014), the red trace simply brings forward consumption by 3 h and increases it by 10 percent to account for round-trip losses of storing "cool." The key advantages of precooling are the reduced slope of the "neck" of the red trace and its lower peak compared to the blue

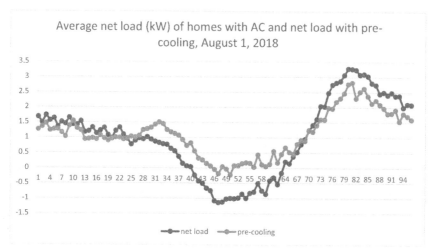

Figure 2.7 Net load and estimate of net load under precooling for houses in Austin, TX. The vertical axis shows the average net load in kW, while the horizontal axis shows the 96 5-min intervals in a 24 h period. *(Source: R. Baldick, 2019. There's More than One Way to Deal With a Duck Curve, Available from: https://rossbaldick.com/theres-more-than-one-way-to-deal-with-a-duck-curve/. (Accessed 26 May 2020), based on data from Pecan Street, Inc.)*

curve. These come at the cost of higher overall energy consumption, but this energy consumption is focused onto times when solar production is higher and, therefore, wholesale prices are lower.

All of these examples could be implemented either through direct response by customers exposed to wholesale tariffs or, perhaps more typically, by retailers having control of the relevant consumer appliances (Silva et al., 2009). The charging and precooling applications can be implemented primarily with software changes to existing control hardware and therefore require essentially no capital investment, which is a key advantage compared to capital intensive storage such as chemical batteries.

There are also a number of higher effort flexible demand applications. For example, dedicated cool storage for buildings can provide significantly more flexibility and higher efficiency than simply precooling the building itself, but requires capital investment and development of appropriate control strategies. Enabling chemical processes to vary production likely involves larger storage facilities than would otherwise be necessary for the end products (Kelley et al., 2018); however, depending on the application, the cost of end product storage could be only a small fraction of the overall cost of the facility. Expanded storage capacity in such cases would add only a small amount to the total cost. As discussed in Seel et al. (2020), the cost-effectiveness of additional storage for processes such as desalination depends on the penetration of renewables. Since such plants are long-lived, planning of such resources should seek to anticipate long-term changes in renewable penetration and resulting prices, including for example, low prices during the middle of the day due to solar penetration and low prices in the night due to wind penetration.

Other industrial demand response, including that of data centers and even Bitcoin miners (PR Newswire, 2020), requires some ability to modulate processing speed, but again the capital investment required in the hardware is likely to be trivial compared to the capital cost of the facility itself. Moreover, such processes may be able to adapt to levels of penetration of renewables without having to explicitly build "storage" capacity.

Finally, new uses for electricity such as electrolysis to produce hydrogen could be designed to have variable production rates but must face the economic implications of capacity factors that may be relatively low compared to other industrial processes (Müller, 2019; IEA, 2019). All of these examples imply higher effort and at least somewhat higher capital cost to enable adaptation of patterns of load; however, they also have the potential to significantly increase the scope of price-based flexible demand.

Although these higher effort applications require some capital investment, a key observation is that most of them are not subject to the same lifetime cycling limitations as chemical batteries, so that capital cost of end-use storage can be amortized over much more energy throughput. Moreover, the capital cost itself of, for example, an insulated tank to store end product or to store a cool liquid tends to be much smaller than the cost of chemical battery storage.

To summarize, the effective capital cost of storage for the various examples ranges from essentially zero to the capital cost of additional end-use storage, which can be expected to be considerably lower than the cost of equivalent chemical batteries. Moreover, in cases such as electric vehicle charging, there is essentially no change in the overall energy used. Other applications, such as precooling of buildings, result in overall increases in energy, but these efficiency losses can be relatively small with well-designed end-use storage. Even with losses, overall energy purchase costs can be lower because purchases are focused into periods of low wholesale prices.

6. Conclusion

This chapter has argued for significantly expanded price-based flexible demand to compensate for the medium timescale variability and intermittency of renewable resources, involving a change in the patterns of load, typically enabled by expansion of end-use storage capabilities for products or services. In regions with storage provided by PSH, demand response would effectively add to medium timescale storage capacity to facilitate deep decarbonization. In regions such as ERCOT without PSH, full decarbonization of the electricity sector may not be achievable at reasonable costs. This might necessitate maintaining some natural gas generation capacity for use during long timescale renewable droughts even with the utilization of flexible demand. Nevertheless, flexible demand can provide for the medium timescale adaptation of patterns of load to renewable variability and intermittency.

There is currently over 20 GW of renewables in ERCOT, primarily wind, providing over 20% of energy annually. Continued strong renewable growth is expected including both wind and solar. At the multi-tens of GW level, and particularly as energy penetration reaches multi-tens of percent, variability and intermittency of renewables poses a huge challenge to an electricity system. With existing flexible thermal capacity, and the ability to "dispatch down" renewables when necessary, ERCOT has so far been able to integrate renewables successfully. But to move to deep decarbonization, the contribution of fossil thermal generation must decrease. With no options for PSH in ERCOT, a way forward is to significantly expand the existing few GW of price-based flexible demand in ERCOT. The required price-based flexible demand resource will eventually need to be multi-tens of GW, but ERCOT's energy-only market has the elements to make it possible.

References

American Wind Energy Association, (AWEA), 2020. Wind Energy in the United States. Available from: https://www.awea.org/wind-101/basics-of-wind-energy/wind-facts-at-a-glance. (Accessed 23 May 2020).

Andrade, J., Dong, Y., Baldick, R., 2018. Effect of Market Changes on the Required Amounts of Frequency Regulation Ancillary Services in ERCOT. Available from: http://users.ece.utexas.edu/~baldick/papers/Regulation_ERCOT.pdf. (Accessed 23 May 2020).

Appunn, K., Haas, Y., Wettengel, J., 2020. Germany's Energy Consumption and Power Mix in Charts. Available from: https://www.cleanenergywire.org/factsheets/germanys-energy-consumption-and-power-mix-charts. (Accessed 25 May 2020).

Baldick, R., 2003. Shift Factors in ERCOT Congestion Pricing. Available from: http://users.ece.utexas.edu/~baldick/papers/shiftfactors.pdf. (Accessed 23 May 2020).

Baldick, R., 2012. Wind energy and energy markets: a case study of Texas. IEEE Sys. J. 6 (1), 27–34.

Baldick, R., 2019. There's More than One Way to Deal With a Duck (Curve). Available from: https://rossbaldick.com/theres-more-than one way-to-deal-with-a-duck-curve/. (Accessed 26 May 2020).

Baldick, R., Niu, H., 2005. Lessons learned: the Texas experience. In: Griffin, J., Puller, S. (Eds.), Electricity Deregulation: Where to from Here? The University of Chicago Press.

Baldick, R., Oren, S., Schubert, E.S., Anderson, K., 2020. ERCOT: success (so far) and lessons learned. In: Glachant, J.-M., Rossetto, N. (Eds.), Handbook of the Economics of Electricity. Edward Elgar.

Bill Blevins, 2012. Voltage Support Service (VSS) for CREZ Region and Baseline Power Angle Ranges Post CREZ. ERCOT. Available from. http://www.ercot.com/services/training/archives. (Accessed 4 June 2020).

Cembalest, M., 2020. 2020 Tenth Annual Energy Paper. J. P. Morgan. Available from: https://www.jpmorgan.com/jpmpdf/1320748699400.pdf. (Accessed 2 July 2020).

Chang, J., Karkatsouli, I., Pfeifenberger, J., Regan, L., Spees, K., Mashal, J., Davis, M., 2014. The Value of Distributed Electricity Storage in Texas. Brattle Group. November, Available from: http://files.brattle.com/files/7924_the_value_of_distributed_electricity_storage_in_texas_-_proposed_policy_for_enabling_grid-integrated_storage_investments_full_technical_report.pdf. (Accessed 29 May 2020).

Cole, W., Will Frazier, A., 2019. Cost Projections for Utility-Scale Battery Storage. National Renewable Energy Laboratory. NREL/TP-6A20-73222, Available from: https://www.nrel.gov/docs/fy19osti/73222.pdf. (Accessed 30 May 2020).

Elliston, B., Diesendorf, M., MacGill, I., 2012. Simulations of scenarios with 100% renewable electricity in the Australian National Electricity Market. Energy Pol. 45, 606–613.

ENTSO-E, 2019. Statistical Factsheet 2018. Available from: https://www.entsoe.eu/Documents/Publications/Statistics/Factsheet/entsoe_sfs2018_web.pdf. (Accessed 24 September 2019).

ERCOT, 2020. Fact Sheet. Available from: http://www.ercot.com/content/wcm/lists/197391/ERCOT_Fact_Sheet_5.11.20.pdf. (Accessed 22 May 2020).

German, A., Hoeschele, M., 2014. Residential Mechanical Precooling. In: Alliance for Residential Building Innovation. NREL. Available from: https://www.nrel.gov/docs/fy15osti/63342.pdf. (Accessed 5 June 2020).

Henze, V., 2019. Battery Pack Prices Fall as Market Ramps up with Market Average at $156/kWh in 2019. Bloomberg New Energy Finance. December, Available from: https://about.bnef.com/blog/battery-pack-prices-fall-as-market-ramps-up-with-market-average-at-156-kwh-in-2019/. (Accessed 30 May 2020).

Hledik, R., Fox-Penner, P., Lueken, R., Lee, T., Cohen, J., 2020. Decarbonized Resilience: Assessing Alternatives to Diesel Backup Power. Brattle Group. Available from: https://brattlefiles.blob.core.windows.net/files/19026_decarbonized_resilience_white_paper_-_final.pdf. (Accessed 3 July 2020).

Hogan, W.W., 2012. Electricity Scarcity Pricing through Operating Reserves. Available from: https://scholar.harvard.edu/whogan/files/hogan_ORDC_042513.pdf. (Accessed 15 May 2016).

International Energy Agency (IEA), 2019. The Future of Hydrogen. Available from: https://webstore.iea.org/download/direct/2803. (Accessed 2 July 2020).

Kefayati, M., Baldick, R., 2012. Harnessing demand flexibility to match renewable production using localized policies. In: Proceedings of the 50th Annual Allerton Conference on Communication, Control, and Computing, pp. 1105–1109. Allerton, IL,

October, Available from: http://rossbaldick.com/wp-content/uploads/electric-vehicle-and-peak-load.pdf. (Accessed 26 May 2020).

Kelley, M.T., Pattison, R.C., Ross, B., Baldea, M., 2018. An MILP framework for optimizing demand response operation of air separation units. Appl. Energy 222, 951–966.

Lee, D., Ross, B., 2020. Wind Variability and Impact on Markets. Available from: http://users.ece.utexas.edu/~baldick/papers/GDFM.pdf. (Accessed 4 June 2020).

Littlechild, S., Baldick, R., 2020. The Texas Competitive Renewable Energy Zone (CREZ) Transmission Expansion Project: Part I the Legislative Framework. Working paper.

Mele, C., 2018. ERCOT Changes and Challenges. Available from: https://www.rmel.org/rmeldocs/Library/Presentations/2018/FALL18/FALL18_Mele.pdf. (Accessed 22 May 2020).

Müller, S., 2019. Green Hydrogen and Sector Coupling. Future Electricity Markets Summit, Sydney, Australia, November.

Potomac Economics, 2019. 2018 State of the Market Report for the ERCOT Wholesale Electricity Markets. Available from: www.potomaceconomics.com. (Accessed 18 September 2019).

PR Newswire, 2020. Layer1 Launches Bitcoin Batteries to Stabilize Energy Grids by Releasing Electricity to Meet Market Demand. Available from: https://www.prnewswire.com/news-releases/layer1-launches-bitcoin-batteries-to-stabilize-energy-grids-by-releasing-electricity-to-meet-market-demand-301063984.html. (Accessed 5 June 2020).

Ruff, L., 2002. Economic Principles of Demand Response in Electricity. Edison Electric Institute. October, Available from: https://hepg.hks.harvard.edu/files/hepg/files/ruff_economic_principles_demand_response_eei_10-02.pdf. (Accessed 5 June 2020).

Seel, J., Mills, A.D., Warner, C., Paulos, B., Wiser, R., 2020. Impacts of High Variable Renewable Energy Furtures on Electric Sector Decision-Making: Demand-Side Effects. Available from: https://eta-publications.lbl.gov/sites/default/files/berkeley_lab_high_vre_impacts-_demand-side_effects_report_2020.05.20.pdf. (Accessed 2 July 2020).

Silva, V., Stanojevic, V., Pudjianto, D., Strbac, G., 2009. Value of Smart Appliances in System Balancing. Intelligent Energy Europe. Available from: https://ec.europa.eu/energy/intelligent/projects/sites/iee-projects/files/projects/documents/e-track_ii_energy_networks_report.pdf. (Accessed 2 July 2020).

Silverstein, A., 2020. Resource Adequacy Challenges in Texas: Unleashing Demand Side Resources in the ERCOT Competitive Market. Environmental Defense Fund. Available from: https://www.edf.org/ERCOTdsm. (Accessed 12 May 2020).

Sixel, L.M., 2020. Texas to get 15 utility-scale battery storage sites. Houston Chronicle. June 10, Available from: https://www.houstonchronicle.com/business/energy/article/Texas-to-get-15-utility-scale-battery-storage-15328684.php. (Accessed 3 July 2020).

Snowy Hydro, 2020. About Snowy 2.0. Available from: https://www.snowyhydro.com.au/snowy-20/about/. (Accessed 22 May 2020).

Sue, K., MacGill, I., Hussey, K., 2014. Distributed energy storage in Australia: quantifying potential benefits, exposing institutional challenges. Energy Res. Soc. Sci. 3, 16–29.

Tesla, 2020. Earn Cash from Your Powerwall and Create a Cleaner, Stronger Grid. Available from: https://www.tesla.com/connectedsolutions. (Accessed 22 May 2020).

Wikipedia, 2020. Electric Car Use by Country. Available from: https://en.wikipedia.org/wiki/Electric_car_use_by_country. (Accessed 30 May 2020).

Ziegler, M.S., Mueller, J.M., Pereira, G.D., Ferrara, M., Chiang, Y.-M., Trancik, J.E., 2019. Storage requirements and costs of shaping renewable energy toward grid decarbonization. Joule 3 (9), 2134–2153.

CHAPTER 3

Rising variability of generation in Italy: The grid operator's perspective

Giacomo Terenzi[1]
Market Analysis, System Strategy, Terna Spa, Rome, Italy

1. Introduction

Global megatrends are radically redefining the structure of political, economic, and social balances at the international level. The growth and development trajectories of the entire planet depend on the answers that institutions, governments, corporate entities, markets, and civil society will provide.

Over the next 3 decades, demographic dynamics, socioeconomic changes, the acceleration of technological innovation, the spread of digitalization, and the growing impact of anthropic activities on the environment and on the exploitation of natural resources, if not properly governed, will transform the world into a place that bears no resemblance to Earth as we know it.

The energy sector is affected by these trends under multiple dimensions and directions. As a consequence of changes in geopolitical balances, energy consumption is gradually shifting from America and Europe to South East Asia. The economic expansion of developing countries, in addition to significantly increasing demand, will bring to the markets an ever-greater share of the billion people who still do not have access to electricity today. Population growth will have an increasing impact on access to scarce resources, including energy resources responsible for the most significant share of climate-changing emissions (IEA, World Energy Outlook, 2019b).

Technological innovation and digitalization can contribute to accelerating the energy transition toward a carbon neutral economy, which has marked the first important steps over the last 2 decades. It is ever more

[1] I would like to thank for supervision and feedback Luca Marchisio, Modesto Gabrieli Francescato, Pierluigi Di Cicco, Andrea Lupi, and Fabio Genoese, all from Terna Spa. Any remaining errors are mine.

Variable Generation, Flexible Demand
ISBN 978-0-12-823810-3
https://doi.org/10.1016/B978-0-12-823810-3.00012-1

necessary, if the planet wants to achieve sustainable, lasting, and inclusive growth, that represents the core of the 17 Sustainable Development Goals (SDGs) adopted by the UN through the 2030 Agenda.

Electricity is at the heart of modern economies and it is providing a rising share of energy services. Demand for electricity is set to increase further as the world gets wealthier, as its share of energy services continues to rise, and as new sources of demand expand, such as digital connected devices, air-conditioning, and electric vehicles. Electricity demand growth is set to be particularly strong in developing economies.

Energy transition will be led by a deep electrification process along with high penetration of RES in the system, and it is clear that flexibility will be a cornerstone of electricity security in a changing power mix.

Cost reductions in renewables and advances in digital technologies are opening huge opportunities for energy transitions, while creating some new energy security dilemmas; if on one hand RES penetration can be considered a success in meeting the European decarbonization targets, on the other it is something that implies many impacts in the daily management of the grid and it obliges the Transmission System Operator to set in place a portfolio of countermeasures to keep the system safe and adequate.

Policymakers, regulators, and TSOs will have to move fast to keep up with the pace of technological change and the rising need for flexible operation of power systems. Terna is deploying a wide set of strategies in this regard, and issues such as new interconnections, market design for storage, demand and flexibility resources, and data management are the backbone of the Terna agenda for the forthcoming years. All these points have the potential to expose the system to new opportunities and risks, topics presented in this chapter.

The chapter is organized as follows:

- Section 2 provides a general overview of the National and European regulation which is supporting and addressing the energy transition;
- In Section 3 the most relevant recent historical trends of the Italian electricity system are presented;
- Section 4 presents the long-term view of Terna's for the electricity system in terms of load, installed capacity, generation, and residual load; and
- In Section 5 the main implications of RES penetration are reported along with Terna's strategies for managing a variable generation on the rise: grid expansions, market evolution, and demand flexibility; followed by the chapter's conclusions.

2. Regulatory framework for the energy transition: European environmental targets

The Paris Agreement[2] is the first with a global reach[3] character in the field of climate change and aims to keep the average increase in temperatures well below 2°C compared to preindustrial levels and to intensify efforts to limit this increase to 1.5°C in order to significantly reduce the risks and impacts of human activities on the environment. These thresholds are considered by the international scientific community to be those, which with a probability of 50% and 66%, respectively, would make the process of raising the concentration of greenhouse gases in the atmosphere reversible.

Nevertheless, reaching these goals would require a radical transformation of the global energy system with the taking of concrete actions in all sectors. To meet the challenge of decarbonization over the next decade and follow up on the commitments made in the Paris Agreement, the European Union has developed the "Clean Energy for all Europeans" Package, also known as the "Clean Energy Package."[4] This is a system of rules that acts simultaneously on all the five dimensions of the Energy Union, which are:

- Energy security;
- Internal market;
- Energy efficiency;
- Decarbonization; and
- Research, innovation, and competitiveness.

The Clean Energy Package has set three new goals to be achieved by 2030 including:

- Reduction of at least 40% of greenhouse gas emissions compared to 1990 levels;
- At least 32% share of final energy consumption represented by renewables; and
- Improvement in energy efficiency, with a reduction of at least 32.5% of primary energy consumption compared to the business-as-usual scenario.

---

[2] https://unfccc.int/process-and-meetings/the-paris-agreement/the-paris-agreement.

[3] It is worth noting that the Paris Agreement per se is on a voluntary base at the international level, while the European Clean Energy Package has a binding character for Member States.

[4] https://ec.europa.eu/info/news/clean-energy-all-europeans-package-completed-good-consumers-good-growth-and-jobs-and-good-planet-2019-may-22_en.

Under the new governance of the Energy Union, by December 2019, each Member State has drawn up a proposal for an integrated National Energy and Climate Plan (NECP[5]) for the 2021—30 period which establishes the necessary policies to effectively reach the targets set at European level.[6] Following the evaluation process envisaged in the legislation, the Commission has provided a series of recommendations to strengthen the commitments, which the countries will have to take into consideration in the final version of the NECP. Furthermore, the European Commission is working on the so-called Green Deal[7] which is a massive wave of investment to focus on the energy transition across Europe in order to boost the process which might slow down if no effective and supportive economic measures will be set in place.

3. The Italian electricity system: historical trends and dynamics

During the last 20 years the Italian electricity load have been presenting two clear phases, as illustrated in Fig. 3.1: pre and post 2008 economic and financial crisis. The first period is characterized by a constant growth of 2.2% on average per year up to 2007, when the load reached its peak at 339.9 TWh. In 2009 the financial crisis hit hard and the load went down by 6%, plunging to 320.2 TWh, which has become some sort of new equilibrium value for the Italian system. In fact, the average value for the postcrisis period is very close to this level; this is particularly evident for the last 3 years.[8] Naturally, great attention is posed to the 2020 outcome, already marked by the global pandemic. First projections are indicating a loss between 7% and 10%, setting back the load around the initial value of the 2000s.

As it is well known, there is a strong positive correlation between electrical load and Gross Domestic Product, as illustrated for Italy in Fig. 3.2.[9] What is worth noting is that if on one hand the diffusion of new efficient technologies and behaviors is lowering the electrical consumption level, on the other there is still not any statistical evidence of the so-called

---

[5] https://ec.europa.eu/info/energy-climate-change-environment/overall-targets/national-energy-and-climate-plans-necps_en.

[6] By mid-May 2020 NEPC from Germany Ireland, Luxembourg and Romania are still missing.

[7] https://ec.europa.eu/info/strategy/priorities-2019-2024/european-green-deal_en.

[8] 2019 provisional data: 319.6 TWh.

[9] Terna internal research note.

Figure 3.1 Italian electrical load (2000—19), TWh. *(Source: Terna, 2019 data provisional.)*

decoupling between GDP and load growth which would imply concentration of the scatter plot in the lower right quadrant of the chart.

In Fig. 3.3 the peak load evolution from 2000 to 2019 is presented; this dimension has an average value of 54 GW across the considered period and it has been increasing in the last 5 years. The peak load is particularly related to the temperature and to the level of electrification (driven by the diffusion of

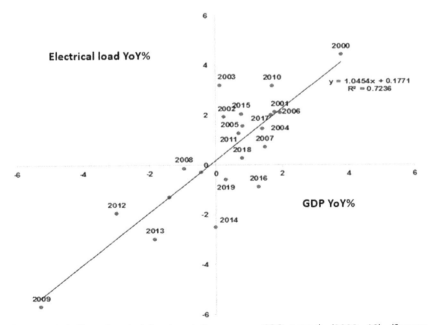

Figure 3.2 Italian electrical load variation versus GDP growth (2000—19). *(Source: Terna and the Italian Institute for Statistics, ISTAT.)*

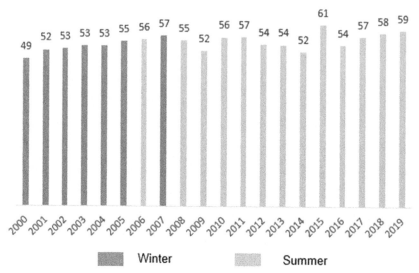

Figure 3.3 Italy's Peak load historical trend by season (2000–19), GW. *(Source: Terna.)*

appliances such as electric air-conditioning/heat pumps). It is important to point out, in this respect, the seasonal shift of the peak from winter to summer starting from 2008. In particular, the maximum value has been observed in July 2015 when a heavy heat wave hit the country for several days.

Moving from the load to the generation, the main feature of the Italian generation capacity mix of the last decade is surely the high level of RES capacity such as photovoltaic and wind heavily pushed by government policies across the years made of strong incentive schemes for this kind of generation.

Those policies reached their goal, can be said, considering the increase in the installed capacity presented in Fig. 3.4 where the PV capacity jumped from basically zero to almost 21 GW, while the wind capacity went up by 7 GW.

The increase in RES capacity has been pushing up also the percentage of electrical load covered by renewable energy,[10] as presented in Fig. 3.5 at different granularities for the year 2019 (hourly, daily, monthly, yearly). The maximum RES penetration in the mix has been achieved, as it usually happens, during the central hours of a spring day (27th April at 2 p.m.), when the load is at the minimum level of the year and there is abundance of solar and wind resources in the system.

[10] Hydro included (Run of river and reservoir).

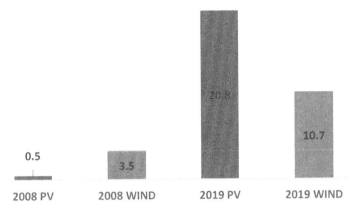

Figure 3.4 Wind and Solar capacity evolution in Italy, 2008 vs 2019, GW. *(Source: Terna.)*

The massive penetration in the market of virtually zero variable cost and heavily subsidized fixed cost RES plants in combination with low carbon policies (coal phase out by 2025 as per Italian National Energy and Climate Plan[11]) have been reducing the thermal generation market share during the last years. In fact, the capacity factor of the thermal fleet has been constantly going down, jeopardizing their capability to recover costs from the market. This phenomenon has been having as main implication the continuous lowering of thermal power plants contribution in the system (Fig. 3.6).

Figure 3.5 2019 maximum RES penetration in the generation mix (hourly, daily, monthly, and annual, as % of total generation). *(Source: Terna.)*

[11] https://ec.europa.eu/energy/sites/ener/files/documents/it_final_necp_main_en.pdf.

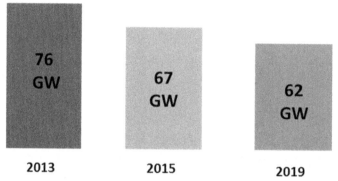

Figure 3.6 Declining thermal capacity in Italy, 2013–19, GW. *(Source: Terna.)*

It is also worth noting that, in terms of future perspectives, as the pressure for meeting the environmental target increases, the trend mentioned above is going to be even more evident, as further described below.

4. Where is the Italian system heading?

In terms of future perspectives of the electrical and energy system, Terna, in cooperation with Snam, the Italian gas TSO, recently issued the Scenario Description Document[12] (DDS 2019) in compliance with the Italian Energy Regulatory Authority resolution 654/2017/R/eel and annex A resolution 627/16/R/eel.[13]

The main focus of this study was to examine the overall primary energy demand in Italy up to 2040 with special attention to electricity and natural gas. The study serves as major input to the national transmission grid development plans for electricity and natural gas sector at the national level.

Many different long-term energy demand scenarios were considered in this analysis under a wide range of hypothesis and assumptions on policy, economics, environment, and technology. For illustrative purposes only two scenarios will be presented here:

- A current policies scenario (Business as Usual—BAU) and
- A very electrified and European decarbonization target compliant scenario (Decentralized—DEC).

---

[12] https://www.terna.it/en/electric-system/grid/national-electricity-transmission-grid-development-plan/scenarios.

[13] https://www.arera.it/it/docs/17/654-17.htm; https://www.arera.it/allegati/docs/16/627-16eng.pdf.

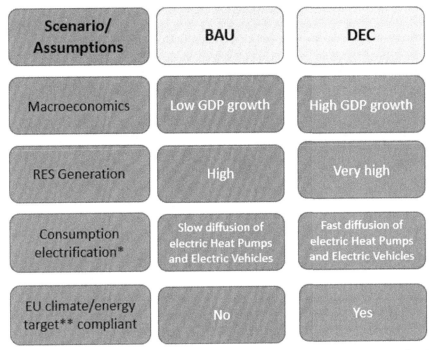

Figure 3.7 *Main scenario assumptions: Business as Usual (BAU) and Decentralized (DEC) scenario.* * a different degree of electrification is expected in the two scenarios according to electric technologies diffusion, especially in residential (heat pumps) and transport sectors (electric vehicles) ** https://ec.europa.eu/clima/policies/strategies/2030_en. *(Source: Terna, 2019b. Scenario Report. https://www.terna.it/en/electric-system/grid/national-electricity-transmission-grid-development-plan/scenarios.)*

The main features of these two scenarios are qualitatively shown in Fig. 3.7.

The above scenarios provide a view of constant growth of the electrical load, as shown in Fig. 3.8. In the Business as Usual scenario, the less electrified scenario, the load increases at 0.71% per year, reaching 340 TWh in 2030 and 371 TWh in 2040, with an electrification[14] rate of 24% and 26%, respectively.

The Decentralized scenario presents a more electrified economy and, in this case, the load increases by 1% per year leading to 356 and 391 TWh of

---

[14] Calculated as electricity final consumption over total final consumption.

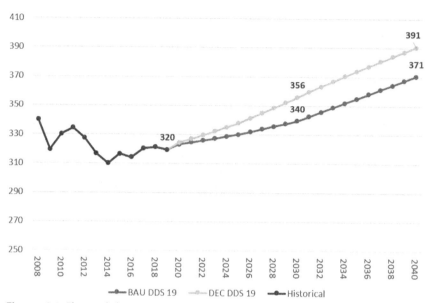

Figure 3.8 Electrical load growth scenarios for Italy—BAU and DEC to 2040, TWh. *(Source: Terna, 2019b. Scenario Report. https://www.terna.it/en/electric-system/grid/ national-electricity-transmission-grid-development-plan/scenarios.)*

load in 2030 and 2040, respectively, implying an overall electrification of the Italian system at 28% and 35%.

The impact on the forecasted peak load is evident in the most electrified scenario, with 72 GW for the summer peak demand by 2040 (Fig. 3.9). It is worth noting that the studies of the peak load are particularly sensitive to the diffusion of electrical appliances and EVs, but also to the assumptions made on the EVs charging/discharging cycle during the day, topics further described in Chapter 8.

Moving to the generation side, RES penetration is expected to keep growing in both scenarios even if at different paces, according to the storylines (Figs. 3.10 and 3.11). In the Business as Usual scenario, in 2040 PV capacity reaches 47 GW (20 GW in 2017) while wind capacity reaches 18 GW (10 GW in 2017).

The Decentralized scenario, which is the most ambitious in terms of electrification and decarbonization, results in even higher RES penetration: 70 GW of solar and 25 GW of wind in 2040. It is worth noting that in both scenarios wind and solar are expected to heavily penetrate in the generation

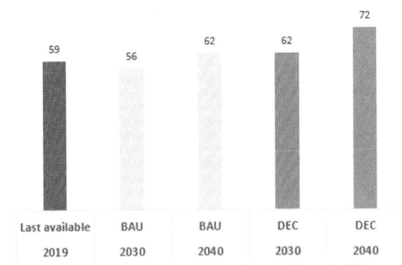

Figure 3.9 Summer peak load in 2030 and 2040—BAU and DEC scenarios, GW. *(Source: Terna, 2019b. Scenario Report. https://www.terna.it/en/electric-system/grid/national-electricity-transmission-grid-development-plan/scenarios.)*

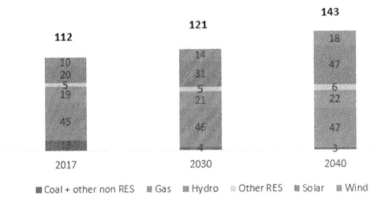

■ Coal + other non RES   ■ Gas   ■ Hydro   ▩ Other RES   ■ Solar   ■ Wind

Figure 3.10 Capacity mix evolution—BAU Scenario, GW. *(Source: Terna, 2019b. Scenario Report.   https://www.terna.it/en/electric-system/grid/national-electricity-transmission-grid-development-plan/scenarios.)*

mix by 2040, reaching 45% in Business as Usual scenario and 55% in the Decentralized scenario, respectively, compared to 26% in 2017.

This evolution is expected to impact the electrical system in man ways. One of the major indicators is the *residual load*, which has different definitions but generally refers to what is left of the load after accounting for RES production as illustrated in Fig. 3.12.

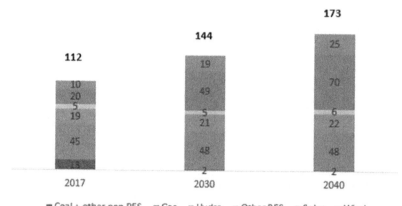

Figure 3.11 Capacity mix evolution—DEC Scenario DEC, GW. *(Source: Terna, 2019b. Scenario Report. https://www.terna.it/en/electric-system/grid/national-electricity-transmission-grid-development-plan/scenarios.)*

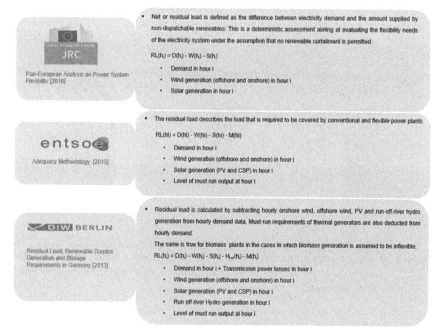

Figure 3.12 Residual load possible definitions. *(Source: Pan-European Analysis on Power System Flexibility (2018); SOAF Methodology Adequacy (2015); Residual Load, Renewable Surplus Generation and Storage Requirements in Germany (2013).)*

The dynamics and consequences of the *residual load*, also called the *net load*, are examined in many reports and studies from many other TSOs and researchers across the world[15] including in Chapter 1 in this book looking at the Californian "duck curve" and the related challenges in managing its fast evolution.

Terna's Scenarios consider the projected RES penetration in the long term and its impact in terms of expected hourly residual load for sample days, as reported in Figs. 3.13 and 3.14, where only the Decentralized scenario is considered. The residual load keeps on dropping over time and varies based on the season. During the spring in 2030 and 2040, for instance, the residual load goes heavily negative during the hours when the RES generation is at its maximum—to −20 and almost −40 GW, respectively. Spring is particularly critical in managing the residual load

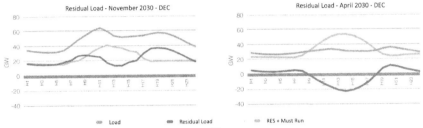

Figure 3.13 Hourly residual load by season[16]—2030—DEC Scenario, GW. *(Source: Terna, 2019b. Scenario Report. https://www.terna.it/en/electric-system/grid/national-electricity-transmission-grid-development-plan/scenarios.)*

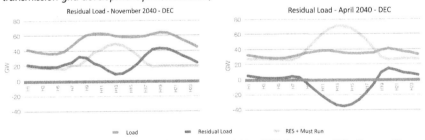

Figure 3.14 Hourly residual load by season—2040—DEC Scenario, GW. *(Source: Terna, 2019b. Scenario Report. https://www.terna.it/en/electric-system/grid/national-electricity-transmission-grid-development-plan/scenarios.)*

[15] As examples consider California Demand Response Potential Study Phase 3: The Potential for Shift Through 2030, Lawrence Berkeley National Laboratory. Feb 2020; and Renewable integration study, Australian Energy Market Operator (AEMO) (2020).

[16] The chart for April 2030 and 2040 simulates a bank holiday for stressing the residual load issue.

because of relatively low load and high level of wind and solar irradiation—nearly identical to the case of California, presented in Chapter 1. This implies that flexible plants have to reduce their output because of the abundance of RES generation. The situation is worse during weekends and holidays when the load is particularly low, the residual load goes negative, the excess RES production must be curtailed, and the need for storage becomes obvious—exactly what happens in the case of CAISO with its duck curve—as illustrated in the right side of Fig. 3.14.

To cope with an increasing number of hours of negative residual load, an estimated 13 and 19 GW of storage—23% and 37% of which are batteries, while the rest is hydro storage—will be required in 2030 and 2040, respectively (Fig. 3.15). The following section describes potential strategies to cope with these challenges including interconnection optimization, flexible demand, and market evolution.

5. RES penetration and TSO's countermeasures for tackling the transition

The Italian electrical system is already moving toward the desired decarbonization targets in a similar fashion to neighboring Spain, covered in Chapter 4. For this transition to succeed, a number of challenges must be addressed including trade-off between the environmental goals and security and adequacy of the electrical system.

As already explained, the combination of reducing the traditional and dispatchable power plants while adding more RES, mostly solar and wind, will lead to the exacerbation of the "duck curve" phenomenon as reported in Chapter 1.

The higher the RES penetration, the more pronounced will be the phenomena, as shown in Fig. 3.16 with steeper morning *deramps* and

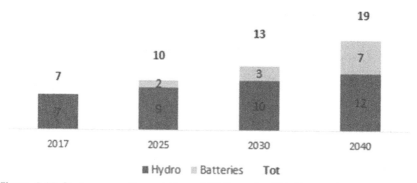

Figure 3.15 Storage capacity evolution—DEC Scenario, GW. (*Source: Terna, 2019b. Scenario Report. https://www.terna.it/en/electric-system/grid/national-electricity-transmission-grid-development-plan/scenarios.*)

evening *ramps* and a bulging "duck belly" with negative residual load. This calls for additional flexible resources on both the supply (gas-fired turbines, pumped hydro storage and batteries, interconnectors) and on the demand side (industrial and households demand response, electric vehicles, power-to-gas, etc.) and market solutions to unlock the needed flexibility.

Fig. 3.16 shows how a typical hourly residual load might look like in 2030 given the assumed RES penetration levels. The residual load profile follows the solar generation—even when the contribution of wind is included. This means that thermal generation must have the flexibility to go down fast—during the morning *deramp*—while being able to quickly restart, during the evening *ramp up*. During the middle hours of the day, storages might be needed to avoid massive RES curtailment—all issues present in California today.

The progressive increase of renewables, which is necessary to deliver the energy transition, is changing the entire architecture of the system and its operation. To be able to maintain the current high levels of security, adequacy, and service quality, while avoiding excessive costs for society, it is essential to understand how to manage the challenges of running a system largely based on variable generation.

Transmission System Operators (TSOs) have the delicate and complex task of balancing electricity production and demand in real time, ensuring consumers a safe and reliable supply of energy. Already today, with the current share of renewables, we can observe significant impacts on the

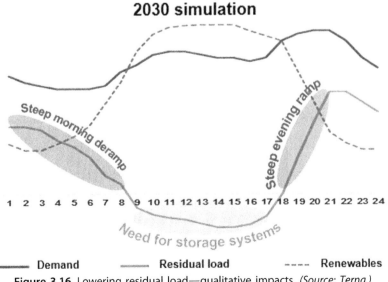

Figure 3.16 Lowering residual load—qualitative impacts. *(Source: Terna.)*

transmission grid mainly due to the technical characteristics, the variability, and the location of renewable energy sources. Managing decarbonization will require huge investments as reported in Box 1 in Chapter 1 as well as Chapter 4.

RES plants usually interface with the network through static machines, which do not provide rotating power, unlike the traditional thermal and hydro generation. Therefore, they do not have the same capability to support frequency and voltage, which are fundamental to keeping the system stable. Frequency and voltage control are essential to cope with perturbations as in the case of a sudden and unexpected loss of generation or other emergencies. With the steady displacement of thermal capacity due to the increasing shares of RES, the inertia of the system is progressively decreasing, making it more vulnerable to frequency deviations. As shown in Fig. 3.17, for example, faced with the unexpected loss of a generator, the frequency decreases more quickly in a system with lower inertia (System B) than in a system with more inertia (System A). In the former case, frequency could drop so fast that automatic load shedding would have to be activated to avoid an uncontrolled blackout. This is similar to the frequency event, which occurred on the 9th of August 2019 in Great Britain's

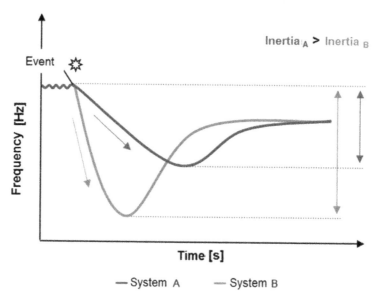

Figure 3.17 Qualitative behavior of two systems with different levels of inertia. *(Source: Terna.)*

electricity system, which led to the disconnection of 5% of the total load after the frequency dropped below 49.2 Hz.[17]

In electricity systems increasingly based on RES, which are subject to weather conditions, balancing production, and consumption, especially during critical periods such as peak demand hours is becoming more complicated. As explained in Chapter 5, the traditional flexibility options will increasingly become unavailable, unsuitable, or expensive, hence the need for new solutions—including a shift toward making demand to follow generation, rather than the other way around.

Furthermore, variable renewables, in particular wind, are often located far from consumption centers, as explained in Chapter 2. The electricity generated therefore needs to be transported over long distances to reach consumers. This can cause a considerable increase in transmission congestions. Moreover, most of RES plants are connected to medium- and low-voltage distribution networks, which have been developed in the past to handle unidirectional rather than bidirectional flows. This suggests that new challenges are emerging in the management of the electricity system at all voltage levels (Fig. 3.18).

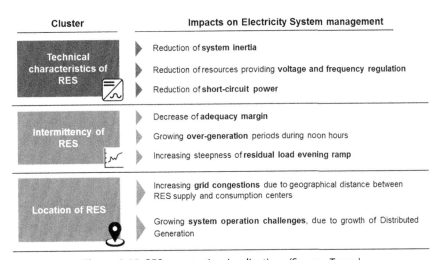

Figure 3.18 RES penetration implication. *(Source: Terna.)*

[17] GB Power System disruption on 9 August 2019— Energy Emergencies Executive Committee (E3C): Final report -Department for Business, Energy & Industrial Energy. https://assets.publishing.service.gov.uk/government/uploads/system/uploads/attachment_data/file/855767/e3c-gb-power-disruption-9-august-2019-final-report.pdf.

In this context, it is of paramount importance that TSOs implement a set of coordinated and coherent strategies to enable this complex transition—as highlighted by Luca Lo Schiavo, in the book's Preface. In particular, investments in electricity network infrastructures, at all voltage levels, will continue to be crucial. However, they will need to be accompanied by other, complementary investments to increase system flexibility and to effectively exploit and manage the flexibility potential of all new resources. Grid operators typically procure flexibility in dedicated ancillary services markets. Consequently, a redesign of ancillary services markets in terms of procurement, remuneration rules, and new resources allowed, such as flexible demand, will be essential ingredient to support the full integration of renewables.

From the TSO's perspective, the major issues that must be dealt with include the following:

- Significant need of ramping up thermal generation in the evening hours to balance the drastic loss of solar output;
- Improved regulation of capacity given the growing share of RES in the generation mix;
- Increasing limited upward reserve margins to cover peak load, following the decommissioning of significant amount of thermal installed capacity;
- Improved management of grid congestions due to the nonhomogeneous distribution of RES across Italy, most notably in the Southern parts;
- Solutions for the increased periods of overgeneration from nondispatchable renewables; and
- Increasing the availability of sources providing voltage regulation (such as reactive power) and frequency regulation (such as rotational inertia against the loss of system stability).

Addressing these critical issues requires further work in the following areas:

- Transmission grid development to optimize power flows among regions, especially considering the geographical concentration of new RES installations in the southern part of Italy;
- Long-term price signals for encouraging investments in efficient and flexible generation;
- Market evolution for unlocking flexibility and enabling new resources; and
- Innovation and digitalization of the assets, processes, and control systems across the entire value chain.

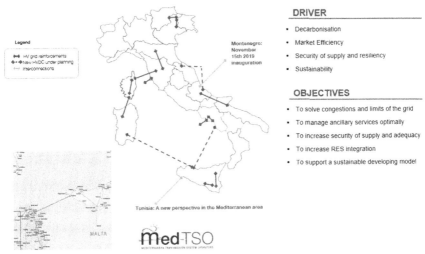

Figure 3.19 Main planned interconnection and reinforcements. *(Source: Terna, 2019a. National Development Plan. https://www.terna.it/en/electric-system/grid/national-electricity-transmission-grid-development-plan.)*

Fig. 3.19 illustrates some of the required transmission grid developments currently planned[18] by Terna, many of which are focused in the central and southern parts of the country where the rise of RES is projected to be higher. In particular, there are plans for the Sardinia—Sicily—Campania interconnection, Tyrrhenian Link, currently under evaluation along with the interconnection with Tunisia, which may be a key enabler for further RES production across regions and countries by connecting the Mediterranean grid.[19]

In terms of long-term price signals, the rising RES penetration in the market has been posing new challenges for the conventional/dispatchable generation. Marginal price reductions and fewer dispatch hours during the day are jeopardizing the ability of these power plants to cover their costs, potentially leading to a suboptimal investment level in dispatchable generation. These kinds of investments are characterized by high level of upfront costs that, considering the volatile and uncertain future market prices, might be perceived as too risky from the investors' perspective.

---

[18] Projects planned for the forthcoming years or still under evaluation. Source: National Development Plan 2019, Terna.
[19] For more details see Mediterranean Master Plan issued by Med-TSO association (2018/2020).

Long-term price signals are considered to be one of the proper instruments for enabling the energy transition which, at the same time, must remain adequate and secure.[20]

The general framework to provide the right incentives for allowing the evolution of the generation mix over time should provide the correct signals while reducing the uncertainty of the revenues. This framework may include:

- Capacity remuneration mechanism for dispatchable generation;
- Power purchase agreement for RES; and
- Long-term contracts for storages and flexible demand.

To address these issues, Terna strongly supported the introduction of the capacity market as a form of market-based capacity remuneration mechanism where Terna buys long-term generation capacity for adequacy purposes.[21]

This market is composed of a series of sequential yearly auctions, which will result in fixed prices, against which the spot price will be compared, generating a positive or negative difference—which will be refunded to/or by Terna through a reliability option mechanism. This market is open to low emissive[22] conventional generators, RES, as well as the demand side, allowing the lowest cost options to prevail. In this market the demand curve is represented by Terna's capacity requirements as illustrated in Fig. 3.20, defined by probability of loss of load or LOLE.[23]

The required energy transition also calls for other market design innovations given the evolving generation fleet and the flexibility needed by TSOs for managing the grid.

In Italy the electricity market is currently organized in a wholesale energy market and an ancillary service market. The energy market is composed of the day-ahead and 7 intraday markets with hourly bidding zone auctions at marginal price. The intraday markets have recently increased from 5 to 7 markets to better approximate real-time conditions

---

[20] It is worth noting that in the long-term decarbonization framework, dispatchable units, that are still of paramount importance for the adequacy of the system, have to be thought either using green fuels or associated with CO_2 absorption mechanism such as CCS.

[21] https://www.terna.it/en/electric-system/capacity-market.

[22] Emission index <550 grCO2/kWh -As per Capacity Market Code (2019).

[23] LOLE represents the yearly hours with EENS occurrence (which is a measure of the amount of electricity demand—in MWh in a given year—expected to be lost when demand exceeds the available generation).

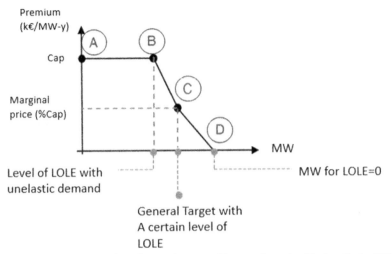

Figure 3.20 Capacity market demand curve. *(Source: Capacity Market Code (2019), Terna.)*

giving more opportunity for market participants, especially RES producers, to adjust their position in the market according the actual production levels.

The ancillary service market is divided into 6 hourly sessions per day with a pay as bid nodal design. The TSO acts as the sole buyer procuring the necessary resources for the management and control of the electricity system up to real time to maintain constant balance between production and consumption while keeping the fundamental network parameters stable.

Different types of ancillary services are secured, some in the competitive market while others are mandatory. These can be grouped in four categories below and shown in Fig. 3.21:

- System Management services ensure a secure and efficient operation and monitoring of the electricity system;
- Frequency Control services guarantee the balance between electricity generation and consumption at any time, in order to maintain a stable frequency, through active power regulation services;
- Voltage Control services maintain the voltage level in the permissible values through reactive power handling; and
- System Restoration services enable grid operators to restore the electricity supply as quickly as possible after a failure (these are last-resort services with mandatory participation).

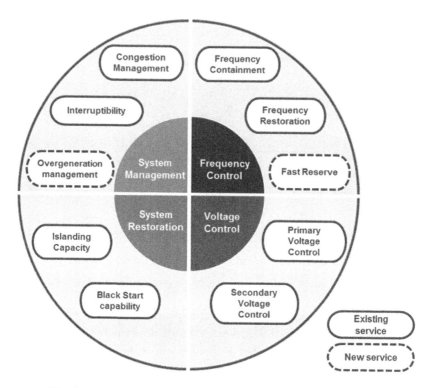

Figure 3.21 Services on ancillary service market. *(Source: Terna.)*

The changing resource mix triggered by the energy transition requires two major changes in the context of designing ancillary services markets:

- Provision of existing ancillary services from new resources to diversify and adapt to the changing generation mix and
- Introduction of new services that are required to ensure a secure system operation but were previously provided implicitly by other resources such as thermal and hydro or simply not needed.

There are services that have always been necessary to manage the electricity system, but that need to be progressively provided by new resources to replace thermal capacity, which is being phased out. Moreover, new services will emerge as the percentage of demand covered by variable RES increases. This includes services for:

- Fast Reserve: a service that delivers a very fast reaction (activation time < 1 s) to frequency variation from its nominal value. It would be capable to contain the rate of change of frequency after a major loss, partly offsetting the decreasing level of system inertia and

- Overgeneration Management (electricity storage): a service capable to face the challenge of structural overgeneration from intermittent renewables by absorbing RES production when it exceeds demand and releasing it again in the form of electricity in daily—weekly cycles.

The first service is essential to manage the consequences of a decreasing level of system inertia, which measures the capability of the system to "resist" or rectify an imbalance between generation and consumption without excessive variations in the frequency. Today, this service is provided "for free" by large power plants with rotating masses. The contribution of inverter-based generation such as wind and solar to system inertia, however, is close to zero, which should be addressed by procuring Fast Reserve explicitly.

A similar argument can be made for managing overgeneration. The expected growth in wind and solar will lead to increased overgeneration. By 2030, for example, electricity production is projected to frequently exceed demand, especially in spring and summer with a high penetration of solar, resulting in renewable curtailment, which is not helpful when the goal is decarbonization. At the same time, to continue operating the network safely, it is necessary to exploit new flexibility resources and this requires a framework that can ensure their development and market participation, while taking into account the different capabilities of each technology, topics covered elsewhere in this volume.

The market, therefore, must play a critical role to enable and unlock new resources, and this, as pointed out by other authors, requires adjusting the market rules and design to let new services—including flexible demand—to be adequately remunerated. During the past few years the ancillary service market has been evolving in response to the National Regulation (Resolution 300/2017/R/EEL), in particular the Resolution 300/2017/R/EEL, which sets the first opening to the ancillary service market to demand side response, RES, and storage.

Supporting these developments, since 2017 Terna has engaged in a few pilot projects to manage new resources, including small ones, to participate in the ancillary market including in a scheme called Virtual Aggregated Mixed Unit or UVAM. UVAM allows for the aggregation of storage, production, and consumption units including e-mobility infrastructures as unique market participant. This is the evolution of two previous pilot projects for consumption and production side only, respectively, as illustrated in Fig. 3.22.

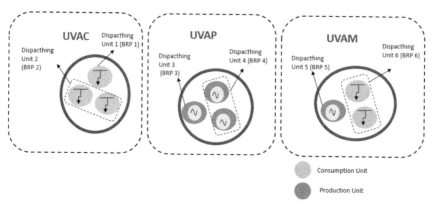

Figure 3.22 UVAM (Virtual Aggregated Mixed Units): composition and perimeter. *(Source: Terna.)*

- UVAC (Virtual Aggregated Consumption Unit) and
- UVAP (Virtual Aggregated Production Unit).

Fundamentally, UVAM consists of a Balancing Responsible Party (BRP) which is the owner of the dispatching unit, either consumption or production, included in the virtual unit and a Balancing Service Provider (BSP) which is responsible for the bidding on the ancillary service market of the entire aggregate.

The ancillary service market evolution is currently the main opportunity to allow the demand side to play an active role in the market. A number of chapters in Part Two of this book are focused on examples of developing and delivering flexible demand. Terna, as demonstrated by the market innovations presented above, supports further development of a market-based flexible demand.

Finally, as parts of its efforts to enable the energy transition, Terna has taken steps to digitalize its own systems to be better able to monitor and control the ever-increasing number of resources that actively interact with the grid. The exponential increase of small-scale nonprogrammable RES production implies a jump in the real-time volatility of the production, putting at risk the security of the grid. Electrification and behind-the-meter production/consumption activities call for new approaches in the real-time grid operations. In particular, a focal point will become the data availability in real time across the transmission and distribution network all the way down to substations where the TSO and DSO interface as schematically illustrated in Fig. 3.23.

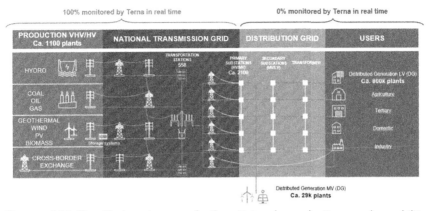

Figure 3.23 Simplified scheme of the Network and Terna observability perimeter. *(Source: Terna.)*

In addition to a deeper level of cooperation between the TSO and the DSO, the necessity to manage real-time data from generation, consumption, and storage instantaneously will require new competencies and tools.

6. Conclusions

The energy transition is posing new hard challenges to cope with. One of the key features of the future energy system will be the integration of increasing RES in the electricity market. Among the consequences will be challenging daily residual load profile, which will be characterized by steep morning deramps and evening ramp as well as negative residual load in the middle hours of the day, the duck curve phenomenon.

Furthermore, the reduction of traditional dispatchable power plants and the increase of nonpredictable RES pose risks for the electrical system adequacy and service quality. This requires the active involvement of the TSO to balance the trade-off between decarbonization and security and adequacy of the system.

As presented in this chapter, as variability of generation in the system rises, the TSO must adjust by implementing adequate countermeasures for safely supporting the transition. Cognizant of these challenges, Terna is focused on the following:

- Development of the transmission grid in the forthcoming years for optimizing the RES power flows across the network while reducing bottlenecks and curtailments including working on several major transmission projects such as the Tyrrhenian Link and Italy—Tunisia interconnection.

- Encouraging the formation of long-term price signals for investments in the generation fleet which has to be balanced between decarbonization and adequacy of the power system.
- Market design evolution for unlocking new grid services needed and for enabling new resources including flexible demand on the ancillary services market.
- Investing in the digitalization of the transmission grid including asset and processes to increase Terna's ability to better monitor and manage the network in real time.

Considering the complexity of the energy transition, to successfully manage the transformation of the system, a coordinated strategy with cooperation from various stakeholders along with a heterogeneous portfolio of solutions will be needed.

Bibliography

AEMO, 2018. Operational and Market Challenges to Reliability and Security in the National Electricity Market. https://www.aemo.com.au/-/media/Files/Media_Centre/2018/AEMO-observations_operational-and-market-challenges-to-reliability-and-security-in-the-NEM.pdf.

Australian Energy Market Operator (AEMO), 2020. Renewable Integration Study. https://aemo.com.au/-/media/files/major-publications/ris/2020/renewable-integration-study-stage-1.pdf?la=en.

DIW Berlin, 2013. Residual Load, Renewable Surplus Generation and Storage Requirements in Germany.

ENTSO-E, 2015. SOAF Methodology Adequacy. https://eepublicdownloads.blob.core.windows.net/public-cdn-container/clean-documents/sdc-documents/SOAF/141014_Target_Methodology_for_Adequacy_Assessment_after_Consultation.pdf.

ENTSO-E, 2018. Ten Years National Development Plan. https://tyndp.entsoe.eu/tyndp2018/.

European Commission, 2015. Energy Union Package, COM/2015/080.

European Commission, 2019a. Clean Energy for all Europeans.

European Commission, 2019b. The European Green Deal, COM (2019) 640 Final.

International Energy Agency (IEA), 2019a. Africa Energy Outlook.

International Energy Agency (IEA), 2019b. World Energy Outlook (WEO).

Integrated National Energy and Climate Plan, 2019. Italian Ministry of Development.

Joint Research Center, 2018. Pan-European Analysis on Power System Flexibility.

Lawrence Berkeley National Laboratory, Feb 2020. California Demand Response Potential Study Phase 3: The Potential for Shift Through 2030.

Med-TSO, 2018. Mediterranean Master Plan. https://www.med-tso.com/masterplan.aspx?f=.

Terna, 2019a. National Development Plan. https://www.terna.it/en/electric-system/grid/national-electricity-transmission-grid-development-plan.

Terna, 2019b. Scenario Report. https://www.terna.it/en/electric-system/grid/national-electricity-transmission-grid-development-plan/scenarios.

Terna, 2019c. Statistical Factsheet. https://www.terna.it/en/electric-system/statistical-data-forecast/statistical-publications.

Terna, 2019d. The Energy Transition in Italy and the Role of the Gas and Power Sector. https://www.terna.it/en/media/news-events/states-general-italian-energy-transition.

CHAPTER 4

Integrating the rising variable renewable generation: A Spanish perspective

Juan José Alba, Julián Barquín, Carolina Vereda, Eduardo Moreda

Endesa, Regulatory Affairs, Madrid, Spain

1. Introduction

The European Union (EU) has started a process that should lead to the European economy's full decarbonization by 2050.[1] Each Member State must develop a plan that outlines the process of transition for that country. The national commitments are spelled out in National Energy and Climate Plans[2] that are to be periodically published and submitted to the European Commission for review.

Spain has presented its first National Energy and Climate Plan (2021−30), called the PNIEC, its Spanish acronym.[3] The plan complies with European objectives and policies and has been welcomed by the European Commission. The central feature of the PNIEC is deep penetration of renewable energy sources (RES). The plan also considers bioenergy and renewable gases for transportation and use in buildings and industrial processes. However, most of the increase under the plan is expected to happen in the electricity sector. The intention is not only to meet current demand with greener electricity but also to electrify end markets currently served by fossil fuels, notably transportation, as well as space heating and cooling.

In the electricity sector, most of the new resources will be renewable and therefore variable, notably solar—both photovoltaic and thermoelectric—and wind. Their increase is expected to be a massive one, as renewables

---

[1] *A Clean Planet for all. A European strategic long-term vision for a prosperous, modern, competitive and climate neutral economy.* COM/2018/773. European Commission, 2018.

[2] The obligation is stated in the *Regulation on the governance of the energy union and climate action* (EU/2018/1999).

[3] Available at the Ministry for the Energy Transition and the Demographic Challenge website, https://www.miteco.gob.es/es/prensa/pniec.aspx.

Variable Generation, Flexible Demand
ISBN 978-0-12-823810-3
https://doi.org/10.1016/B978-0-12-823810-3.00017-0

85

become the generation backbone of the system. The challenge of variability must be addressed by using a number of flexible energy resources, including storage technologies, sector coupling, and demand response.

The variability challenge grows as electricity decarbonization advances. Decarbonization of electricity is linked to a parallel electrification process, which provides an opportunity to harness variability. For instance, storage technologies are useful for electrifying sectors where the potential for electricity use is limited, as batteries for transportation and hydrogen for industries requiring high temperature heat. Demand-side technologies, such as high-efficiency low-temperature heat pumps, are also a form of electrification that furthers decarbonization. The full deployment of the technologies needed to reach zero net carbon requires the consideration of energy sector-coupling mechanisms, as well as alignment of regulation and taxation among all energy sectors.

The chapter is organized per below:
- Section 2 describes the Spanish Climate and Energy Plan 2021—30;
- Section 3 examines the period after 2030 when deep decarbonization kicks in and the need for long-term storage becomes critical;
- Section 4 discusses the role for demand flexibility after 2030;
- Section 5 examines sector coupling, linking electricity, natural gas, and oil; and
- Section 6 examines some regulatory issues to ease the transition to and after 2030; followed by the chapter's conclusions.

2. The Spanish energy and climate targets

The Spanish targets are spelled out in the National Energy and Climate Plan, PNIEC, which is updated every year. The numbers quoted below are taken from the 2020 version[4] and shown in Fig. 4.1. The plan envisages the share of RES in final energy consumption in 2030 to be 42% (33.2 Mtoe), up from 18% (15.3 Mtoe) in 2017. The RES share increase reflects a strong decrease in the consumption of oil and oil derivatives, from 43 Mtoe in 2017 to 29.3 Mtoe in 2030, a slight increase in natural gas consumption from 13.5 to 13.8 Mtoe during the same period, and further reductions of an already relatively small coal consumption from 1.8 to 1.4 Mtoe.

---

[4] See previous footnote.

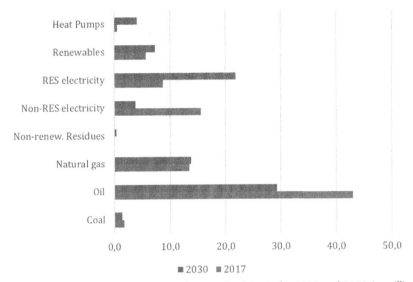

Figure 4.1 Final Energy consumption for Mainland Spain for 2017 and 2030, in million tons of oil equivalent (Mtoe). *(Source: Spanish National Energy and Climate Plan, 2020.)*

Renewable energy, both as electricity generation and in direct final use (mainly biomass), rises significantly. Heat pumps are closely related to the increase in renewable energy and merit a specific comment. They work by harnessing electrical energy in order to funnel an even greater amount of heat energy (up to 3 times or even more) from the outside environment into the buildings to be heated. These heat flows into buildings, according to European regulations, count as renewable energy as shown in Fig. 4.1.

Most of the reductions in oil consumption occur in the residential and transportation sectors due to:

- An increase in energy efficiency in final uses (e.g., better insulation and more efficient car engines) and
- A shift from fossil fuels to electricity (e.g., heat pumps and electric cars).

Mainland Spain's electricity demand[5] is forecasted to grow from 253 TWh in 2017 to 263 TWh in 2030. This modest increase in demand reflects the expectation that energy efficiency in final uses will improve as a

---

[5] This chapter's data and graphics are for mainland Spain. The Balearic and Canary Islands must also decarbonize their economies by 2050, possibly earlier. However, they present quite specific problems and solutions related to their small size and the nature of their economies.

Table 4.1 Electricity generation mix for Mainland Spain in TWh.

	2017	2030
Hydro	18,4	32,4
Pumped storage generation	2,2	13,8
Nuclear	55,5	22,0
Coal	42,4	0,0
CCGT	33,6	27,6
Wind	47,5	109,5
Photovoltaic	8,0	65,2
Concentrated solar power	5,3	19,8
Other RES	6,8	12,1
Cogeneration	28,2	18,4
Battery generation	0,0	3,3
Generation	**248,1**	**324,0**
Pumped storage consumption	−3,6	−18,4
Battery consumption	0,0	−3,7
Net int'l trade	8,0	−39,3
Demand	**252,5**	**262,7**

Source: Spanish National Energy and Climate Plan, 2020.

result of electrification, which is significantly more efficient than the fossil fuels they replace.

The overall changes in the electricity mix from 2017 to 2030 are enormous as shown in Table 4.1.

When comparing the 2017 and 2030 generation mix, the most striking element is the huge increase in RES generation, from 68 to 208 TWh,[6] which requires an even larger increase in RES capacity. Wind capacity, for example, is expected to grow from 23 to 49 GW, concentrated solar power (CSP) from 2 to 7 GW, and photovoltaic from 4 to 38 GW. The average annual investments are projected to be more than 6 GW.[7]

To minimize the required investments, variable generation must be used to the maximum extent possible. Storage, as discussed below, provides a way to avoid curtailment (spillage) and make economic use of the generated power. Exports might provide another venue, and the plan forecasts a huge increase in exports. However, currently the Spanish plan is not coordinated

---

[6] 2017 was a dry hydraulic year and 2030 is taken as an average hydraulic year. That largely explains the difference in hydro generation for those 2 years.
[7] Similar challenges arise in other systems transitioning toward decarbonization. In particular, see in this book Chapters 1−3, where the California, Texas, and Italy cases are examined.

with other national plans, and in particular the French one. So, trade might not be so favorable to Spain as expected under the plan.

Currently, there are interconnectors with Portugal, France, and Morocco. The Portugal interconnection is very strong. Both systems operate within the same electricity market (MIBEL). The day-ahead price is the same in both countries more than 90% of the time. No significant changes are expected before 2030. The interconnection with France is significantly weaker, as reflected in much greater price differences. Finally, the Morocco interconnector is relatively small. The PNIEC assumes that interconnection capacity with France will have doubled by 2030 to 8000 MW. The Morocco interconnector capacity will remain at 600–900 MW. In 2017 net electricity imports into Spain amounted to 9.2 TWh. The PNIEC assumes that in 2030 Spain will export 31.8 TWh. The 41 TWh shift is problematic, as it seems that the huge RES investments in Portugal, France, and other European countries that are included in National Energy Plans have not been considered. RES production in these countries is correlated with the Spanish production. So, trading balances are likely not to be as favorable as assumed by the Spanish Government (Fig. 4.2).

Given these limitations, and in order to manage this amount of variable generation, large investments in storage facilities are planned. They include 3.5 GW of additional pumping (pumped storage) stations, 5 GW of new

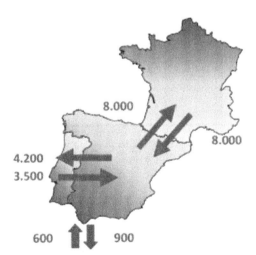

Figure 4.2 Mainland Spain interconnections, current intertie capacity in MW. *(Source: Spanish National Energy and Climate Plan, 2020.)*

CSP stations incorporating thermal storage, and 2.5 GW of new batteries. However, the Plan is unclear about how these investments translate into storage capacity (GWh) and about other specifics.

The plan envisages about 5 million electric vehicles in 2030, 3.5 million of which are electric cars.[8] As described in Chapter 8, an electric car can be conceptualized as a "battery on wheels." Total associated battery capacity is not specified in the plan and is highly uncertain, but assuming 20 kWh average per vehicle it might amount to 100 GWh. Power to grid technologies embodied in these vehicles might provide a significant fraction of the required batteries' capacity. For instance, if 5% of parked cars' batteries can be used as a system storage asset on a daily cycle, it would imply something of the order of 5 GWh of energy flow in and out of batteries per day, or about 1.8 TWh per year. This figure is more than half the 3.3 TWh for batteries in Table 4.1.

Obviously, the 5% of battery capacity can correspond to different cars each day. In any case, "smart" schemes to manage the charging and possibly the discharging of the electrical fleet will be critical. These schemes are actually required in order to integrate electric vehicles without excessive network investments. But if electric vehicles are to participate actively in providing storage and other services to electricity systems, Vehicle to Grid (V2G) capabilities must be developed, as discussed in Chapter 8.

It is noteworthy that, under the Plan, no new technologies are required, nor huge changes in consumers or industries behavior. For example, the Plan does not consider green hydrogen (using renewable electricity to split water into hydrogen and oxygen through the process of electrolysis), marine renewable sources, or massive deployment of low carbon sea transportation.[9] It rather assumes that massive deployment of new technologies will not start until the 2030s or even later. Likewise, consumers

---

[8] Electric vehicles, as used in the Plan, encompass all kind of cars, trucks, buses, bikes, etc. as well as electric and hydrogen fuel cells—powered vehicles. It is unclear about hybrid electric vehicles. Nonetheless it strongly stresses the need for recharging infrastructures, suggesting that at least plug-in hybrid vehicles are considered also as electric vehicles.

[9] Sea transportation is quite important when discussing the energy transition in the Balearic and Canary archipelagos. Possibly, sea transport decarbonization should start with domestic traffic, as international shipping decarbonization requires an international legal framework that does not yet exist. In any case, measures intending ports' decarbonization (also important from a local air quality point of view) are already being considered.

and industries are assumed to operate largely as they do now until 2030. So, for instance, efficiency assumptions suggest that online work from homes will grow but not become dominant, and that industries will not dramatically change the timing of production to adapt to renewable electricity production.

As further explained in the following section, deep decarbonization not only requires massive investments in RES but also fundamental changes in how society functions and the economy operates, including new ways of making better use of the inherent variability of renewable generation, topics covered elsewhere in this volume.

3. Deep decarbonization and storage

According to the current plan, in 2030 there will still be a significant capacity of natural gas—fired CCGTs and cogeneration. Gas-fired generation output will be much lower than now but will be needed to supply electricity during hours of low output from renewable resources (e.g., during the night or under low-wind climatological conditions). New RES investments in electricity will be increasingly inefficient. As solar and wind generation are correlated across the country, eventually the output of additional RES investments exceeds demand and increases curtailment. This is the well-known cannibalization effect. As variable renewable generation increases in hours of already-high renewable generation, not only does it not cover hours when generation is needed, it also further depresses prices and market income for already existing renewable generation assets.

An obvious solution is to store the variable RES generation. Dramatic decreases in batteries' costs have sparked a marked optimism in storage possibilities. And, indeed, batteries can be an extremely efficient way to address daily variability. For instance, it is nowadays usual that PV installers offer, together with residential solar panels, batteries to make use of midday energy after sunset.[10] Hybrid PV and battery farms (or hybrid wind and battery parks) might also be an efficient solution. An advantage is that network access can be sized to be significantly lower than the maximum

---

[10] Residential PV is more expensive, on a €/kW basis, than industry scale PV. However, it can lead to significant savings in network development, if properly placed alongside optimized battery storage. That highlights the importance of well-designed network tariffs and charges. In other words, well-designed network tariffs and charges are essential for efficient PV and storage deployment.

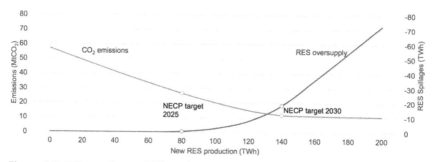

Figure 4.3 RES curtailment (TWh, right axis) and carbon emissions (millions of tons of CO_2, left axis) as a function of additional RES investments. *(Source: Endesa's own simulations.)*

generation output. This is because energy generated during peak hours (e.g., summer middays in the PV case) can be stored locally and fed later into the system. Network savings might be significant and, in addition, the hybrid plants can be efficiently controlled in order to provide other system services (e.g., secondary regulation).[11]

Fig. 4.3 shows the results of a simulation computing the carbon emissions and expected curtailment as a function of new RES investments. Generation capacity of all technologies, with the exception of wind and PV, is assumed to be the target 2030 capacity. Wind and PV capacities are increased from 2017 values. The capacity ratio for new wind and PV capacity is fixed in order that assumed 2030 target capacities are met. Electricity demand is the 2030 target demand and the hydro year is assumed to be the average one.

The horizontal axis shows the generated electricity from the new, post 2017, wind and PV capacities. It runs from 0 (no new capacity) through 140 TWh (2030 target) to a maximum of 200 TWh. With no new RES investments (and 0 new RES production) carbon emissions are slightly below 60 MtCO2 per year, and spillage is negligible. When the 2030 target RES production (140 TWh) is reached, carbon emissions are greatly reduced to about 10 MtCO2 per year. However, wind and PV spillage are about 20 TWh per year. The Plan does not consider RES production over the target of 140 TWh before 2030. However, further increasing wind and PV investments do not significantly decrease carbon emissions, but they increase curtailment.

---

[11] The increasing popularity of hybrid plants in the United States is reviewed in "The Rise of Hybrid Generation", *EEnergy Informer*, August 2020.

It should be noted that the simulation does not consider exports or imports (mainland Spain is assumed to be an isolated system). Interconnections might allow the export of generation otherwise curtailed in Spain. On the other hand, energy otherwise curtailed in other countries might be imported. Therefore, curtailment levels should be reduced when interconnectors are considered, although the amount and its sharing between Spain and the foreign systems strongly depends of the correlation of RES production between Spain and the neighboring countries.

The simulation also assumes that storage capacity is fixed at the values forecast by the PNIEC in 2030. However, if additional storage is allowed, increasing RES investments can lead to even lower carbon emissions as well as less curtailment. In particular, batteries should be considered.

Currently, batteries are an efficient technology to manage variable RES generation on a daily basis. Even so, significant investments may be needed. For instance, Fig. 4.4 shows simulation results for a fully decarbonized electricity system, like the one expected for 2050. The target is to move excess solar energy from the midday hours (the yellow hills) to serve the demand during the rest of the day (the black line). From the simulation it can be concluded that about 345 GWh of battery capacity would be required. Even at a cost of 100 €/kWh, that entails an overnight cost of about €34.5 billion, a staggering figure. But not unknown when total power investments during a number of years are considered.

To further move toward a deeply decarbonized system, we must face the problem that daily storage, as provided by batteries, is not capable of providing flexibility over a longer period of time, from weeks to months.

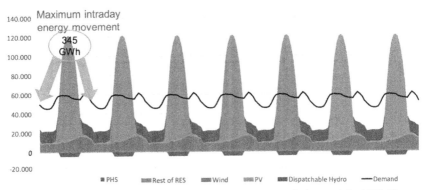

Figure 4.4 RES daily cycle storage. System generation and demand, in MW. *(Source: Endesa's own simulations.)*

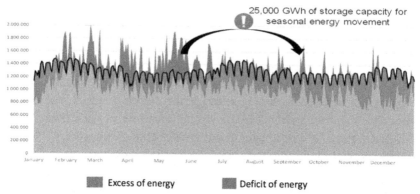

Figure 4.5 RES seasonal cycle storage, in GWh. (Source: Endesa's own simulations.)

In order to build a RES-based system, energy must be moved from the summer (when solar generation is higher) or spring (when wind generation is higher) to the wintertime. While batteries may be a reasonable solution for shifting limited amounts of energy for a few hours, they are not suitable for shifting energy across the seasons. For instance, in Fig. 4.5, a similar simulation to that of Fig. 4.4 is presented. However, the energy transfer is now seasonal, mostly from spring to winter.

About 25,000 GWh are moved from spring to winter. This figure might seem small when compared with the 345 GWh per day, or about 126,000 GWh per year of the previous simulation (that of Fig. 4.4). However, there is a fundamental difference. 345 GWh on a daily cycle requires 345 GWh storage capacity. 25,000 GWh on a yearly cycle require 25,000 GWh of storage capacity. The relevant figure is not the stored energy but the stored energy during a load−unload cycle. The required storage capacity for seasonal cycles is so large as to make batteries a clearly unfeasible option.

Traditionally, seasonal storage has been used to manage hydro systems. Countries with large hydro reservoir capacity, as in Norway, Brazil, or Colombia, can manage large seasonal energy transfers. This eases the penetration of other variable RES. In particular, hydro generation can be easily ramped up and down, balancing variable PV and wind generation. There might also be other significant synergies.

For instance, Brazil meets most of its electricity demand from hydro generation in normal years. The challenge is how to handle multiyear dry years. Supply crises are not related to lack of generation (MW) capacity but to lack of energy (i.e., water to generate electricity) during very dry spells.

Wind and PV resources follow other climatological rhythms than hydro inflows. Moreover, in some regions dry years are correlated with windy or sunny ones, providing a natural hedge.

However, other countries including Spain are not so blessed with hydro resources; other sources of long-term storage are needed.[12] Hydrogen and other power-to-gas options are currently the leading candidates. Hydrogen from water electrolysis, powered by RES electricity, is an old technology whose performance and costs are fast improving. Hydrogen can power gas turbines or fuel cells to generate electricity. Storage is cheap, either short to medium term in tanks or long term in salt caves and other geological formations. As discussed below there are also a number of additional uses for hydrogen that are outside the electricity sector.

4. Deep decarbonization and demand-side flexibility

Demand-side flexibility is an energy resource that has been largely neglected until recently. It can provide cost-effective ways to balance an increasingly variable and inflexible generation system. Moreover, increasing variable RES generation implies that there will be long periods of low electricity prices. Investing in demand flexibility not only facilitates variable RES penetration but also enables demand to profit from lower average electricity costs.

Demand-side flexibility is similar to storage in the sense that it rarely implies the complete foregoing of the energy demand but rather that the delivery is either delayed or shifted. So, system-wide energy consumption falls during peak periods and increases during off-peak periods. Actually, that is how storage works, as shown in Figs. 4.4 and 4.5.

Most discussions on demand-side flexibility focus on explicit mechanisms, under which consumers are instructed to lower and/or shift their consumption away from peak demand hours. The rationale for this is that explicit instructions can provide significant benefits to the system and that the required technical, entrepreneurial, institutional, and regulatory structures for other mechanisms are not yet ready. However, implicit

---

[12] Even so, the PNIEC forecasts 5 GW of additional pumping capacity by 2030. This is a great challenge. There are not many locations, and most of the technically possible sites have significant environmental issues. In any case, this capacity is mostly intended for daily cycles, and does not meet the requirements for seasonal management.

demand-side flexibility, where consumers react to well-designed price signals, could be even more important.[13]

In residential and commercial sectors, distributed PV panels coupled with batteries provide, from the system viewpoint, services essentially identical to demand management. Electric appliances can also be turned on and off, without loss of comfort, topics extensively covered in Chapter 7. As electricity penetration increases with electrification of heating and transport, the potential for demand flexibility increases. For instance, efficient electric heat pumps can be installed together with hot water tanks that effectively store energy as heat to be released as needed. Resistance heating refractory bricks, albeit much less efficient, provide the same service at a much lower capital cost. They could absorb RES electricity that would otherwise be wasted. Even more basically, buildings have significant thermal inertia that can be harnessed by smart automatic management, for instance heating in advance of need, if the conditions (e.g., high RES generation) are favorable. Note that all these measures tend to reduce curtailed energy and, consequently, improve variable RES economics.

Industrial loads can also provide flexibility, as described more fully in Chapters 7 and 16 of this book. An advantage over residential demand is that the required control systems, which do not scale or scale very weakly with facility size, have a much smaller relative cost (€/MW or €/MWh). This may be particularly relevant if explicit demand side schemes are introduced, as they typically require sophisticated metering and control systems. On the other hand, digitalization is reducing monitoring and control costs. That should facilitate participation by small commercial and residential in the provision of flexibility services.

As the energy system is decarbonized and electricity penetration increases through electrification, demand–side flexibility will grow. Processes that can profit from extended periods of low or very low electricity cost will gain relevance. Optimal technologies will combine high operational flexibility with relatively low capital costs; the latter will not be the only parameter to be considered when deciding on investments. For instance, flexible water desalinization facilities have both high capital costs and high energy consumption. Long spells of low electricity prices tilt the balance toward more flexible designs with higher capital costs, as the precise

.

---

[13] The issues are explored in much more detail in other chapters in this book. For a general overview of demand flexibility, see Chapter 5.

moment in which desalinated water is fed into the water supply system is not of great relevance.

There are also measures on the operations side that will sometimes imply out-of-the-box thinking. For instance, electricity cost is likely to be lower in summer, when PV generation is higher. In some industries, summer holiday might become winter holiday, to take advantage of this spread. In any case, industries are very different, and there will be a number of specific and sometimes surprising solutions.

5. Deep decarbonization and sector coupling

Electricity consumption currently represents a small part of final energy consumption: 24% in 2017 for Spain.[14] Most energy consumption to be decarbonized happens outside the electricity sector. As the transition advances, linkages between energy sectors—up to now almost disconnected—will grow. This is particularly relevant for heat and for transportation.

Generally speaking, sector coupling between the electricity, natural gas, and oil sectors has been relatively weak until recently. In most European countries, electricity generation was based on hydro, nuclear, or coal. There was little overlap in the uses of these primary energy sources and of oil derivatives or natural gas. Consequently, electricity, oil, and natural gas were considered as three essentially independent energy sectors, often subject to very different regulatory and taxation regimes. Increased penetration of natural gas—fired generation caused electricity price to be dependent on natural gas prices, although the causal relationship in the opposite direction was weak. As modern electricity-based technologies, such as electric vehicles or heat pumps, directly compete with their traditional fossil fuels counterparts, regulation and taxation of the different energy sectors must be better aligned to avoid economic inefficiencies, as noted by Luca La Schiavo in the book's Preface.

Moreover, most RES are actually renewable electricity sources, with the important exceptions of biomass and solar thermal heating. However, and especially in highly populated regions as Europe, biomass can only supply a fraction of the total energy needs. Solar thermal heating is, in any case, a very specific technology. Therefore, the share of renewable electricity in

---

[14] Consumo de energía final, IDAE. http://sieeweb.idae.es/consumofinal/bal.asp?txt=Consumo%20de%20energ%EDa%20final&tipbal=s&rep=1. Accessed July 20, 2020.

primary energy supply can only grow. Renewable electricity will supply an increasing fraction of final energy, either directly or through intermediate vectors (e.g., hydrogen).

Currently, heating needs in Spain are mostly met by burning natural gas, although oil derivatives are still quite relevant. There are two kinds of heat: low temperature mostly intended for space heating, but also important for some industrial processes; and high temperature that is needed mainly for other industrial processes. In addition, smaller amounts of fossil fuels are needed as raw materials for the chemical industries.

Low temperature heating shows a marked seasonal character.[15] In 2017, natural gas demand by residential and small consumers amounted to 67 TWh, mostly for space heating. Gas consumption peaks at about 575 GWh per day, more than three times the daily average. As mentioned above, electrical heating using heat pumps can be much more efficient than the alternatives. On the other hand, the additional electrical demand for heating comes at times when the "traditional" demand is already high (cold winter days). In other words, electricity demand is likely to become "peakier" than today, further increasing the attractiveness of storage and demand flexibility. In any case, an outcome of the decarbonization process might be the disappearance of nonelectrical low temperature heating, with the exception of few niches.[16]

The situation is quite different in the industrial sector. Electrical high temperature heating is not so efficient or cheap as electrical low temperature heating. In addition, some industrial process (e.g., ironworks) use fossil fuels not just as an energy vector but also as a reducing chemical agent.[17] Hydrogen and other electro-gases can be harnessed to play both roles.

Green hydrogen obtained from renewable electricity through water electrolysis is a promising technology.[18] As opposed to blue hydrogen from fossil methane, it does not require carbon capture and sequestration processes, which are quite unpopular in Europe and whose development has

---

[15] See ENAGAS, "Boletín estadístico del gas". https://www.enagas.es/enagas/es/Gestion_Tecnica_Sistema/Seguimiento_del_Sistema_Gasista/Boletin_Estadistico_Gas. Accessed July 20, 2020.

[16] See A. Bloess, *Impacts of heat sector transformation on Germany's power system through increased use of power-to-heat*. Applied Energy 239 (2019).

[17] See Material Economics, *Industrial Transformation 2050, Pathways to Net-Zero Emissions from EU Heavy Industry*, University of Cambridge, 2019.

[18] See I. Petkov and P. Gabrielli, *Power-to-hydrogen as seasonal energy storage: an uncertainty analysis for optimal design of low-carbon multi-energy systems*. Applied Energy 274 (2020).

remained essentially stagnant for many years. In addition, whenever carbon capture and sequestration are involved, carbon emissions are merely low, as opposed to being essentially nil in the case of green electrolytic hydrogen. On the other hand, and as opposed to biological hydrogen, the resource base for green electrolytic hydrogen is almost unlimited. Consequently, energy bioresources might be better employed for other purposes, for instance as biofuels in difficult-to-decarbonize transportation.

Green hydrogen might be further processed into liquid fuels, albeit at significant energy cost. Hydrogen can be reacted with air nitrogen, in order to synthetize ammonia.[19] Ammonia is easy to store in liquid form, has a number of applications in the chemical industries, and can be burned to propel vehicles. However, attention must be paid in order to avoid the release of nitrogen oxides, as they are inter alia strong greenhouse gases and local pollutants.

Hydrogen can be also reacted with a carbon source to synthetize liquid and gaseous hydrocarbons.[20] However, the carbon source should not have fossil origin in order to avoid greenhouse gas emissions. Biogenic carbon or carbon dioxide from air might be valid alternatives. As noted above, biomass resources can be quite limited. Carbon dioxide from air is unlimited, but the capture process and processing to synthetize hydrocarbons involve a significant amount of energy.

Regarding green hydrogen, electrolizers are quite small for industrial standards, so they might be best placed close to industrial sites, minimizing the need for gas transportation infrastructure. Current gas infrastructure might mostly become, like fossil power plants, a stranded asset. As stated above, hydrogen could double as an industry fuel and a long-term medium for storage. These synergies will bring lower hydrogen prices to industrialists. This is important because the likely outcome is a significant wedge between prices for low-temperature and high-temperature heat. Both will come from renewable electricity, but in the first case costs will be low due to the very high efficiency of heat pumps, whereas in the second case, costs of heat will be higher due to the limited energy efficiency of electrolizers, unlikely to climb above 85%.

---

[19] D. Miura and T. Tezuka, *A comparative study of ammonia energy systems as a future energy carrier, with particular reference to vehicle use in Japan.* Energy 68 (2014).

[20] S. Brynolf et al., *Electrofuels for the transport sector: A review of production costs.* Renewable and Sustainable Energy Reviews 81 (2018).

Oil derivatives markets will also be linked to the electricity sector. These derivatives are used both in heating and in transportation. A heating sector discussion can be carried out in similar terms to the one for the gas sector. Transportation coupling can lead also to system benefits.[21] As discussed above, short-term storage for the power sector might be mostly carried out by batteries. There are at least three ways in which transportation batteries can provide demand flexibility in the power sector:

- Firstly, car batteries can be directly harnessed, either by smart charging or by providing balancing services such as V2G technologies.

- Secondly, old car batteries can enjoy a second life as stationary batteries embedded in the electrical network, since less demanding performance is adequate.

- An often-overlooked option is that high capacity public chargers will be fitted with significant battery capacity, even if only to avoid oversized feeders.[22] The owners are likely to be familiar with electrical market operations and keen to obtain additional revenue streams, topics further examined in Chapter 8.

6. Deep decarbonization and regulation

The transformed energy system that will emerge after the transition will require proper regulation given the scope of changes required, as described in broad terms by Luca La Schiavo in the book's Preface. Avoiding economic inefficiencies is extremely important, as the energy transition cost, no matter how needed it may be, will be enormous. At present, it is only possible to suggest some general ideas. Indeed, regulation usually follows technological developments, and the specific nature of these is still unclear. However, it is worth thinking about a number of ideas with regulatory relevance to ensure the transition is as efficient as possible.

- First, short-term economic signals—short-term prices—are going to be more important than now. Marginal generation costs will change from essentially zero during excess energy periods to reach very high values when backup energy (e.g., electricity from hydrogen fuel cells) is needed. The timing of these events, as that of intermittent RES energy itself, is highly uncertain. In addition, highly variable flows throughout

---

[21] See Chapter 8 in this book.

[22] For instance, a 1 MW charging station can be supplied by a 500 kW feeder plus 500 kW batteries, charged during idle periods.

the network require short-term network tariffs.[23] The variability is needed so that consumers have incentives to move their demand to times when the system is uncongested. So, moving away from short-term electricity markets is the wrong move. Indeed, market prices need to move closer to real time. Current regulatory reforms to shorten the bidding periods in Europe to 15 min and generally moving market operations closer to real time are positive developments.

- Second, the associated short-term markets should clear, as in any market, at the point where supply and demand meet each other. Traditionally electricity demand was considered rigid and electricity generation flexible. For instance, planning studies traditionally considered the hourly demand curve of previous years, possibly affected by forecasted GDP growth and other economic indices, and then computed the best generation fleet to meet that demand. We move now to a situation in which the generation is more inflexible, as intermittent RES generation is largely independent of prices, whereas demand must become much more flexible.

- Third, increased short-term price volatility creates a greater need for hedging structures that allow agents to lock in a price. The discussion is not a new one: it is central to discussions about capacity remuneration mechanisms. Regulatory authorities, and especially European ones, must cease to consider these mechanisms as a subsidy, possibly to justify subsequent regulatory interventions that have unclear support in European Treaties. It is also very important to avoid ill-designed mechanisms, such as feed-in-tariffs, which National authorities often justify as a way to deal with investment risk premia. These feed-in-tariffs typically undermine the fundamental role of short-time prices to provide efficient signals for operation. Reliability options and other market-friendly capacity remuneration mechanisms can provide useful templates not only to hedge prices but also to collect revenue in order to support the transition, as discussed below.

- Fourth, it is likely that the energy transition will require extra support besides markets. In the past, RES facilities required significant economic support, sometimes many times over the market revenues, in order to justify investment. As PV and wind generators have matured, this is no longer the case. However, storage technologies as well as a number

---

[23] See S. Haro at al., *Toward Dynamic Network Tariffs: A Proposal for Spain,* in Innovation and Disruption at the Grid's Edge. Elsevier, 2017.

of final use electrical technologies are still relatively immature. Past errors must be avoided.

- Finally, taxes and levies must be harmonized across energy vectors. In Spain, RES investments in electricity—which account for most of the RES investments—have been almost solely funded by levies included in electricity prices. General ad-valorem taxes (VAT and a specific excise tax) add an additional 25% mark-up on the final price. As a consequence of these levies and taxes, more than half of the electricity price paid by households and small businesses is unrelated to the supply cost. Gasoline and diesel excise taxes are also very high. There is no legal relationship between the amount of these hydrocarbon taxes and the transportation infrastructure cost (mostly roads) although, on an average basis, both figures are similar. Natural gas is, by comparison, very lightly taxed. That can be seen as a heritage of the 1980s, when it was public policy to substitute oil imports by natural gas imports. However, this structure does not make sense any longer.

The lack of harmonization avoids effective interfuel competition and discourages investment in flexible demand. For instance, highly efficient heat pumps fitted with hot water tanks can be better than natural gas boilers from the point of view of energy system cost as well as for security of procurement, besides providing much needed flexibility. However, heat pumps will be uncompetitive as long as the electric customer faces much higher prices due to relatively higher taxes and levies for electricity than for natural gas.

In closing, the harmonization of taxes and levies based on externalities (carbon emission, local pollutants, road financing, security of procurement costs, etc.) must substitute the current fiscal structure. Costs not related to the energy supply cost and its externalities come from societal decisions and, therefore, are ideally financed through general taxation. This is the case of the energy transition support costs. A second best is to charge all energy vectors. Energy content taxation is allowed by European regulations and might provide an alternative.

7. Conclusions

Climate targets are linked to two parallel processes: enormous investments in variable renewable electricity generation and much higher electrification levels. Renewable electricity is required because most of the renewable energy is actually renewable electricity. Furthermore, electrification (using renewable electricity) is required because electric appliances are generally much more energy efficient than the nonelectric alternatives and are required to balance a less flexible generation system.

Technically, electricity system balancing is increasingly provided by storage and demand-side technologies. Short-term (e.g., batteries) and long-term (e.g., hydrogen) storage are both needed. Interestingly these technologies also play an important role outside the electricity sector, in particular to decarbonize transportation and industries, providing an avenue for fossil fuel phaseout in the economy.

During the transition period, these new storage and demand technologies have to compete with traditional fossil technologies. Regulation and taxation must be aligned in the electricity and nonelectricity sectors to avoid economically inefficient outcomes that would raise the costs of decarbonization. Revenues needed to finance specific energy infrastructure costs (including roads and other transportation infrastructure) should be recovered through sector-specific regulation. On the other hand, energy transition costs (e.g., support to nonmature technologies) should preferably be financed by the government budget.

PART 2

Flexible demand

CHAPTER 5

What is flexible demand; what demand is flexible?

Fereidoon Sioshansi
Menlo Energy Economics, Walnut Creek, CA, United States

1. Introduction

Forecasting system peak demand—as it is often called—and how to meet it has kept generations of demand forecasters, system planners, grid operators, economists, rate design specialists, statisticians, and modelers gainfully employed. The task is not a trivial one. Those in charge of running electricity networks anywhere in the world are responsible to keep the lights on at all times, making sure there is ample generation, transmission, and distribution capacity available to meet the peak demand—the highest level of coincident demand experienced in the course of a year. A lot of planning, modeling, and scenario analysis typically goes into forecasting how big the peak demand is likely to be and when it can be expected.

Since building electricity infrastructure is capital intensive and takes time, the planners usually look a decade or more into the future. And since lots of variables affect the peak demand, they consider alternative scenarios with different assumptions—e.g., higher rates of economic growth, rising temperatures, varying levels of appliance ownership and utilization rates, to name a few—in coming up with alternative projections. And since nobody has a crystal ball, and those in charge do not wish to be blamed for shortages—which can result in blackouts or brownouts—they typically overinvest a bit just to be on the safe side. There is massive literature on topics such as resource adequacy (RA), reserve margins, loss-of-load probability, etc.

Another important consideration in deciding how much capacity may be needed is due to the highly *asymmetric* penalty and reward system for grid operators, utilities, regulators—even the politicians. This has always been, and still applies universally. As long as the lights stay on, a little—or even a lot—of overcapacity is generally tolerated. If, however, the lights go out because those in charge did not plan ahead and/or did not build sufficient

Variable Generation, Flexible Demand
ISBN 978-0-12-823810-3
https://doi.org/10.1016/B978-0-12-823810-3.00004-2

capacity to meet the peak demand would find themselves in grave trouble. Numerous governors,[1] politicians, regulators, utility CEOs, and those running the grid have been summarily fired and/or have had to explain what happened and why they did not see it coming. This asymmetry has resulted in a culture where everything in the long electricity value chain is modestly—or grossly—oversized, overbuilt, and/or is gold-plated.[2]

Making matters worse, historically the rate of return regulation, under which virtually the entire industry operated up to fairy recently, incentivized the stakeholders to overinvest in infrastructure—everything and anything upstream of customers' meters. The more they had in their asset base, the more they could earn through regulated tariffs. This perverse and pervasive incentive added to the asymmetry in the previously mentioned penalty and reward system—virtually guaranteeing that when in doubt everyone in the chain of command would rather overinvest, oversize, and gold-plate—as long as they could get away with it. Which, they generally did, because as a rule, regulators were (and still are) understaffed and underresourced. These topics may be found in the history of rate of return regulation—not the topic of this chapter.

What is important is that this deeply rooted culture of overplanning and overinvesting has led to another outcome, which still prevails in the industry, even in places where the underlying rate of return regulation no longer applies. And that legacy is the ingrained belief that customer demand should be taken as a "given." It is what it is. Generations of engineers, forecasters, planners, and modelers did forecast the maximum demand and built sufficient infrastructure to meet it plus a little extra, just to be safe. Many within the industry have yet to change that mindset.

That, as the following chapters of this book explain, is what has led to the fact that historically very little was ever attempted to modify, adjust, manage, shape, shift, or shed customers' demand, especially to reduce peak demand.

Why is this important? Because meeting peak demand at all times and at any cost more than anything else determines how expensive a system we end up with and ultimately how much customers collectively have to pay for it.[3]

---

[1] Gray Davis, Governor of California at the time of the 2000—01 electricity crisis, for example, lost the reelection mostly due to his mishandling of the crisis and the ensuing power outages.

[2] The term usually refers to overinvestment in the network, especially the regulated poles and wires segment.

[3] One study, for example, concluded that roughly a quarter of the costs of operating and maintaining a modern electricity system may be attributed to having sufficient spare capacity to meet the peak load during the highest 100 h of the year.

This chapter begins by asking how much of the customer demand may in fact be flexible, and if so, how much of the peak demand can be shaped, shifted, or shed.[4] The remainder of the chapter is organized as follows:

- Section 2 describes the historical context that has given rise to the current double problem of peak *and* inflexible demand, which is getting worse as more nondispatchable variable generation is added to the system;
- Section 3 examines the implications of three long-held myths that all kWhs are the same, that all customers are alike, and that customer demand should be treated as a *given*;
- Section 4 returns to the chapter's title, namely, what is flexible demand; what demand is flexible?
- Section 5 speculates on how big is flexible demand and how can it be delivered; followed by the chapter's conclusions.

2. The legacy of taking customer demand as a "given"

As outlined in the preceding section, the electric power systems around the world evolved over a century with vertically integrated utilities under rate of return regulation. The emphasis, especially in the early rapidly growing years of the industry, was to provide sufficient investment to build, operate, maintain, and expand the infrastructure, from coal mine to the customers' meters.

During decades of rapid growth, the focus was on exploiting the massive economies of scale and scope—which explains the tendency to favor ever larger vertically integrated monopolies that could capture these economies. This tested and successful formula was fine-tuned during years of double digit demand growth leading to bigger and more efficient infrastructure, which resulted in lower prices, which in turn led to even higher electricity demand and lower prices. This period, sometimes referred to as the golden age of the power sector, was truly magical.

As illustrated in Fig. 5.1 for the United States, in the decade after World War II, electricity generation grew steadily before plateauing in recent years.[5] During the rapid postwar expansion of the industry between 1950

---

[4] Refer to California demand response potential study, Phase 3, Lawrence Berkeley National Lab, Feb 2020 at https://energycentral.com/c/em/lbl-says-there-more-flexible-demand.

[5] Note that the persistent growth came to a virtual halt in 2010, a phenomenon called *decoupling*—which refers to continued growth in GDP essentially with a flat demand for energy, further described in a McKinsey report at https://www.mckinsey.com/industries/electric-power-and-natural-gas/our-insights/the-decoupling-of-gdp-and-energy-growth-a-ceo-guide.

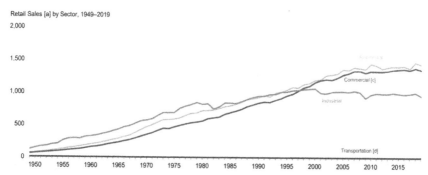

Figure 5.1 *Retail US electricity sales, 1950–2019,*[6] *in billion kWhs.* (Source: Energy Information Administration, Retail Sales, Table 7.6 at https://www.eia.gov/totalenergy/data/monthly/index.php#electricity.)

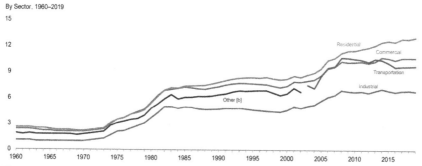

Figure 5.2 *Average retail US electricity prices in cents/kWh, 1960–2019.* (Source: Energy Information Administration, Avg. retail prices, Table 9.8 https://www.eia.gov/totalenergy/data/monthly/#prices.)

and 1970, the average retail prices were flat or slightly declining, as reflected in Fig. 5.2. Similar trends were experienced nearly everywhere around the globe.

The falling cost of electricity led to rising demand and the rapid proliferation of appliances and electricity using devices in all sectors of the economy, which in turn resulted in even lower per unit costs. In such happy times, why would anybody care about the efficiency of end use devices or how, when, and why did customers use electricity. The industry's modus operandi was to take demand as a *given* and build more capacity to meet it. It was simple and, for a while, it worked.

A memorable and frequently (and apparently mis-) quoted statement attributed to Lewis Strauss, the chairman of the US Atomic Energy

Commission in 1954, was that nuclear power—a favorite large-scale generation technology at the time—would soon be "too cheap to meter.[7]"

For much of its history, the industry was exclusively focused on the *supply side*, the upstream infrastructure—bigger and more efficient power plants, transmission, and distribution lines. Who would want to bother with customer demand? Why would anybody waste any time or effort on the *demand side* if electricity would soon be too cheap to meter. For anyone convinced of such a scenario, you could probably argue why bother with an electric meter at all?

The end result of all this was that aggregate demand—that is the sum of all electricity using devices on the network—evolved over time to resemble a load curve such as the one shown in Fig. 5.3, or a load duration curve such as the one shown in Fig. 5.4.

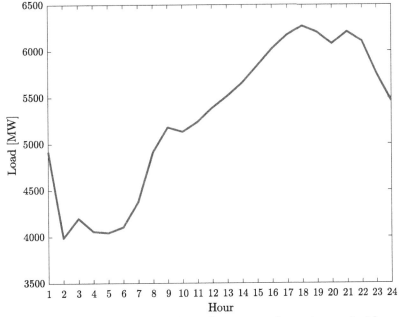

Figure 5.3 *Typical daily load curve shows the amount of capacity required for each hour of the day.* (Source: Author.)

---

[6] 2019 data are preliminary, 2020 figure is expected to show a significant decline due to the pandemic. But setting that aside, the all-time US peak is not expected to be reached or exceeded any time soon.

[7] Refer to https://public-blog.nrc-gateway.gov/2016/06/03/too-cheap-to-meter-a-history-of-the-phrase/.

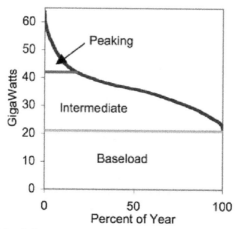

Figure 5.4 Typical load duration curve showing the portion served by baseload (bottom), intermediate (middle), and peaking (top) plants. *(Source: https://www.researchgate.net/profile/Stanton_Hadley/publication/247852903/figure/fig2/AS:670033948704794@153 6759950705/Load-Duration-Curve-and-different-power-plant-classes.png.)*

At the time, engineers who were in charge—that is before accountants, lawyers, and MBAs became CEOs and began to examine the numbers—thought nothing odd about the daily and seasonal demand cycles. Demand, as far as they were concerned, was a "given." Their job was to meet it at all times, at whatever cost it took. Monopoly utilities could simply pass on the costs of investing in necessary infrastructure to hapless, passive customers—they were called *ratepayers* in those days reflecting the fact that they had no options and no choices.

Oddly, the regulators were broadly OK with this arrangement. The regulated volumetric tariff, the cents/kWh applied to what each customer used, was believed to reflect a simple-to-understand and explain, fair, reasonably adequate, and equitable way to allocate the costs of upstream infrastructure among customers.[8] Since everyone used all of the infrastructure built to serve them—there was no opportunity to bypass, self-generate, or go off-grid—the concept of charging everyone a uniform and regulated "bundled service" made perfect sense.[9] So long as the rates were "just and reasonable"—and falling—few ratepayers complained.

---

[8] Reference Principles of Public **Utility** Rates, James C. **Bonbright**, at https://www.amazon.com/Principles-Public-Utility-Rates-Bonbright/dp/0910325235.
[9] In the age of *prosumers* and *prosumagers*, this is no longer the case.

Moreover, in most industrialized countries, electricity bills represent 1%–2% of the typical customer's disposable income.

In the mean time, resourceful engineers devised a clever way to deal with the resulting varying demand patterns. A vertically integrated utility could easily justify investing in a portfolio of different generation technologies to follow the daily and seasonal cyclical demand patterns with the costs passed on to the ratepayers with the regulator's consent.

The result was that each vertically integrated utility would build, maintain, and operate a fleet of plants to cover the different portions of the demand:

- Baseload plants—typically coal or nuclear-powered—would operate 24/7 covering the minimum load hours of demand;
- Intermediate plants—with capability to cycle up or down—would cover the remaining portion of the load; and
- To ensure reliability during extremely high demand hours called peak demand—which occur infrequently and do not generally last for many hours—utilities would build, maintain, and operate a fleet of *peaking plants*, typically smallish gas-fired units capable of coming on line quickly and able to adjust their output to fit the variations in load.

The resulting portfolio is illustrated in Fig. 5.4.

As far as the engineers were concerned the problem of meeting variable demand was successfully solved. Demand was taken as a *given*. It could be whatever it wanted to be. Customers, large and small, could literally flip the switch of any device at any time at will. That was their God-given right. And utilities had to have sufficient capacity to meet it. Generation would be adjusted to follow demand. QED. There was no need to manage, shift, shape, shed, or modify demand.

Another noteworthy consequence of the engineers' mentality of taking demand as a *given* was the total lack of attention to the *efficiency* of energy use, namely how and why customers consumed all the kWhs they were using. The prevailing belief was that the price of electricity would continue to drop as demand continued to grow. Terms such as sustainability or climate change were not in the industry's vocabulary.

As the ownership and use of appliances proliferated with the rising income levels in the industrial countries, they also got bigger and more electricity hungry. Refrigerators, for example, increased in size and features with no efficiency standards.[10] Central air conditioning became standard

---

[10] California was the first state to introduce appliance energy efficiency standards in the 1970s, years before the US federal government followed.

adding to the volume of demand while making the shape of the daily load cycles far more problematic without any efficiency standards—nor were there any standards or building codes minimizing heat loss or gain.

The result of this dual neglect became evident in even more cyclical—read expensive to serve—load shapes. Partly as a result of the lack of building codes plus the rapid proliferation of air conditioning loads, many summer peaking systems ended up with grotesque demand shapes with steep but infrequent peaks that are notoriously expensive to serve.[11] Commercial and residential AC loads on some summer peaking systems account for a disproportionate percentage of peak demand. Fig. 5.5 shows the typical daily load shape for a warm but not extremely hot day in California.

Still, the engineers saw nothing wrong or unusual about this outcome. More peaking units were built to cover the spiky peaks even though these units are notoriously polluting, inefficient, and unprofitable to maintain and operate. But the bundled regulated tariffs masked the cost of serving the expensive peaks since they were spread among millions of hapless customers throughout the year.

Another amazing feature of the prevailing regulation, which still endures, is the uniform flat tariffs.[12] Customers pay the same price at all times,

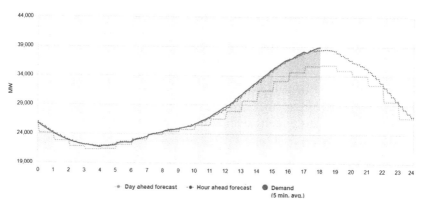

Figure 5.5 CA daily load shape June 10, 2020. *(Source: CAISO at http://www.caiso.com/ TodaysOutlook/Pages/default.aspx.)*

---

[11] In the case of California, unusually high peak demands are experienced every decade or so during extended periods of extreme and dry heat waves.

[12] For discussion of the advantages of time-of-use (TOU) tariffs or dynamic pricing refer to https://www.utilitydive.com/news/6-reasons-why-california-needs-to-deploy-dynamic-pricing-by-2030/578156/.

for any and all kWhs used regardless of where they are or their pattern of consumption. A customer with a massive house and a large AC load who exerts high costs on the network by soaking up expensive kWhs during the peakiest hours of the year pays the same cents/kWh as one who has a flat load profile and no AC.[13]

When these issues first surfaced, the industry's response was to say, with some justification, that "we do not have smart meters to distinguish between different customers with different load profiles.[14]" It was convenient not to mention that nobody had bothered with smart meters or communication infrastructure necessary to measure and bill customers by time-of-use. For a sophisticated industry that could build and operate complex nuclear power plants, it was a lame excuse. Still more amazing is that regulation were lagging to implement time of use (TOU) or real-time pricing even if advanced metering infrastructure (AMI) technology were available.

This ingrained mentality began to be challenged in the 1980s and 1990s by those who argued that:

- We must understand *why* customers use so many kWhs and *when* they use it in the first place;
- Second, we must explore ways to *reduce* overall consumption by encouraging efficient use of electricity[15]—at all hours but especially at the times of system peak; and
- We must shape, modify, and manage the pattern of consumption, especially discouraging electricity consumption when it is expensive to generate and deliver.[16]

When first introduced, these were *radical* ideas—unknown, unpopular, and unwelcomed to old-timers within the industry. It was also unpleasant and awkward for regulators—who because of their neglect to face the issue earlier through innovative rate design—were essentially *complicit* in allowing

---

[13] As further explained in Chapter 8, this becomes even more pressing as more customers buy electric vehicles over time.

[14] Refer to chapters in this volume and a prior one by Carlo Stagrnao and Simona Benedettini.

[15] The energy efficiency "gurus" such as Amory Lovins have been advocating these issues for years as further covered in Negawatt Revolution at https://rmi.org/insight/negawatt-revolution/.

[16] For a comprehensive discussion refer to California demand response potential study, Phase 3, Lawrence Berkeley National Lab, Feb 2020 at https://buildings.lbl.gov/demand-response.

demand to be taken as a *given* for decades. They had to catch up with the new realities of the business.

Still, it could be argued, that nothing would have materially changed had it not been for the dramatic rise of variable renewable generation on many systems during the last decades plus the indications that this trend will continue as more systems march toward higher renewable penetration levels. The variability of renewable generation, the fact that it is non-dispatchable and not totally predictable—for example during extended periods when there is little or no wind and/or solar generation—means that the engineer's paradigm of running baseload, intermediate, and peaking units to match variable demand is no longer practical nor sustainable. As described in Chapters 1—4, the grid operators now routinely experience periods when there is *overgeneration,* that is generation in excess of load. Hitherto, this phenomenon was unknown to them simply because it never happened before. Historically, plants would only be dispatched when there was need for their output. But as we know, renewables generate power based on the prevailing wind and solar conditions, not any commends or dispatch instructions from the grid operator.

Other strange phenomena now routinely experienced include *negative prices*—again something that never happened before because generators were only dispatched when needed. Here again, the old-timers did not know what to do or how to handle negative prices—the notion that you had to *pay* customers or neighboring utilities to *take* surplus capacity was totally alien to them.

With the exception of times when hydro reservoirs were full and the excess water had to be spilled, grid operators were dismayed when forced to resort to *renewable curtailment,* the equivalent of spilling excess solar and wind energy because it could not be used, stored, or exported.

The day of reckoning, one might say, had finally arrived. It was obvious that the emperor was wearing no clothes. Electricity prices were no longer falling. Nuclear power plants were expensive and took a very long time to build. Electricity was not too cheap to meter. The time had arrived to meter not just the kWhs but when and where they were being used. As time went on, concerns about the use of natural resources—including water—and climate change were added to what is the best way to meet customer demand.

3. Overcoming three big myths

Over many decades, the combination of the factors outlined in the preceding section led to three deeply ingrained beliefs within the insular electric power sector, which traditionally was dominated by engineers who knew how to build big power plants and could string cables and wires that would transmit the power to customer meters and sockets, but not much else.

The three myths were:

- All kWhs are the same;
- Ditto for customers—one size fits all; and
- For any network demand, both its peak and its shape, is a *"given."*

Today, almost anybody who knows anything about the power sector agrees that all kWhs are *not* the same. They are different depending on when, where, and how they are generated and by whom and—even more important, when, where, and how they are consumed and, again, by whom.

A kWh coming from a wind turbine, a utility-scale solar plant, a coal-fired plant, a nuclear one or a rooftop solar PV panel, have different attributes, different costs, and different characteristics. The ones from renewable sources (and nuclear) have zero (or low) carbon footprint, which may make them more desirable to some customers and/or grid operators. Their location—far or close to the load center—makes a huge difference. When and how much they produce and whether their output can be adjusted also matters a great deal.

Similar arguments apply on the consumption side. A kWh used to run a critical load, such as a medical facility, an airport, or a data center has different value than a kWh used to run the pump on a backyard hot tub or an electric hot water heater with a storage tank. When and where kWhs are used also makes a huge difference in terms of the costs imposed on the system—both to generate and deliver.

Consider a retirement community in Palm Springs, California. During the long, hot, and dry summer months when midday temperatures reach 120°F and everyone's air conditioners are running at full blast, the community exerts enormous strain, and cost, on the system. Since peak demand is partially met by inefficient and polluting peaking plants, it also exerts high environmental costs. By contrast, kWhs used on a cool sunny day in the spring generated from nearby solar plants may be essentially free, and with negligible carbon content. Clearly, all kWhs are *not* the same.

Yet, the great majority of customers continue to pay flat tariffs that do not vary by time or location. That much for the first myth.

The second myth has to do with customers. As opportunities to self-generate, store, share, and/or potentially trade with others continue to proliferate, the notion that all customers are, and can be treated the same, no longer applies. Nor does the flat bundled regulated tariff. These issues are further covered in the preceding volumes by the same editor and published by Academic Press, namely,

- Behind & behind the meter: Digitalization, aggregation, optimization, monetization, 2020;
- Consumer, prosumer, prosumager: How service innovations will disrupt the utility business model, 2019;
- Innovation and Disruption at the Grid's Edge, 2017;
- Future of Utilities, Utilities of the Future, 2016; and
- Distributed Generation & its implications for the utility industry, 2014.

In countries such as Australia, where nearly one in four customers have solar roofs, the transition from consumer to *prosumer*—the second from top in the above list—is well under way. In other countries including Japan and Germany, a growing number of people installing rooftop solar PVs are also investing in distributed storage, which suggests that there will be more *prosumagers* in many parts of the world.

Clearly, *prosumers* are different than consumers; *prosumagers* even more so. Not only do they buy fewer net kWhs from the network, their expectations of what they need and expect from the distribution and retail company diverges from those of consumers who remain totally reliant on the incumbent providers and continue to pay regulated bundled tariffs, topics extensively covered in the same volumes mentioned above.

The third myth, central to this chapter and book, argues that neither the size nor the shape of demand is *preordained* or is a *"given"* as has long been assumed even if not explicitly stated. Which leads to the question why has this myth persisted for as long as it has?

Historically, the explanation was that three major obstacles had to be overcome before there was any hope for adjusting or modifying customer demand:

- Smart meters;
- Smart prices; and
- Smart (connected) devices.

The basic idea is that we need smart prices to be delivered to smart devices before we could meaningfully manage demand because

- Until and unless we have AMI it is not possible to measure or charge customers based not only on how much they use but when (and possibly where) they use it;
- Even if and when we get smart meters, we do not have TOU rates or dynamic pricing that can be applied to the bulk of customers; and
- Based on empirical research, customers cannot be expected to meaningfully engage in managing their total usage let alone consumption of individual devices even if price signals could be delivered and responses measured. To manage individual loads, especially for small consumers, we need to automate things, essentially bypassing the customers.[17]

These are serious and persistent obstacles that are being resolved to varying degrees in different parts of the world. But at least, now there is growing recognition of the fact that demand is *not* necessarily a *given*, that both its shape and its peak can in fact be adjusted only if we have the necessary *functionality* and can apply the proper tools using advanced technology, the topic of a number of following chapters in this book.

4. What is flexible demand; what demand is flexible?

After discussing the historical context and dismissing the three big myths about customers and demand, let's focus on the title of the chapter, namely:

- What is flexible demand and
- What demand is flexible?

Flexible demand refers to any customer load that need not be on or totally served at all times. For example, take any electric water heater with a storage tank, commonly found in millions of households and commercial establishments the world over. The water stored in the tank can be heated more or less at any time—assuming that it has sufficient capacity and has proper insulation to prevent heat loss. It makes absolutely no sense to heat up the water in any given tank during times when demand on the network is high and so are the prices. While shifting the time when a single water tank is heated has a negligible effect on the system peak, doing the same on millions of water tanks can have an appreciable effect[18] as further described in Chapter 7.

---

[17] For example, refer to Chapter 18 on limitations of customer response even when subjected to TOU tariffs.

[18] For example, see Voltalis at https://corporate.voltalis.com/?lang=en or Emulate Energy at https://www.emulate.energy.

As for what demand is flexible, virtually all customer demand has *some* flexibility. Very few electricity using devices need to be on all the time or at full capacity. Take pool pumps, air conditioners, ice makers, or washing machines. Nobody needs or expects them to operate 24/7. And once again, even though cycling a single air conditioner has negligible impact on system peak, cycling thousands or millions of them can make a difference.[19]

Additionally, there is the terminology issue[20]—the fact that many terms are used in the following chapters to describe loads that can be modified or may be flexible—hopefully any nuanced differences can be deciphered from the context of the discussion.

The puzzle of flexible demand, of course, is why aren't we doing more with millions of devices and applications with significant demand flexibility potential? The short answer is that up to now, the engineers running the systems virtually everywhere did not have to. They preferred adjusting generation to follow the load rather than managing the load itself. They believed that it would be easier, more reliable, and less expensive to adjust generation.

Perhaps they were correct in the past, but the relative economics of adjusting generation versus load have been dramatically altered in recent years in favor of the latter for a number of reasons including the following:

* Digitalization and connectivity of end use assets—An increasing number of electricity using devices are smart and increasingly "connected," which means they can be remotely monitored and managed at rapidly falling costs;

* Aggregation and automaton of behind-the-meter (BTM) assets using software[21]—Large numbers of such connected smart devices can be monitored and remotely controlled using sophisticated software that rely on artificial intelligence and machine learning; and

* Rising share of nondispatchable renewable generation—As noted in Part One of the book, as the resource mix of many systems moves toward higher levels of renewable generation, it becomes increasingly difficult if not impossible to adjust generation to balance supply and

---

[19] Southern California Edison Company, as an example, runs one of the largest air conditioner cycling programs in the United States.

[20] The author is indebted to Ross Baldick for pointing out this issue.

[21] Behind and beyond the meter volume, 2020, covers many of these issues in more detail.

demand. Renewable generation is variable by nature and neither solar nor wind, the two biggest sources of future growth, can be dispatched. Over time, an increasing number of system operators must resort to new tools and schemes to balance supply and demand in real time. As the cost of adjusting flexible demand drops, it will become one of the many options needed to maintain grid operability while minimizing the costs.

5. How big is flexible demand and how can it be delivered?

It is probably not much of an exaggeration to say that today, flexible demand is where aviation was when the Wright brothers "flew" their "plane" near Kitty Hawk, North Carolina, in Dec 1903. The Wright flyer weighed a mere 600 lbs., could barely carry a single "pilot," and had a top speed of 30 MPH.

Compare that to today's modern planes that fly nonstop for 15 + hours while carrying 300 + passengers across oceans. Even more astonishing is how the costs of air travel has dropped over time while safety and performance has improved.

This analogy is not to suggest that it will take over a century for demand response (DR) to reach maturity and deliver results at low cost but rather to emphasize the skepticism and disbelief that Wright brothers faced in trying to advance their idea into a practical and useful product or service. But once the many advantages of air travel were widely recognized, the business rapidly grew.

DR, no doubt, has long ways to go but it is fair to say that it has a great potential.[22] Multiple studies of DR potential conducted by the Federal Energy Regulatory Commission (FERC)[23] and others over the years in the United States and elsewhere have consistently concluded that as much as 20% of the coincident peak demand on many networks can potentially be "managed" through DR—consistent with the results presented by Hledik and Lee in Chapter 9. One, of course, always has to read the fine print, namely the assumptions that go into such predictions and every study of the subject points to the lack of ubiquitous smart meters (Fig. 5.6) and smart tariffs as the two most important obstacles.

---

[22] Refer to Chapters 10 and 11 for further coverage.
[23] Reference to 2019 assessment of Demand Response and Advanced Metering, FERC, Dec 2019, at https://www.ferc.gov/legal/staff-reports/2019/DR-AM-Report2019.pdf.

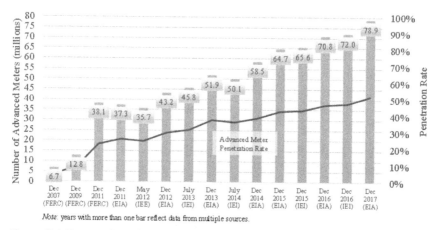

Figure 5.6 *Penetration of smart meters in the United States.* (*Source: 2019 Assessment of Demand Response and Advanced Metering, FERC, Dec 2019 at https://www.ferc. gov/legal/staff-reports/2019/DR-AM-Report2019.pdf.*)

Moreover, these studies have consistently concluded that managing peak demand by DR is less costly and certainly far less polluting than the alternative, which is to take demand as a "given" and meet it with expensive and highly polluting natural gas–fired peaking plants—which is usually the alternative.

A number of studies looking at the cost of meeting the "uncontrolled" peak demand on typical systems have concluded that it is *exorbitantly* high. One such study concluded that as much as 25% of the operation costs of a "typical" network may be attributed to having sufficient capacity to meet load during the 100 highest demand hours in a year, an astonishing fact. The explanation, however, should be intuitively obvious. A lot of spare capacity—fuel, generation, transmission, and distribution—is needed to serve the peak load in just a few hours in a typical year. Peaking plants, for example, have to be staffed, fueled, maintained, and ready to go at a few moments' notice 24/7, just in case. And since they are rarely utilized, they are notoriously unprofitable for the operators.[24]

This is not to say that DR is a cheap or *free* resource, but it should handily compete with a much more expensive option—namely peaking plants. The same, of course has been said about energy efficiency—that

---

[24] In California, for example, a number of critical peaking units are "artificially" kept viable through out-of-market payments. Many are designated as "reliability must run" plants, which qualifies them to receive payments merely to be "available" if and when needed. They may sit idle for very long periods of time in between short periods of operation.

negawatts are always cheaper than the *megawatts*—a phase coined by energy efficiency guru Amory Lovins.[25]

As further explored in other chapters of this book, DR has to be gradually cultivated, encouraged, supported, and developed to reach its full potential. The key enablers include:

- Supportive policies that encourages full-scale implementation of smart meters with capabilities to administer time-of-use and/or more exotic pricing schemes to broad range of customers;
- Supportive polices, standards, guidelines, and regulatory clarity that encourages more smart devices to receive smart prices and can be remotely monitored, controlled, managed, aggregated, and optimized resulting in favorable conditions for monetizing the value of DR inherent in a large portfolio of electricity using devices;
- Tariffs that better reflect the variable costs of consuming or not consuming energy/capacity by time and potentially location;
- Supportive regulatory policies that allows aggregation and more active participation of DR in wholesale and retail markets; and
- Policies that identify who should be responsible for developing and delivering DR—the retailers, the DSO, the regulated distribution companies, private aggregators with new business models that can deliver services profitably at scale, or a combination of the above.

In his latest book, Power after Carbon, Peter Fox-Penner[26] speculates about new players or possibly existing ones assuming new roles and responsibilities to provide useful services to customers including aggregators, energy service companies (ESCOs), and different types of retailers or distribution companies with new business and service models.

Supportive regulations are sorely needed for these types of business models to flourish. A number of chapters in this book highlight the current lack of clarity on these issues as serious obstacles to further development of flexible demand.

Needless to say, the way forward is to identify where the low-hanging opportunities lie in different sectors of the economy and develop the means and the functionalities required to capturing them starting with the most cost-effective ones first.

- In the industrial sector, which may be a good place to start, energy-intensive processes with inherent flexibility should be identified and targeted;

---

[25] For example, see https://www.latimes.com/archives/la-xpm-1990-05-27-mn-160-story.html.

[26] Power after Carbon, Peter Fox-Penner, Harvard Univ. Press, 2020.

- Water and sewage processing are huge users of electricity with significant opportunities in pumping, as an example;
- In the commercial sector, which is typically dominated by lighting and space conditioning loads, there is significant flexibility to preheat and/or precool buildings to avoid peak demand hours without appreciable impact on comfort levels; and
- The residential sector, probably the most challenging sector due to the small size of individual customers, nevertheless is virtually unexplored to date hence offering significant potential for DR. Recent advances in smart, connected devices and smart wireless thermostats and home energy management systems offers promising opportunities as well.

One specific device with huge potential is electric vehicles (EVs) and electrified transport, topic covered in Chapter 8. There are expectations that millions of EVs will be in use within a decade[27]—and in some places—the sale of internal combustion engines (ICEs) is being banned as early as 2030s. EVs not only use a lot of electricity to charge their batteries but can potentially discharge some of their stored energy back to the grid at times of high prices and peak demand.

Aggregating the inherent flexibility to charge and potentially discharge millions of EV batteries allows DR opportunities that could dwarf all other applications.[28] If EVs are charged during off peak hours—say when there is excess solar generation at low or possibly negative prices—and discharged back during peak demand hours, the shape of the daily demand curve can be appreciably modified.

6. Conclusions

This chapter described what is meant by *flexible demand* and explored how much demand may in fact be *flexible* and hence subject to being developed and delivered to scale. It also explained why this significant resource has not been explored to date in significant scale. Clearly, a lot more can be expected from flexible demand over the next couple of decades.

---

[27] State of California, for example, expects to have as many as 5 million EVs by 2030.

[28] One estimate, for example, put the aggregate storage capacity of the EVs in the United States at 28 GW within a decade. Needless to say, many innovative companies are developing the tools to turn this into a profitable business with large scale.

CHAPTER 6

Who are the customers with flexible demand, and how to find them?

Carlo Stagnaro[1], Simona Benedettini[2]
[1]Observatory on the Digital Economy, Istituto Bruno Leoni, Milano, Italy; [2]Pwc, Rome, Italy

1. Introduction

Can small customers participate in electricity markets? Are they willing to? And do they? Under which circumstances?

Historically, the role of small customers in electricity markets has been largely passive—further described in Chapter 5. This resulted from a variety of reasons, including (but not limited to) the industry's structure, regulation, and transaction costs. Things remained virtually unchanged until the late 1980s, when a major transformation began, that led to the vertical disintegration of the industry, on one hand, and the liberation of the customer, on the other hand (Kiesling, 2009). Technological development also increased the consumer's reliance on electricity to meet her energy needs and expanded the array of small-scale devices to generate, use, and store power. Thanks to these progresses, *consumers* are evolving into *prosumers* or *prosumagers* (Sioshansi, 2019). As such, they are no longer mere recipients of electricity but also producers and traders who may provide services to the grid, such as shifting load, storing power which is either self-generated or received from the grid, or supplying power to the grid. A new wave of assets is being installed behind the meter that may further encourage the consumer's ability to be an active part of the system, by creating private as well as social value (Sioshansi, 2020). These include rooftop solar panels, smart devices, and batteries. Other components are equally important: *hardware* and *software*, i.e., devices such as Amazon's Alexa or Google Home, and applications, may be employed to manage the generation, storage, and consumption of energy behind the meter while shaping it in a way that contributes to accommodate the needs of the grid. In other words, the

Variable Generation, Flexible Demand
ISBN 978-0-12-823810-3
https://doi.org/10.1016/B978-0-12-823810-3.00016-9

complex interaction between assets behind the meter, the industry's upstream, and potential intermediaries (such as traditional utilities, demand aggregators, or third parties) may maximize the social welfare under the constraint of maximizing the customer's own utility.

The fact that this is *technically feasible*, though, does not necessarily mean that opportunities for private and social value creation are also *actually exploited*.

This chapter is concerned with three main questions:

- What kind of flexibility services can be provided by small consumers of electricity?
- Do small consumers exploit this opportunity?
- What changes should occur in order to encourage small consumers to become more engaged in flexibility services?

The chapter is structured as follows:

- Section 2 after this introduction provides a taxonomy of flexibility services which can be provided from distributed energy resources (DERs) at both the global level, i.e., Transmission System Operator (TSO) level, and the local level, i.e., the Distribution Service Operator (DSO) level. Given the increasing penetration of variable energy resources at the local level, Section 2 will provide a major focus on local flexibility services and the evolving role of DSOs with respect to the procurement of these services. To this aim, examples from European electricity markets will be provided;

- Section 3 illustrates the state-of-the-art situation with respect to the provision of flexibility services. It will review the cases as well as the circumstances under which flexibility services are supplied (or at least considered) by small customers. As one might argue, the mismatch is large, between the amount of flexibility services that might *theoretically* be provided and the amount which is *actually* supplied.

- Section 4 digs deeper into the causes thereof. The short answer is simple enough: if it can be shown that there is money on the table but no one collects it, then the blame is likely to be placed upon transaction costs Coase, (1960b). This is, in fact, the case: two sets of reasons are identified, namely *regulatory* barriers (Stigler, 1971) and *behavioral* barriers (Thaler, 2017).

- Section 5 summarizes and concludes, also drawing some policy implications.

2. A taxonomy of flexibility services

In the European Union, the mandatory 2030 Climate and Energy targets, and the subsequent EU Communication "A Clean Planet for All" to achieve net zero GHG emissions in 2050, assign to variable renewable energy sources (VRES) for electricity generation a significant role for the decarbonization of the European energy sector. Figs. 6.1 and 6.2 provide an illustration of the incidence of VRES in several EU Member States.

To accommodate in a cost-effective and secure mode the increasing penetration of VRES, flexibility services becomes crucial. To this aim, the 2019 EU Electricity Directive[1]—which is bound to be transposed in the national legislations of the member states by December 31, 2020—affirms the necessity to allow DERs to contribute providing such services and to promote an active role of DSOs in their procurement. According to the Directive, DSOs are allowed to procure flexibility services through market-based procedures and consistently with transparent and nondiscriminatory rules. In addition, the Directive establishes that Member States' regulatory framework shall incentivize the procurement of flexibility services from DSOs and implement effective methodologies for the remuneration of the investments performed by distributors to the aim.

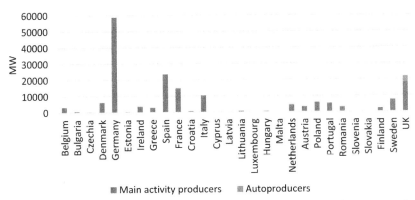

Figure 6.1 Installed Wind generation capacity (MW) in 2018 in the EU member states. *(Source: Own elaboration on Eurostat data.)*

---

[1] Directive EU 2019/944 of the European Parliament and of the Council of 5 June 2019 on common rules for the internal market for electricity and amending Directive 2012/27/EU.

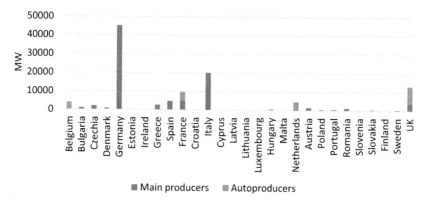

Figure 6.2 Installed Solar PV generation capacity (MW) in 2018 in the EU member states. *(Source: Own elaboration on Eurostat data.)*

Flexibility services can be defined as services provided within the energy system as an effect of the modification of generation and consumption patterns in reaction to an external signal (price signal or activation).

A review of several European countries suggests that DERs can provide the following flexibility services at both the global and local level. Fig. 6.3 summarizes.

A sample of EU member states is reviewed in Table 6.1. They show a significant heterogeneity with respect to the types of DERs allowed to provide flexibility services at both the local and global level as well as with respect to the types of services that DERs can provide (Table 6.1).

With respect to the role of DSOs in the procurement of flexibility services from DERs the selected EU Member States highlight the emergence of the following models.

At the global level:

- DSOs as neutral facilitator of local flexibility services: DSOs facilitate the TSOs' procurement of flexibility services by acting as facilitator, e.g., providing real-time information regarding the potential unavailability of DERs;

- DSOs as validator: DSOs not only act as facilitator but they have the possibility to prevent TSOs from the activation of a given DER if such activation may impact efficient and secure operations on the distribution network.

At the local level:

- DSOs procure flexibility services through market-based procedures;

Table 6.1 Flexibility services provided by DERs at global and local level by EU member states.

	Global level									Local level					
	Frequency regulation			Congestion management			Voltage regulation			Congestion management			Voltage regulation		
Member state	DG	D	S	DG	D	S	DG	D	S	DG	D	S	DG	D	S
Belgium	X	X	X				X	X	X	X	X	X			
France	X	X								X	X	X			
Germany	X	X	X	X	X	X	X	X	X	X	X	X	X	X	X
Great Britain	X	X	X		X			X	X	X	X	X		X	
Greece	X	X	X							X	X		X	X	
Ireland		X		X	X	X	X	X	X	X	X	X		X	
Netherlands	X	X	X	X	X		X	X	X	X	X				
Portugal							X								
Spain	X			X			X	X		X	X		X	X	
Sweden	X	X		X	X		X	X	X	X	X		X	X	X

Note: D, Demand; DG, Distributed generation; S, Storage.

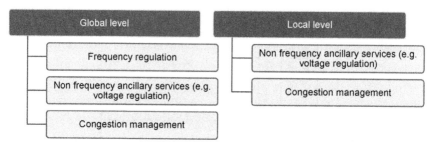

Figure 6.3 Flexibility services provided by DERs at global and local level in the EU.

- DSOs procure flexibility services through alternative mechanisms with respect to market-based procedure, e.g., through flexible connection contracts.

  Fig. 6.4 provides an overview of the models in place in the selected EU Member States at both the global and local level. European initiatives for the provision of flexibility services have been developed only recently due to the input given by the 2019 Electricity Directive. Only time will tell what works and what doesn't, and why.

Figure 6.4 *Models to allow the provision of flexibility services from DERs in a sample of EU member states.* Note: hatching means that two models coexist within the same country.

3. Do customers with flexible demand exist?

The fact that something is technically and legally feasible does not necessarily mean that it happens in practice. While DERs are increasingly providing flexibility services, the involvement of small generators becomes more complex as the size of the generator itself decreases. The same applies on the demand side. Demand flexibility, especially at a small scale, critically depends on a number of enhancing conditions, such as a sufficient degree of electrification, smart devices, or other assets that may be used to generate, store, or shift load, smart meters, or similar devices, demand aggregators, etc. While the number of small customers who own these characteristics is still small, it is growing and becoming significant. Hence one would expect to see a corresponding increase in experiences of market participation. Unfortunately, actual applications are still scarce (see also Chapters 5 and 7, on what demand is potentially flexible and what role can aggregators play, respectively).

To be fair, the history of demand participation and flexibility is as old as (wholesale) electricity markets themselves. Since the very first experiences with market liberalization in the 1980s, the underlying idea was that an efficient allocation of resources could be achieved by letting demand and supply interact with each other, coordinated by the price system. Although technicalities may differ, this acknowledgment lies at the basis of power exchanges all over the world (Cretì and Fontini, 2019). In wholesale markets, energy consumers (as well as demand aggregators or traders) make an offer in day-ahead markets showing their willingness to pay a given amount of money for a given volume of energy to be delivered at a given point in time, and subsequently adjust their positions in the following market sessions by reacting to exogenous changes as well as price signals. Several papers showed that large customers—such as energy-intensive industries—are very keen to react to price signals, by increasing their demand when prices are low, reducing it as prices increase, and shifting load when the system benefits from, and pays for, such adjustments (Rassenti et al., 2002).

Moreover, an entire stream of literature is dedicated to assess the potential for flexibility on the demand side. To what extent, under which conditions, and for how long electricity consumers may reduce, shift, or adjust their load request in order to provide flexibility services to the system? And what do they expect to get in exchange? The aggregate potential may be large. Gils (2014) performs an assessment on demand response potential in 40 European countries, finding a minimum load reduction capacity of 61 GW and a minimum load increase capacity of 68 GW, under conservative assumptions, although the available capacity shows large variations across time, countries, and sectors. These results are confirmed by other reviews on the potential in various industrial sectors, particularly those where the productive processes are relative energy-intensive and the size of firms medium-to-large (Shoreh et al., 2016; Helin et al., 2017; Grein andPehnt, 2011). Hledik and Lee argue that the potential for load flexibility in the United States is as high as 198 GW of peak reduction, or 20% of peak (see Chapter 9).

Commercial and residential buildings show a potential for flexibility, too. That is mainly the result of the gradual electrification, that increases the demanded load as well as the ability of energy consumer to shift their energy requests. While the potential for flexibility in the industry sector relies on production processes, when it comes to less energy-intensive activities it mostly lies with the management of buildings. Research has found a large potential from commercial buildings (Yin et al., 2016), residential and small business activities (Martínez Ceseña et al., 2015), and residential buildings (Lopes et al., 2016).

While the literature on the *technical* and *economic* potential for demand-side flexibility is large, the amount of evidence on this turning into reality is scarce. Experiments and surveys fill the gap by inquiring into the actual willingness of households and entrepreneurs to get enrolled in some flexibility program. Before a few results are illustrated, an important point should be made. Providing flexibility services to the system entails some sort of loss of utility for those who stand on the demand side: changing the temperature indoor in order to reduce energy needs, advancing or postponing an activity (such as a productive step within a business or a washing cycle inside home), or shutting down a machinery, all require individuals to give up part of their utility, at the very least because they need to change their programs. In order to convince them to do so, four main obstacles need to be overcome (see also Chapters 7 and 18):

- The required adjustment should be technically feasible (technical barriers);
- The required adjustment and the underlying trade must comply with the relevant legal and regulatory framework (regulatory barriers);
- The so-called divestiture aversion, i.e., the failure to give up something people already have as opposed to their willingness to pay for the same object, should be small (behavioral barriers); behavioral barriers also

include the legitimate concern that personal data, once obtained by third parties, might be misused or transferred without an explicit consent; and

• A compensation should be awarded that is at least as large as the utility loss, at the margin (economic barrier).

This chapter is not concerned with technical barriers, which are assumed to be given. What is or is not technically feasible is by and large a function of the technology, i.e., it changes (and presumably increases) over time. Once a given kind of flexibility service can be technically provided by a small energy-user, it becomes interesting to understand whether, and at what cost, she would be willing to provide it. On top, as Sioshansi argues in the Introduction and in Chapter 5, a number of technological barriers have already been removed—for example, households and small and medium enterprises are increasingly equipped with smart meters, smart devices, and smart grids. The increasing reliance on electrified appliances increases both the peak demand from small customers, and the load that may be shifted easily and/or at a limited cost. Access to real-time metering, pricing, and billing is the most important condition that must be met in order to enable demand-side response and other forms of demand participation (Nordic Council of Ministers, 2017). Significant progresses are taking place virtually everywhere so it is safe to assume that these constraints are being removed (Stagnaro and Benedettini, 2020).

On top, there are reasons to believe that people are willing to participate in demand flexibility schemes—at least in principle. For example, a survey on 835 respondents in the Netherlands found that 11% would be willing to enroll in a flexibility program. This rate is remarkably high, if one considers that 60% had never heard about "smart grids," although younger respondents were more familiar with the concept than older ones (Li et al., 2017). Another survey on 2020 British households (Nicolson et al., 2017) found that about a third would be willing to switch to a 3-tiered smart time-of-use tariff, but it also found a considerable degree of loss aversion that, as shown in the following section, contributes to explain the mismatch between a generally high interest in flexibility programs and their relatively low level of take-up where they are actually deployed. On the other hand, electric vehicle owners were significantly more willing to join such programs than nonowners. In California, about 2.2 million customers are already receiving electricity on time-of-use rates, a figure that is expected to increase to 10 million customers by 2025: the spread of time-of-use tariffs is expected to go hand in hand with a wider diffusion of electric vehicles and rooftop solar panels (Faruqui, 2020).

The evidence from experiments is broadly in line with what emerges from surveys. For example, a large pilot test in Belgium involved 239 households for 36 months. It turned out that—under an appropriate set of incentives—a maximum increase of 430 W per household could be realized at midnight, while a maximum decrease of 65 W per household could be realized in the evening, mostly thanks to smart wet appliances, i.e., washing machines and washers (D'hulst et al., 2015). Kubli et al. (2018) ran an experiment that involved 7216 individual decisions from 902 participants in Switzerland. They found that current and potential electric car and solar PV users are more willing to participate in flexibility programs than heat pump users. Klaassen et al. (2016) equipped 77 households in the Netherlands with solar PV and a connection with district heating and introduced a dynamic tariff for electricity. They found that i) the largest contribution to the flexible load within households comes from wet appliances and ii) the load shifting (from the evening to midday) tended to persist over time rather than responding to short-run price variations. The authors conclude that simple and transparent, dynamic tariffs are the most important driver of demand response.

Table 6.2 summarizes.

4. Transaction costs: where art thou?

In short, the literature shows that there is a growing technical capability for demand-side response and other flexibility services. From a systemic perspective, it would be desirable to mobilize the potential which is sleeping behind the meter: making an efficient use of decentralized production and storage as well as of smart devices would create social as well as private value. Small customers themselves claim they are willing to become engaged, and the share of those available for enrollment increases when adequate information is provided. So, why doesn't this translate into a broad participation of small customers in electricity markets?

The standard answer that economists provide to such a question is: transaction costs (Coase, 1937, 1960a; Stigler, 1966; Williamson, 2008). In other words, there must be reasons that explain the failure to take advantage from what might appear as a mutually profitable exchange. Transaction costs can take many forms. Before getting into an attempt of classifying the various kinds of transaction costs, a general caveat should be raised:

Table 6.2 Evidence from surveys and experiments.

	Potential for flexibility	Surveys	Experiments
D'hulst et al. (2015)			Households willing to provide max 430 W load increase at midnight and 65 W load decrease at evening
Gils (2014)	Min load reduction of 61 GW and max load increase of 68 GW in 40 European countries		
Faruqui (2020)	10 million customers on time-of-use rates in California by 2025		
Hledik and Lee (2020) (Chapter 9)	198 GW load flexibility at peak time in the US		
Klaassen et al., (2016)			Simple and transparent tariffs is the main driver to get people switch to time-of-use rates
Kubli et al. (2018)			EV and PV owners more willing to participate in flexibility programs
Li et al. (2017)		11% willing to become enrolled in flexibility program	
Nicolson et al. (2017)		Ab. 1/3 customers willing to switch to time-of-use	
Shoreh et al. (2016), Helin et al. (2017), Grein and Pehnt (2011) Yin et al. (2016), Martínez Ceseña et al. (2015), Lopes et al. (2016)	Large potential from energy-intensive industries and medium and large enterprises Large potential from residential and commercial buildings		

technology is a powerful ally to reduce and overcome transaction costs. Sioshansi provides several examples in Chapter 7 regarding how IT, software, and artificial intelligence may be used to address transaction costs. "Technology" also refers to institutional innovation, in the sense that rules and regulations—that were designed within a different environment—oftentimes limit the economic agents' ability to trade and find innovative solutions to the problems they face. In order to liberate the potential for flexibility, not just new products or processes should be developed and deployed, but also new uses for the existing products should be invented and, a fortiori, should be allowed.

In the following, transaction costs are grouped into two categories:
- Transaction costs stemming from regulatory choices and
- Transaction costs stemming from behavioral biases.

Their features and potential solutions are described below.

4.1 Regulatory barriers

Regulatory barriers refer to all those circumstances, embedded in the legal or regulatory framework, that either impede or make more costly market participation on the part of small customers. The existing regulatory barriers can be often understood with the benefit of hindsight: electricity has historically been a strongly concentrated industry, with high coordination costs and strong incentives to overinvest in reliability. Despite the major changes intervened in the past few decades, this is still true to a considerable extent. Regulation has been designed and has evolved over time having this kind of context in mind; hence it has overlooked, if not willingly excluded, the contribution from smaller agents, whose participation could increase the costs of coordination as well as the probability of a failure. Technological and institutional innovations have significantly reduced both risks, but regulators (as individuals) and regulation (as a body of norms) still pay tribute to a secular legacy of dealing with the "Big Guys." On top, the entry of smaller subjects in the market may also threaten the position and margins of the incumbents. As Stigler (1971) wrote, more often than not "regulation is acquired by the industry and is designed and operated primarily for its benefit." In dealing with the regulatory obstacles that hinder small customers' participation in electricity markets, a peculiar effort needs to be performed, in order to separate those regulations (or parts thereof) whose benefits outweigh the costs, from those that result primarily in limiting competition without pursuing any meaningful general interest.

Removing regulatory barriers is necessary, although by no means sufficient, to get small customers engaged in flexibility services (see also Chapter 5).

A first issue is related to the very design of the markets for services, and the role of DSOs and TSOs—including how vertical coordination is achieved and how, and to what extent, the grid activities are unbundled within vertically integrated companies. The main challenge to—and the main resistance from—grid operators lies in the margins and cash-flow shift that is implied in a greater involvement of small customers. Grid regulation has historically been predicated upon the principle that grid operators would recover their costs mainly through levies that are charged according to the volume of energy withdrawn from the public grid and, to lesser extent, as fixed charges. The increasing volumes of self-generated, or stored, energy (which is a consequence of low-cost renewables and batteries) and the flattening of the load profile (that may stem from a greater engagement of small customers in flexibility markets) may result in lower revenues, hence a greater difficulty in recovering costs—or, alternatively, in the need to increase the levies, all else being equal. Moreover, network companies may also become exposed to risks that they are not supposed, and not equipped, to deal with. At the very least, investment priorities and market conduct should be updated to take into account the new array of (potential) market participants and business conducts (Rossetto et al., 2019; Shittekatte et al., 2018). Grid operators may be more or less willing to adapt to this changing environment, but the choice is not really theirs: it depends also on the kind of evolution that regulators envision. If the approach to investments and grid management does not develop, though, grid operators will keep doing business as they have always done in the past decades, perceiving small consumers as a threat—or nuisance—rather than an opportunity for the whole system.

A second source of obstacles to demand participation in flexibility markets comes from the market design itself. In particular, three barriers have been identified (Koliou et al., 2014; Eid et al., 2016):

• Minimum bidding volumes;
• Minimum bid duration; and
• Binding up and down bids.

These requirements reflect both the supply-side bias of electricity regulation (Sioshansi, 2020) and the underlying idea that large generators matter the most. In fact, balancing the grid is easier (and possibly less costly) if bids are relatively large, if they need to be finalized well in advance, and if up and down bids are binding. Each of these requirements have a rationale and was perfectly acceptable in a supply-driven electricity system. Yet, they prevent, or make more costly, the entry of new actors, and come increasingly at odds with a trend toward distributed generation and active demand.

This leads us to a third issue. Small consumers—both households and SMEs—can provide little individual contribution to balancing the grid. They may turn into a powerful force, though, through aggregation. But this presupposes that demand aggregation is possible, in the first place. An aggregator acts as a third-party intermediary between end consumers and power generators. As such, a demand response aggregator need not be the supplier as well, although there are obvious synergies between providing energy to an end consumer and managing her load for the sake of grid balancing. If compared with traditional utilities, aggregators have both strengths and weaknesses. For example, they tend to be flexible and run on relatively low operational costs, but the cost for acquiring new customers may be high and the underlying financial risk may be also high (Ponds et al., 2018). More importantly, the sustainability of their business model relies critically on two features of the market design: (i) relatively low barriers to entry in the wholesale markets and (ii) the customers' trust in the market functioning. Sioshansi provides several examples of workable—and successful—business models for aggregators in Chapter 7.

Hence, the design of *retail* markets is equally important. For small customers to trust aggregators and become enrolled in a demand-side response, there must be a widespread understanding of what a consumer's rights are (He et al., 2013) and where opportunities are to be found. The role of retail competition is often overlooked but in this case it may be crucial. First, consumers must feel that they are not going to be cheated and, even more important, that switching supplier or becoming part of a demand response program will not be detrimental to the quality of service. Again, well-functioning smart grids and smart meters, on which rapid and effective switching procedures depend, emerge as a precondition (CEER, 2018). Secondly, consumers should understand that electricity is a service rather than a commodity, and that there is wide scope for product differentiation in the industry. This implies a major departure from the traditional view of the industry and from traditional price regulation or price-setting mechanisms (Stagnaro et al., 2020). Thirdly, such features as real-time metering and dynamic pricing should be technically and legally available (CEER, 2020; Schreiber et al., 2015).

Beyond all that, consumers must be made aware that providing flexibility services is technically feasible, socially desirable, and potentially profitable. It is not easy to convey such information in a context where the opposite has been true for generations. Retail competition and freedom to choose may provide the infrastructure whereby profit-oriented retailers or aggregators make the "dirty job" of nudging customers toward a more active participation in electricity markets—a sort of application of Say's Law that supply creates its own demand.

4.2 Behavioral barriers

Even if all the above barriers were removed, a rapid take-up of small customers' participation in flexibility markets would not be obvious to follow. Other, more subtle barriers exist, that entail the *perceived* costs and benefits from demand-side response.

To begin with, people must be aware that they may draw financial benefits, while producing positive externalities to the network as well as the environment, by contributing to balancing the grid. The degree of awareness is today relatively low, particularly among older people (Li et al., 2017). Especially in the initial stages, electricity users cannot be expected to seek information by themselves. Hence informational campaigns (both institutional and advertising by aggregators and other intermediaries) may be needed to catch the attention of small customers. A fortiori, this suggests that aggregators or other facilitators do not only provide a *technical* service by managing the available load in a way that is beneficial to the system. A big chunk of their contribution—and a large share of the value they create—lies in their *commercial* operations, that include the task of providing information to the interested party, acquiring customers, and being a reliable intermediary among so different subjects as small customers and grid operators.

A strictly connected issue is that of trust: for customers to be willing to engage in load-shifting and similar activities, a high degree of trust is needed toward those in charge of managing the load request within houses and SMEs. According to a survey performed on behalf of the European Union (EC, 2016), even though households have been free to switch electricity suppliers since 2007 in most member states, still more than one-third do not know or believe it is not allowed. Yet, about 80% of EU households either agree or strongly agree that electricity companies respect rules and regulations. The share is higher in countries where the market is more mature (86%) and among those who are highly informed about their own energy use (91%). The most successful intermediaries found their value proposition upon reputation and trust (see, for example, *EEnergy Informer*, 2020).

Trust is also positively related to the perception of the gains from trade, and negatively related to the perceived difficulties of joining a demand-side response program (including the time and effort needed to become enrolled). Nicolson et al. (2018) show that—despite a relatively high interest stated in surveys—the take-up rate of time-of-use tariffs (a basic kind of demand-response setting) can be very low unless specific steps are taken to close what they define the "intention-action" gap. If significant effort is made to provide guidance and ease switching or enrollment, then the share of those actually involved may increase up to as much as 43%. Consumers also seem to prefer simple tariffs—for example they would rather have time-of-use rather than real-time pricing—although there may be a trade-off between simplicity and effectiveness. While it may be desirable, from a systemic perspective, to nudge consumers into shifting loads in time (say, from late evening to midday), that is based on ex ante observations. Prompting real-time responses from consumers, in order to contribute to balancing the grid when and if needed, might deliver much larger benefits.

A major issue to overcome behavioral barriers lies in how commercial offers are presented to small customers. As it can be expected, opt-out enrollment is much more effective than opt-in, but a case can hardly be made for *requiring* customers to participate markets unless they decide otherwise. In drafting proposals, aggregators or other market agents should consider that energy consumers display loss aversion to a significant degree, and loss-averse consumers are less likely to switch to time-of-use tariffs or other forms of market participation (Nicolson et al., 2017). In other words, the compensation these customer are willing to accept in order to give up something they already have (say, the right to run their dish-washer at 6 p.m.) is larger than the amount they would be willing to pay in order to run their dishwasher at 6 p.m. if they had not had that right in the first place.

The same papers and surveys that documented this phenomenon—also known as "endowment effect" in behavioral economics—show that small consumers are more likely to become enrolled in sophisticated tariff designs if they expect a higher compensation. Which leads us to a final question: How large can reasonably be the financial rewards for inducing small customers to participate the markets? Unfortunately, the scope for making a business out of providing flexibility services to the market is still limited, although it is reasonable to expect them to grow over time (Nolan and O'Malley, 2015). To make things even worse, it is not easy to tell in advance how profitable it might be to become enrolled. This—perhaps unavoidable—lack of transparency goes to detriment of trust and what we have called "willingness to accept." Studies show that demand-side flexibility may be worth about 3% of the annual bill for a typical household (Tveten et al., 2016; Feuerriegel and Neumann, 2016).

Table 6.3 provides an estimate of the potential savings for European customers under a conservative estimate (3% of their annual expenditure for

Table 6.3 Estimate of potential savings from demand-side flexibility services in the EU member states.

	Yearly consumption	Price	Yearly expenditure	Potential savings (conservative scenario)	Potential savings (optimistic scenario)
	kWh/y	Euro/ kWh	Euro/year	Euro/year	Euro/year
Romania	1639	0.14	233	7	15
Lithuania	1963	0.13	250	8	17
Latvia	2022	0.17	351	11	23
Poland	2124	0.15	312	9	21
Italy	2840	0.23	665	20	44
Slovakia	3262	0.16	517	16	34
Hungary	2912	0.11	319	10	21
Estonia	2993	0.14	422	13	28
Netherlands	3063	0.21	629	19	42
Portugal	3114	0.22	679	20	45
Germany	3694	0.29	1061	32	70
Czech Republic	3645	0.18	645	19	43
Denmark	4102	0.29	1199	36	79
UK	4553	0.22	1006	30	66
Bulgaria	3698	0.10	354	11	23
Belgium	4198	0.29	1201	36	79
Spain	3790	0.24	907	27	60
Luxembourg	3876	0.18	697	21	46
Croatia	4138	0.13	548	16	36
Slovenia	4280	0.17	713	21	47
Greece	4471	0.16	693	21	46
Ireland	5205	0.22	1150	34	76
Malta	4555	0.13	594	18	39
Austria	4620	0.21	958	29	63
Cyprus	4841	0.22	1082	32	71
France	5425	0.17	944	28	62
Finland	7848	0.15	1195	36	79
Sweden	9770	0.18	1723	52	114
EU average	**4023**	**0.18**	**752**	**23**	**50**

Source: Own elaboration on data from Eurostat, Odyssee database.

electricity) and under an optimistic one (6%, under the assumption that innovation and the gradual increase in rooftop solar panels, small storage facilities, and smart appliances will eventually augment the number of people willing to become enrolled, and that as a consequence power demand would increase by 10%).

As it can be seen from Table 6.3, the potential savings are relatively limited, as they range between 23 and 50 euro per year at the EU level. At the country level, potential savings would be between 7 and 52 euro/year in the conservative scenario and 15—114 euro/year in the optimistic one. Flexibility services would be more profitable in the countries where either consumption or prices are the highest—or both. Perhaps not surprisingly, the country where the expected gains are the highest, i.e., Sweden, is the one with the highest consumption per household (more than twice as much as the EU average). Still, one may question whether people would be willing to give up to some extent the control over their own consumption behavior, if the remuneration is so low.

On the other hand, an increase in demand-side participation might deliver relatively large benefits to the whole grid (a positive externality) and to renewable generators in particular, with lower power curtailment and more revenues (Tveten et al., 2016; Kubli, 2018). The take-up rate might increase under optimal tariff design (Schreiber et al., 2015), innovative business models (Lombardi and Schwabe, 2017), and an effective targeting of more inclined categories of consumers, such as those living in detached houses rather than apartments (Torstensson and Wallin, 2015) or the owners of electric vehicles (Kaschub et al., 2016; Nicolson et al., 2017).

The good news is that at least some of these barriers are being overcome as technology progresses and markets mature. As time passes, everything suggests that a number of things will change: the cost of rooftop solar, storage, and electric vehicles will decreases, leading to a greater diffusion of these devices; a larger share of total energy consumption by households and SMEs will be covered by electricity rather than other forms of energy; smart devices and appliances will gain traction; consumers will become more skilled in, and likely to, switch to a more sophisticated tariff; consumer protection regulation will evolve in a way that is predicated upon promoting the engagement of small customers, rather than limiting product differentiation. All of the above, and more, will translate into (i) a larger expenditure on electricity per household or per SME and (ii) greater trust in electricity suppliers, aggregators, and other intermediaries. This, in turn, will lead to greater savings or other kinds of financial rewards from, and less hesitation in, becoming enrolled in flexibility programs. More sophisticated software, the use of artificial intelligence, and smarter devices will provide the enabling condition for flexibility services to be supplied even by small customers (under aggregation schemes), but ultimately knowledge, awareness, and more active market behavior will make the difference.

5. Conclusion and policy implications

The provision of flexibility services from small customers is reality, not science fiction, but its scope is still limited. This is due to a number of obstacles that can and should be removed. This can often be done relatively easily, as long as it entails minor regulatory changes. Examples are the reduction of the minimum volumes for bids in balancing markets, the reduction of minimum bid durations, and introducing more flexibility with regard to potential changes in up and down bids.

Other barriers are harder to remove. This is particularly the case of behavioral barriers, that depend on *information* asymmetries and *trust*. If consumers do not even know that they may provide services for which they would be remunerated, or they do not trust those who would like them to become enrolled in flexibility programs, they are unlikely to grow more active. In some instances transaction costs—including the cost of information and trust—can be reduced by regulatory choices, but ultimately it is up to market operators (such as traditional utilities, aggregators, or third parties) to overcome these barriers. In other cases, the problems do not lie in *costs* but in *prices*, i.e., in the customer's willingness to accept a change, however little, in her consumption choices. For the time being, there seems to be limited scope for large gains. Our review of EU member states showed that, even under favorable assumptions, the potential savings do not exceed a few dozen euro per year.

At the same time, it is clear that a greater demand-side participation would bring a real, sizable, and growing value to the system. Suppliers, aggregators, and potentially new entrants from the digital arena might play a critical role in promoting demand-side flexibility, for example by embedding these services in broader offers. As the AC/DC would sing, "It's a long way to the top."

References

CEER, 2018. Roadmap to 2025 Well-Functioning Retail Energy Markets. Council of European Energy Regulators, C17-SC-59-04-02.

CEER, 2020. Recommendations on Dynamic Price Implementation. Council of European Energy Regulators, C19-IRM-020-03-14.

Coase, R.H., 1937. The nature of the firm. Economica 4, 386—405.

Coase, R.H., 1960a. The problem of social cost. J. Law Econ. 3, 1—44.

Coase, R.H., 1960b. The problem of social cost. J. Law Econ. 56, 837—877.

Creti, A., Fontini, F., 2019. Economics of Electricity. Cambridge University Press, Cambridge.

D'hulst, R., Labeeuw, W., Beusen, B., Claessens, S., Deconinck, G., Vanthournout, K., 2015. Demand response flexibility and flexibility potential of residential smart appliances: experiences from large pilot test in Belgium. Appl. Energy 55, 79—90.

EC, 2016. Second Consumer Market Study on the Functioning of the Retail Electricity Markets for Consumers in the EU. Report prepared by Ipsos-London Economics-Deloitte consortium for the European Commission Directorate-General for Justice and Consumers.

EEnergy Informer, 2020. First PVs, then EVs, now VPPs. EEnergy Informer 30 (7), 10—11.

Eid, C., Koliou, E., Valles, M., Reneses, J., Hakvoort, R.A., 2016. Time-based pricing and electricity demand response: existing barriers and next steps. Util. Pol. 40, 15—25.

Faruqui, A., 2020. 6 reasons why California needs to deploy dynamic pricing by 2030. Util. Drive. May 19, 2020. https://www.utilitydive.com/news/6-reasons-why-california-needs-to-deploy-dynamic-pricing-by-2030/578156/.

Feuerriegel, S., Neumann, D., 2016. Integration scenarios of Demand Response into electricity markets: load shifting, financial savings and policy implications. Energy Pol. 96, 231—240.

Gils, H.C., 2014. Assessment of the theoretical demand response potential in Europe. Energy 67 (1), 1—18.

Grein, A., Pehnt, M., 2011. Load management for refrigeration systems: potentials and barriers. Energy Pol. 39, 5598—5608.

He, X., Keyaerts, N., Azevedo, I., Meeus, L., Hancher, L., Glachant, J.-M., 2013. How to engage consumers in demand response: a contract perspective. Util. Pol. 27, 108—122.

Helin, K., Käki, A., Zakeri, B., Lahdelma, R., Syri, S., 2017. Economic potential of industrial demand side management in pulp and paper industry. Energy 141, 1681—1694.

Kaschub, T., Jochem, P., Fichtner, W., 2016. Solar energy storage in German households: profitability, load changes and flexibility. Energy Pol. 98, 520—532.

Kiesling, L.L., 2009. Deregulation, Innovation and Market Liberalization. Routledge, London.

Klaassen, E.A.M., Kobus, C.B.A., Frunt, J., Slootveg, J.G., 2016. Responsiveness of residential electricity demand to dynamic tariffs: Experiences from a large field test in the Netherlands. Appl. Energy 183, 1065—1074.

Koliou, E., Eid, C., Chaves-Ávila, J.P., Hakvoort, R.A., 2014. Demand response in liberalized electricity markets: analysis of aggregated load participation in the German balancing mechanism. Energy 71, 245—254.

Kubli, M., 2018. Squaring the sunny circle? On balancing distributive justice of power grid costs and incentives for solar prosumers. Energy Pol. 114, 173—188.

Kubli, M., Loock, M., Wüstenhagen, R., 2018. The flexible prosumer: measuring the willingness to co-create distributed flexibility. Energy Pol. 114, 540—548.

Li, R., Dane, G., Finck, C., Zeiler, W., 2017. Are building users prepared for energy flexible buildings? — A large-scale survey in the Netherlands. Appl. Energy 203, 623—634.

Lombardi, P., Schwabe, F., 2017. Sharing economy as a new business model for energy storage systems. Appl. Energy 188, 485—496.

Lopes, R.A., Chambel, A., Neves, J., Aelenei, D., Martins, J., 2016. A literature review of methodologies used to assess the energy flexibility of buildings. Energy Procedia 91, 1053–1058.

Martínez Ceseña, E.A., Good, N., Mancarella, P., 2015. Electrical network capacity support from demand side response: techno-economic assessment of potential business cases for small commercial and residential end users. Energy Pol. 177, 222–232.

Nicolson, M.L., Huebner, G.M., Shipworth, D., 2017. Are consumers willing to switch to smart time of use electricity tariffs? The importance of loss-aversion and electric vehicle ownership. Energy Res. Soc. Sci. 23, 82–96.

Nicolson, M.L., Fell, M.J., Huebner, G.M., 2018. Consumer demand for time of use electricity tariffs: a systematized review of the empirical evidence. Renew. Sustain. Energy Rev. 97, 276–289.

Nolan, S., O'Malley, M., 2015. Challenges and barriers to demand response deployment and evaluation. Appl. Energy 152, 1–10.

Nordic Council of Ministers, 2017. Flexible Demand for Electricity and Power: Barriers and Opportunities. Tema Nord, Copenhagen.

Ponds, K.T., Arefi, A., Sayigh, A., Ledwich, G., 2018. Aggregator of demand response for renewable integration and customer engagement: strengths, weaknesses, opportunities, and threats. Energies 11, 2391.

Rassenti, S.J., Smith, V.L., Wilson, B.J., 2002. Using experiments to inform the privatization/deregulation movement in electricity. Cato J. 21 (3), 515–544.

Rossetto, N., Dos Reis, P.C., Glachant, J.-M., 2019. New Business Models in Electricity: The Heavy, the Light, and the Ghost. EUI Policy Brief, RSCAS 2019/08.

Schittekatte, T., Momber, I., Meeus, L., 2018. Future-proof tariff design: recovering sunk grid costs in a world where consumers are pushing back. Energy Econ. 70, 484–498.

Schreiber, M., Wainstein, M.E., Hochloff, P., Dargaville, R., 2015. Flexible electricity tariffs: power and energy price signals designed for a smarter grid. Energy 93, 2568–2581.

Shoreh, M.H., Siano, P., Shafie-khah, M., Loia, V., Catalão, J.P.S., 2016. A survey of industrial applications of demand response. Elec. Power Syst. Res. 141, 31–49.

Sioshansi, F. (Ed.), 2019. Consumer, Prosumer, Prosumager. Academic Press, London.

Sioshansi, F. (Ed.), 2020. Behind and Beyond the Meter. Digitalization, Aggregation, Optimization, Monetization. Academic Press, London.

Stagnaro, C., Benedettini, S., 2020. Smart meters: the gate to behind-the-meter? In: Sioshansi, F. (Ed.), Behind and Beyond the Meter. Digitalization, Aggregation, Optimization, Monetization. Academic Press, London, pp. 251–266.

Stagnaro, C., Amenta, C., Di Croce, G., Lavecchia, L., 2020. Managing the liberalization of Italy's retail electricity market: a policy proposal. Energy Pol. 137, 111150.

Stigler, G.J., 1966. Theory of Price. Macmillan, New York.

Stigler, G.J., 1971. The theory of economic regulation. Bell J. Econ. Manag. Sci. 2, 3–21.

Thaler, R.H., 2017. From Cashews to Nudges: The Evolution of Behavioral Economics. Nobel Prize Lecture, December 8, 2017.

Torstensson, D., Wallin, F., 2015. Potential and barriers for demand response at household customers. Energy Procedia 75, 1189–1196.

Tveten, Å.G., Bolkesjø, T.F., Ilieva, I., 2016. Increased demand-side flexibility: market effects and impacts on variable renewable energy integration. Int. J. Sustain. Energy Plann. Manag. 11, 33–50.

Williamson, O.E., 2008. Transaction cost economics. In: Ménardé, C., Shirley, M.M. (Eds.), Handbook of New Institutional Economics. Springer, Berlin, pp. 41–65.

Yin, R., Kara, E.C., Li, Y., DeForest, N., Wang, K., Yong, T., Stadler, M., 2016. Quantifying flexibility of commercial and residential loads for demand response using setpoint changes. Appl. Energy 177, 149–164.

CHAPTER 7

How can flexible demand be aggregated and delivered?

Fereidoon Sioshansi[1]

Menlo Energy Economics, Walnut Creek, CA, United States

1. Introduction

If customer demand is inherently *flexible*, as claimed, then why is it difficult to deliver demand flexibility on a large scale in practice? That is the proverbial $64,000 question.

Since Chapter 5 in this volume provided the background and context for the demand flexibility paradox, this chapter presents empirical evidence, anecdotal examples, case studies, and evolving business models of a number of innovative companies that have succeeded and/or are trying to do so focusing on particular segments of the market, in different parts of the world and taking different approaches with different degrees of success.

As will be described, it is fair to say that the state of the art in what works and is profitable is still evolving as existing players and innovative newcomers experiment and fine-tune their approaches. The main challenge is to identify, capture, and monetize value in a way that is win–win–win: for the participating customers, for the aggregator and for the network, the DSO, and/or the grid operator.

Another challenge is that not all regulators and policymakers necessarily recognize the full value of flexible demand nor do they actively encourage further development and participation of demand response (DR) schemes in electricity markets. Despite these challenges, as experience is gained in delivering low-cost, reliable products and services, DSOs and grid operators are gaining confidence in incorporating demand flexibility into markets making it easier for the nascent business to expand.

[1] The author has benefitted from original content as well as useful references and case studies provided by Daniele Andreoli and Paul Troughton of Enel X and Pierre Bivas of Voltalis. Moreover, Daniele Andreoli of Enel X, Pierre Bivas of Voltalis, Stephen Marty of KiwiPower and Michele Governatori representing the European Retail Electricity Retailers participated in a symposium in Florence, Italy on 19 Feb 2020, whose efforts are gratefully appreciated.

Variable Generation, Flexible Demand
ISBN 978-0-12-823810-3
https://doi.org/10.1016/B978-0-12-823810-3.00014-5

The balance of the chapter is organized as follows:
- Section 2 features Enel X, currently the biggest and best known firm with a global and expanding footprint primarily focused on large commercial and industrial (C&I) segment of the market;
- Section 3 features Voltalis, an innovative French enterprise primarily focused on managing electric space and water heating loads in Europe;
- Section 4 features OhmConnect, a San Francisco-based start-up currently focused on the residential and small commercial sector;
- Section 5 briefly describes a number of other noteworthy start-ups with promising business models including CPower, Leap, Olivine, Voltus, EnergyHub, and Enbala plus others active in integration or implementation of DR and distributed energy resources (DERs); followed by the chapter's conclusions.

It must be noted that this is *not* an exhaustive list nor are the companies presented in any particular order. Moreover, mentioning the companies by name is not to be construed as an endorsement of their approach or potential. Finally, this chapter does *not* include companies engaged in electric vehicle charging business—which is covered in the following chapter.

2. Enel X[2]

Enel X, recognized as the biggest and best known in the field, is an established firm with a global footprint. It is part of Enel Group of companies, which has been rapidly expanding its market reach with the recent acquisition of EnerNOC and eMotorWerks in North America. The company claims to have 6.3 GW of demand under management in 15,000 sites across the world. It is mainly but not exclusively focused on the large commercial and industrial (C&I) segment of the market, where the marketing, customer acquisition, and transaction costs are low relative to the potential to generate and deliver value.

The company's DR business model—Enel X does many other things besides DR—may be summarized as identifying and approaching C&I customers with large energy bills *and* demand flexibility as further described in Box 7.1. Moreover, Enel X is most active in markets where demand flexibility is appropriately valued—for example, in places with a capacity

---

[2] The assistance of Paul Troughton and Daniele Andreoli from the London and Rome offices of Enel X, respectively, who provided helpful content and reviewed an initial draft of the chapter, is hereby acknowledged.

BOX 7.1 [3]How does demand response work in practice

As already explained in preceding chapters, demand flexibility is becoming increasingly important for all stakeholders including individual consumers, energy suppliers, and Transmission System Operators (TSOs).

Virtually any facility can reduce and/or shift demand to some extent. For an aggregator such as Enel X, any customer with at least 100 kW of demand flexibility is a good candidate for Demand Response (DR) participation. Examples may include schools and college campuses, hospitals and healthcare, manufacturing, food processing and cold storage facilities, commercial property buildings, grocery stores, retail malls, government buildings, and many more.

Customers' ability to be flexible about when they use energy represents an opportunity that can be monetized whenever there is a utility or grid-sponsored DR scheme allowing participants to get paid for offering that flexibility. Nearly all DR programs provide financial payments to customers who agree to adjust their energy consumption in response to signals. Such schemes are gaining traction across the globe and if you look at the benefits, it's easy to see why.

By participating in DR programs, customers can earn substantial payments for being on call to modify nonessential energy use when the electric grid needs support—a win-win for customers looking to boost their bottom line and the grid operators who avoid incurring significant capital expenditures.

Most large DR providers focus mostly on commercial and industrial (C&I) customers by managing their energy-intensive assets to avoid peak demand periods while using energy efficiently. In exchange for being on call to reduce energy, participants receive *availability payments* and also receive additional *energy* payments based on how much they can reduce consumption when called on. The amount of payment varies depending on the specifics of the schemes and the incentives offered by the utility or the grid operator. DR providers work with customers to ensure that they receive the maximum financial benefits for their participation.

According to Enel X, a typical DR engagement consists of three steps:

- Prepare;
- Respond; and
- Restore.

In the first step Enel X liaises with the customer to identify ways to reduce nonessential energy without adversely affecting business operations, comfort, or product quality. After examining customer's energy intensive operations, Enel X comes up with a strategy that delivers maximum value with the minimum operational impact. Energy reduction measures are customized for each facility and may include turning off lighting, air-conditioning, pumps, and other nonessential equipment.

Moreover, many C&I customers can shift energy-intensive processes by a few hours to maximize the value of their participation in the DR program based on the available incentives. In some cases, customers may increase their impact by switching to backup generation and/or on-site storage, further reducing

Continued

BOX 7.1 [3]How does demand response work in practice—cont'd

demand on the grid. DR aggregators work with customers to create a customized strategy that works and delivers the best value.

An integral part of this step is to install the necessary monitoring, communication, and metering devices at the customer's facility, for example, a gateway device to enable communication with Enel X's Network Operations Center (NOC). This ensures that Enel X can monitor energy-intensive devices in real time during DR dispatches while allowing the customer to access the same.

The second step enables the customer to receive and acknowledge dispatch notifications from the NOC. When the utility or grid operator needs the participants' response, it notifies the DR provider, who in turn sends a notification to customers alerting them when a DR episode will begin and how long it will last, typically between 15 min and 4 h. Once the customer acknowledges the receipt of the notification, they will be instructed to reduce electricity usage according to the predetermined plan, usually conducted automatically by the NOC using sophisticated software.

The last step is to return to normal operations once the DR dispatch is over. The goal of the scheme is to ensure that participants achieve their reduction targets and receive the highest level of financial payments.

[3] This box is based on 9 May 2020 post by Daniele Andreoli at https://www.linkedin.com/pulse/unheard-what-once-cost-now-new-revenue-stream-how-daniele-andreoli/?trackingId=NPV1ZW%2BsR%2B2ErMNWzNTPhA%3D%3D.

---

market. If there is nobody willing to pay for flexibility, there is no point in looking for customers with demand flexibility.

The next phase of engagement typically includes gaining a detailed understanding of the customers' operations and operational flexibility—as further described in the chapter by Lobbe et al.—followed by discussions on how much of this flexibility can in practice be managed in a way that generates value, typically in the form of lower energy bills and/or demand charges.[4] The conditions to proceed to the next step typically include:

- Sufficient scope for savings for the customer relative to the costs;
- Agreement on the protocols on how to deliver the demand response and create value in the process; and

[4] Some DR players such as Stem primarily focus on avoiding or reducing monthly demand charges, which can be significant, further details at https://www.stem.com.

- How the resulting savings, net of costs—both Opex and Capex—will be captured and shared.

The bottom line is that the expected savings must exceed the effort, the investments, and/or any inconvenience or service degradation that may ensue—for example, requirements to shift certain energy-intensive operations to off-peak hours, change processes, invest in new equipment including storage devices that can shift loads, and so on.

While these may not seem onerous, getting to the "yes"—i.e., an executed agreement to cooperate—is not trivial. Typical challenges include:

- Most customers are initially skeptical and reluctant to allow an outsider to monitor and manage their energy consumption;
- Many are reluctant or unable to invest in technology that may be needed to deliver savings;
- Since the estimated savings cannot be guaranteed unless the necessary investments are made and operations are rearranged, there is a risk that may prevent some customers from participating;
- There may be disagreements on who does what to deliver value and who gets what portion of the value generated; and
- Since necessary investments are usually needed up-front and savings accrue over time, the parties must agree on how to share the risks and rewards—both of which are uncertain.

Enel X people with hands-on experience dealing with customers explain that the central issue is *not* that you usually need to make investments to generate savings, but rather how to deal with the uncertainties about the savings that may be accrued following the investment. Some of this uncertainty has to do with adjustments in the customer's operations, but usually more relates to the prices and available incentives in a given market. For example, in a market with an established capacity payment scheme, there may be a few years of known price visibility, but typically not enough to fully derisk the investment. In other cases, there may be little price certainty or track record, hence more risk exposure, as captured in the final bullet above.

Typically DR customers don't have to make any up-front financial investment, with Enel X paying for any equipment that requires installation and monitoring. Still, getting customers to participate in the DR scheme requires staff and management effort.

Talking with people involved in a typical engagement reveals that these and many other issues need to be identified, negotiated, and agreed before proceeding to implementation and operations, which explains why transaction costs tend to be high and why smaller customers with small demand flexibility are less attractive to companies like Enel X. Box 7.2 provides a few examples.

BOX 7.2 Case study: Kimberly-Clark is rewarded for its demand flexibility[5]

A typical example of Enel X engagement with C&I customers is its successful project with Kimberly-Clark, which has a major plant in Ontario, Canada, operating 24/7 producing a range of paper products including Kleenex, Scott, Huggies, Pull-Ups, Kotex, etc., which are sold globally.

In 2010, the paper mill decided to participate in Ontario Power Authority's (OPA) DR Program to reduce its energy bills. Initially, the customer was thinking of offering 500 kW of flexible load in OPA's DR scheme. But after Enel X visited the site to evaluate the flexibility of customer's energy use, the two agreed that a much bigger portion of the facility's 7 MW load could potentially participate in the program. After a successful trial, it became clear that as much as 5.3 MW of the load, representing the most energy-intensive processes at the plant, could in fact be shut down for up to 4 h to maximize the savings from the DR program. This was possible because the plant had the flexibility to continue uninterrupted operations by processing product inventory on hand—such as packaging and shipping.

Subsequently, in Nov 2011, Kimberly-Clark entered into an agreement allowing Enel X to take advantage of its demand flexibility to reduce its energy bills. Over time, the client has developed a culture of flexibility around being able to respond quickly to a DR dispatch and make the most of its 4-hour period of downtime to perform required maintenance on the tissue manufacturing machine—the most energy-intensive load at the plant. This has been an exceptional success story because no capital investment had to be made and no major changes in operations were necessary—what is called "an empty wallet approach."

At the same time, the successful DR program has been a clear win-win-win:

- For Kimberly-Clark, who receives $325,000 per annum by participating in the DR scheme;
- For the Ontario's power grid, which is spared over 5 MW of load during peak demand periods; and
- For Enel X who has been able to assist in the successful implementation of the scheme.

Since 2011, offering its flexible demand has resulted in more than $2.3 million in DR payments to the plant.

[5] This box is based on Daniele Andreoli's presentation in a symposium in Florence, Italy on 19 Feb 2020 as well as Enel X Customer Highlight, *Kimberly-Clark's Huntsville Mill Reduces Cost, Improves Performance with Enel X Demand Response* at https://www.enelx.com/n-a/en/resources/case-study/cs-kimberly-clark.

The above case study illustrates what it takes in a typical engagement. Enel X has a portfolio of similar success stories working with customers with flexible demand across the world including a DR program with Glenwood properties in New York City, which offers load reduction during the critical summer peak period.

In this case, Enel X helped Glenwood to optimize its existing distributed energy resources (DER) consisting of 6.3 MWh of battery storage and 2 MW solar-plus-storage in its real estate properties in New York City resulting in roughly 3.5 MW of flexible demand—which allows the client to make it through the peak demand hours during summer months.[6] According to Josh London, Vice President of Glenwood Management,

We are expanding our partnership with Enel X by adding demand response to the portfolio of solutions that we are leveraging to reduce our energy costs, improve the resiliency of our buildings for our tenants, and support the stability of the power grid.

Other successful examples include an agreement with the Dublin Airport (DAA) to offer 11+MW of flexible capacity in Ireland's ancillary services market and numerous others.[7] Enel X, which acquired EnerNOC, serves more than 3,600 customers in nearly 11,000 sites in North America.

3. Voltalis

Voltalis is a Paris-based flexible demand innovator, which originally started managing electric space and water heating loads in the residential sector in France, now expanding its footprint in Europe and beyond.

Both electric water and space heating are ideal loads with built-in flexibility. Since most electric water heaters come with a storage tank, it doesn't much matter *when* the water is heated. What matters is to have a sufficient supply of hot water when needed. This means that electric water heating load is truly flexible. There is no reason to heat up the water during peak demand hours—when supplies may be limited and prices tend to be high. Space heating has similar characteristics since thermal inertia of the

---

[6] Based on "Enel X expands partnership with New York luxury residential real estate Group Glenwood to include demand response services," 17 May 2018 at https://www.enelx.com/en/news-and-media/press/2018/05/enel-expands-partnership-with-new-york-luxury-residential-real-e.

[7] Refer to https://www.enelx.com/en/news-and-media/news/2018/08/demand-side-response-Dublin-Airport.

space allows brief curtailments of such loads without any impact on the occupants' comfort. The other major electricity using device with similar flexible demand characteristics are electric vehicles with their massive batteries, the topic of the next chapter.

Since both electric water and space heaters are widely used in Europe, the potential scale of the aggregated load is enormous. Managing this relatively easy to control load is not new. For some time, France has had a two-part tariff to shift some heating load to off-peak evening hours. The two-part tariff is frequently mentioned as an example of a successful demand management scheme. In fact the International Renewable Energy Agency (IRENA) credits it for modifying the daily shape of demand curve in France over time.[8]

Voltalis,[9] founded in 2006, initially focused on controlling the aggregated load of large fleets of electric space and water heaters, primarily in apartment buildings that are operated by large property owners/developers in urban areas. This reduces marketing and transaction costs since once the manager of the property is convinced, all the water heaters in the buildings can be fitted with the necessary tools to adjust when water is heated and stored.

Once agreement is reached with the property owner/manager, Voltalis installs the monitoring devices at customers' premises (Fig. 7.1) and manages them free of charge. Customers share the *flexibility* of their consumption— which Voltalis aggregates and operates at scale without inconveniencing them. It sells these services to the grid operator and/or in the wholesale market, which provides a revenue stream. It is a win–win–win.

In describing the company's business model, its founder, Pierre Bivas,[10] says participating consumers essentially get a *free* energy monitoring and management service and save on electricity bills, by up to 15%. There is no up-front cost and no risk. They do not have to do anything—Voltalis does it for them from its control center (Fig. 7.2) where it monitors generation, loads, and prices on the network to decide when to adjust the water and/or space heaters without affecting occupants' comfort.

---

[8] Refer to Demand-side flexibility for power sector, IRENA, 2019 based on data from EDF R&D at https://www.irena.org/publications/2019/Dec/Demand-side-flexibility-for-power-sector-transformation.

[9] For further details visit https://www.voltalis.com.

[10] The assistance of Pierre Bivas is greatly acknowledged. Moreover, Bivas participated in a symposium in Florence, Italy, on 19 Feb 2020 providing valuable insights.

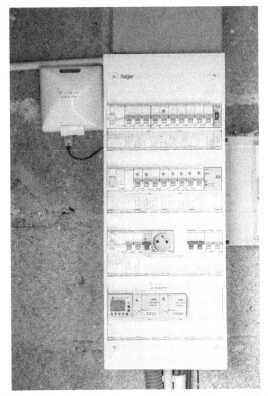

Figure 7.1 Typical Voltalis control device. *(Source: Voltalis)*

Figure 7.2 Voltalis control center. *(Source: Voltalis)*

Moreover, Bivas emphasizes that the scheme not only benefits the participants but also nonparticipants because it reduces overall operational costs of the network.

Companies in the generation business should also benefit from flexible demand participating in the market, as far as these companies adapt their range of assets and strategies. Indeed, DR participation in the market is a way to avoid volatility of prices—which is a risk for all market participants—and also, physically, a way to smoothen load curves, so that generators are used more efficiently. In addition to contributing to the reliability of the power system, it will be easier for generators to get rid of their old, expensive and polluting power plants.

The company has attracted roughly 100,000 customers amounting to 300 MW of inexpensive capacity aggregated from roughly a million electrical appliances mainly in homes, but also in commercial premises, offices, and various public and municipal buildings. It claims to account for more than 80% of the DR volumes delivered in the French market in 2018. More recently, Voltalis has received the support of the European Investment Bank and the European Commission to install its technology in some 150,000 additional sites in France.[11]

Getting the business started, however, was not easy. Initially, the French network operator, Réseau de Transport d'Électricité (RTE), wanted evidence of the reliability of the scheme. It was convinced after a successful proof-of-concept trial where Voltalis demonstrated that it could effectively and reliably deliver balancing services to the grid. Once RTE gained confidence in the solution and the underlying technology, Voltalis was allowed to expand its business.

A critical feature of all DR schemes is that they benefit *nonparticipating customers*, since the intervention *reduces* electricity costs for *everyone*. Indeed, offering DR is far less expensive than paying generators to provide peaking capacity or flexibility services to the grid while there are typically other benefits such as less pollution or the need for grid reinforcements.

As is always the case in starting a new business, there were obstacles including skepticism and resistance from incumbents as well as regulatory uncertainties.[12] Yet Bivas is optimistic that Voltalis and others offering

---

[11] Refer to 23 Jan 2020 EIB Press Release, France: EIB, with the support of the European Commission, is financing the deployment of 150 000 smart boxes to actively manage electricity demand.

[12] For further details refer to Chapter 3 in Behind an beyond the meter in F. Sioshansi (Ed.), Academic Press, 2020.

similar services will prevail because they offer a superior service—flexible demand—at lower cost to the alternative, such as expensive, polluting peaking plants. In recent years, regulators in Europe and the United States have issued orders forcing organized market operators to acknowledge the growing role of DR and—more recently—storage by allowing them to actively participate in ancillary services as well as in wholesale markets.[13]

Moreover, Bivas explains that after working mostly on demand control and reduction, Voltalis has extended its solution to include control of various distributed resources including PVs and EVs to allow more renewable generation to be integrated into the network and provide what he calls "an overall optimization solution"—where flexible demand plays a more active role in balancing supply and demand.

4. OhmConnect[14]

OhmConnect is a San Francisco—based start-up currently focused on the residential and small commercial sector in California, but the company's basic business model is applicable anywhere with a "California duck" curve phenomenon, namely large variations in demand with commensurate swings in prices.

The company, founded in 2014, encourages its customers to avoid using electricity during peak demand hours when it is expensive to supply.[15] Growing from 70,000 customers in 2016, it claims to have over 500,000 in early 2020 and is planning to expand to a million.

Customers who sign up and actively participate, according to Ohm-Connect, can save an average of $300—500 per annum on their electricity bills—the savings vary from customer to customer. The company keeps the participants engaged with regular updates, keeping track of their performance, and offering them points—similar to frequent flyer miles offered by airlines—as illustrated in Fig. 7.3.

During a 3-month period in the summer of 2018—a hot one—OhmConnect says that it awarded more than $2.5 million to participants

---

[13] Refer to FERC Order 745 at https://www.ferc.gov/EventCalendar/Files/20110315105757-RM10-17-000.pdf and the European Commission at https://eur-lex.europa.eu/legal-content/EN/TXT/?uri=CELEX%3A32019L0944.

[14] Refer to https://www.prnewswire.com/news-releases/ohmconnect-names-new-ceo-after-hitting-major-milestones-300924315.html.

[15] Refer to https://www.ohmconnect.com/impact/gamification-for-good-how-to-level-up-your-bank-account-your-relationships-and-your-impact.

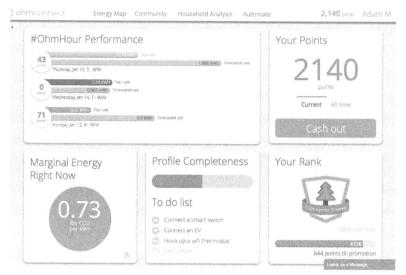

Figure 7.3 OhmConnect: Save energy, Get paid.[16] *(Source: OhmConnect at https:// assets.website-files.com/571f8d1539ec57464276d9ff/5c5f39b3d885ee1e20cf9505_tips-and-tricks-ohmconnect-gamification-for-good-how-to-level-up-your-bank-account-your-relationships-and-your-impact.jpg.)*

while saving 500 MWhs during the peak demand hours from over 100,000 participants.[17]

The company's approach is rather simple. Customers who have signed up are encouraged to shed load during peak demand hours, the neck of the "duck curve," by an easy-to-use online alert system, called OhmHour. And if they do, they get rewarded. OhmConnect says its efforts have resulted in approximately 150 MWs of energy reduction during peak demand hours rewarding its customers with more than $10 million.

OhmConnect pays its participants who are customers of the three major California utilities—Pacific Gas and Electric (PG&E), Southern California Edison (SCE), and San Diego Gas and Electric (SDG&E). Participants receive a text or email notifying them to save energy for a specified period, which are called OhmHour[18] events (Fig. 7.4).

---

[16] Refer to https://www.ohmconnect.com/impact/gamification-for-good-how-to-level-up-your-bank-account-your-relationships-and-your-impact.

[17] Refer to https://www.solarpowerworldonline.com/2018/10/ohmconnect-awards-more-than-2-5-million-california-users/.

[18] Refer to https://www.ohmconnect.com/how-it-works/what-is-an-ohmhour.

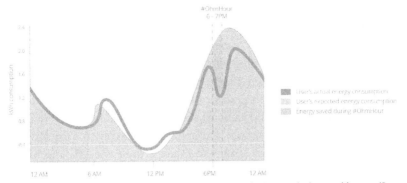

Figure 7.4 Participants are paid to avoid using energy during peak demand hours. *(Source: OhmConnect at https://assets.website-files.com/53cda9eccbc8e0894bcf7766/5c58fee83e262c-c2e81ac9c1_what-is-an-ohmhour-ohmconnect-energy-consumption-min.png.)*

OhmConnect's main success may be attributed to its low-cost marketing and customer acquisition business model, which is totally paperless, online, and requires virtually no labor or effort.[19] Participants sign up and receive electronic messages, prompting them to respond to signals. They have the option to directly connect their wireless thermostats—and possibly other devices—to signals hence making the process automatic. Since the individual savings tend to be small, transaction costs must be minimal to make it a viable business. The scheme seems to work well currently in California; it is not clear whether it would work in other parts of the world.

5. Other means of aggregating demand flexibility

The three companies highlighted in the preceding sections have developed particular approaches to engage customers and aggregate and monetize the resulting flexible demand. There are plenty of others who are trying a variety of approaches to do the same. A few of them are described below, not in any particular order.

One example of an entrepreneurial company trying to capitalize in aggregating and monetizing flexible demand is Voltus.[20] It offers its services

---

[19] Refer to http://calenergycommission.blogspot.com/2019/03/ohmconnect-gamifies-energy-use-toshift.html?utm_source=feedburner&utm_medium=email&utm_campai gn=Feed%3A+CaliforniaEnergyCommissionBlog+%28California+Energy+Commis sion+BLOG%29.

[20] The material is excerpted from the May 2020 issue of EEnergy Informer available at www.eenergyinformer.com.

to customers usually by working with distribution utilities. It offers customers some 150 DR options to choose from and says that participants can reap $500,000/MW-yr. in value by *stacking the revenue streams* of various schemes, depending on their operational flexibility and the types of tariffs they are on.

The San Francisco—based company says it can better manage energy through simple, risk-free programs, generating cash for participants who allow Voltus to maximize the value of their operational flexibility in energy markets. The basic idea is simple and somewhat similar to OhmConnect. Customers sign up and allow Voltus to manage their energy consumption, both when and how much is used, and receive a bonus—a portion of the savings that Voltus can generate. There is no up-front costs, no risk, and little or no discernible loss in service.

As with all DR service providers, Voltus focuses on reducing and/or shifting loads from the peak demand hours, which typically compromise a significant part of the bill for large commercial, industrial, and institutional customers who face exorbitant demand charges. As noted at its website,

The top 10% of grid-wide electricity demand lasts less than 1% of the hours of the year. Yet, these hours represent as much as 40% of your electricity bill. By reducing your demand from the grid during critical hours you can generate significant cash flow and savings. With the Voltus technology platform, we make it easy.

It says typical participant could see savings ranging from 5% to 30% of their annual electricity bill, depending on customer's operational flexibility as well as prevailing tariffs and other variables. It notes that sophisticated customers can generate additional savings by taking advantage of "stackable" DR programs.

Founded in 2014 in Baltimore, Maryland, CPower[21] is a demand-side energy management company that helps its customers save on energy costs while earning revenues through energy curtailment by providing services that enhances the reliability of the grid.

Its business model is similar to that of Enel X. At the outset, CPower examines the intricacies of each client's operations—how, why, when, and how much energy and capacity is utilized in a given facility and how much flexibility may be available. Using analytical tools and software, CPower reconfigures schedules, processes, and operations to optimize energy use while minimizing costs. The company claims to have more than 1400 commercial and industrial customers at nearly 9,000 sites in the United States and Canada.

---

[21] Refer to https://cpowerenergymanagement.com.

Another noteworthy company is Leap.[22] It aggregates and enables distributed energy resources (DERs) to participate in energy markets. Its main *forte* is to automate real-time trading of virtually any energy using, producing, or storing asset regardless of its capacity or location. In other words, it allows a collection of DERs to perform as a virtual power plant (VPP), delivering useful services to the grid. Its business model is more holistic, maximizing the utilization and participation of all assets, not just flexible loads.

As noted on its website, "When the electric grid is stressed, we send a signal to our partners and they reduce load through their connected homes and business – thermostats, EV chargers, and the like," as illustrated in Fig. 7.5. As with everyone else in this business, Leap relies on sophisticated software that optimizes and monetizes without human involvement.

The business model is, once again, a shared saving scheme. As described on its website,

Combined with other participants, those resources form a virtual power plant. Grid operators pay Leap's virtual plant just as they do for traditional power plants, and Leap pays its partners for participating.

Additional value is derived from each resource by slicing and dicing the portfolio of assets generating synergies that may not be obvious to individual customers. It notes that local load is more valuable than regional, and

Figure 7.5 Participants form a virtual power plant. *(Source: Leap at https://leap.energy/img/vpp.png.)*

[22] Refer to https://leap.energy.

individual customers rarely qualify to participate in wholesale markets, among services that Leap delivers.

Another noteworthy company in this space is Olivine.[23] It says its technology and unique service offerings help implement DR and DER programs and provide effective efficiency options to unlock greater value streams. It offers services for

- Utilities and retail energy service providers;
- Aggregators; and
- Distributed energy resources.

Its services encompass market entry analyses such as revenue opportunity assessments, ongoing operational management, and administrative support. Olivine, a registered Demand Response Provider (DRP) and Scheduling Coordinator (SC) in California, has a strong presence in California and claims to have fully integrated its DER Management Platform with those of the grid operator, the California ISO. On its website, it says,

"Olivine helps aggregators unlock the full value of distributed energy resources (DERs), such as renewable generation, energy storage, electric vehicles and demand response, by integrating these assets into wholesale markets," adding "Our operations team manages asset registration, energy and capacity bids, and settlements in the market on a daily basis."

Also active in this space is EnergyHub,[24] which notes that

Over 200 million distributed energy resources (DERs) are on track to come online globally by 2030. To keep up with this dynamically evolving grid topology, a majority of North American utilities have initiatives that aim to upgrade or deploy third-party developed advanced operational and informational systems—mainly through the deployment of ADMS (advanced distribution management system) and DERMS (distributed energy resource management system) solutions.

It adds,

The definitions of these two systems, however, remain amorphous, resulting in the pressing need for some clarity around the line of demarcation between them. It is equally critical, however, that utilities also understand the synergies between these two systems. With this clarity, utilities can then identify the optimal configuration between the ADMS and the DERMS to achieve evolving mission-critical objectives in the network operator's control room.

---

[23] Refer to https://olivineinc.com.
[24] Refer to https://www.energyhub.com.

Energyhub has developed the Mercury DERMS platform, which uses advanced machine learning—based artificial intelligence to manage fleets of DERs, leveraging the unique characteristics of different assets to deliver vital grid services and achieve mission-critical outcomes for utilities.

Whether it is delivering more predictable load reduction during demand response events, or optimizing electric vehicle charging in real time to match renewable generation or locational pricing, Mercury allows utilities to extract maximum value from every type of DER, from the grid edge to the substation.

Relying on the strength of Mercury's deep-learning capabilities, the platform generates detailed DER load and capacity forecasts. These forecasts increase visibility into DER activity behind the meter and enable utility operators and program managers to assess expected DER behavior and better inform grid-edge management decisions.

According to EnergyHub, "When combined with meter data and other utility telemetry, Mercury's analytics provide utilities with unprecedented situational awareness from the substation to behind-the-meter."

Enbala,[25] like other players in this space, offers software and a platform for managing distributed energy resources (DERs), it enables VPPs to operate using real-time communications to monitor, control, coordinate, and manage energy assets connected to the network. On its website, it says,

The portfolio built with these aggregated DERs becomes a single dispatchable resource that can provide grid services, such as capacity, operating reserves or regulation service.

Moreover, it says, these tools

… open up opportunities to strengthen customer relationships by providing a way for your customers to achieve even more value from their DER investments.

Enbala claims to have experience bidding VPP capacity into wholesale markets allowing customers to gain additional revenue streams by bidding their DER assets in a variety of schemes.

In addition to these companies, and numerous others that are not mentioned, there are a number of other players active in various aspects of DR integration or distributed energy resources (DERs). Some focused on narrow niches such as distributed solar PVs, storage, or EV charging

[25] Refer to https://www.enbala.com.

including well-recognized names like Sunrun[26] and Tesla[27] or smaller ones like Extensible Energy[28] and Emulate Energy[29]—start-ups that add intelligence to distributed solar + storage investments and monetize the value of storage, respectively.

Moreover, there are yet other players such as Oracle—which acquired Opower—a company engaged in behavioral DR as well as players who are primarily engaged as implementers of utility DR programs such as Energy Solutions, CLEAResult, ICF, Honeywell, Willdan, and Nexant, to name a few.

6. Conclusions

The preceding discussion highlighted a few companies active in developing and delivering flexible demand following variations of the same theme. One way or another, participants with flexible demand must be identified and recruited to join a given scheme, typically by offering them a portion of the saving that can be derived from their contribution to the program. The larger the participants' load and flexibility, the larger the rewards.

As a general rule, customer acquisition, enabling customers to participate in DR schemes and keeping them engaged are frequently identified as major challenges to profitability and scale. Another major obstacle is that the nascent business, which is trying to bring flexible demand to participate in electricity markets traditionally dominated by generators, is not broadly recognized for what it can offer nor adequately rewarded for the services delivered.

Despite these challenges, there are hopeful signs that many innovative companies are trying to identify and develop promising niches—assisted by the falling cost of sensors, telecommunication, data crunching, and artificial intelligence—allowing successful companies to scale up while fine-tuning their business models.

Finally, supportive regulations and policy clarity is often mentioned as a critical requirement for the success of companies engaged in developing and delivering demand flexibility.

---

[26] Refer to https://www.sunrun.com.
[27] Refer to https://www.tesla.com.
[28] Refer to https://www.extensibleenergy.com.
[29] Refer to https://www.emulate.energy.

CHAPTER 8

Electric vehicles: The ultimate flexible demand

Fereidoon Sioshansi[1]
Menlo Energy Economics, Walnut Creek, CA, United States

1. Introduction

Electric vehicles (EVs) have been around for a *very* long time, including the ubiquitous carts on golf courses the world over. Auto companies have been flirting with the technology for decades but, until recently, their heart was not in it—perhaps because they saw EVs as competition to internal combustion engines (ICEs). Why would anyone with a commercially successful product wish to cannibalize its own business by introducing a competing technology[2]? But the rising pressure to decarbonize economies[3] and move away from exclusive reliance on fossil fuels in the transport sector has reached a point where EVs are increasingly seen as the future of personal road mobility.

Even though currently EVs represent a small fraction of global car sales, they are expected to move mainstream within a decade, eventually replacing ICEs in ever increasing numbers. A few countries, notably China, France, and the United Kingdom, and a few major cities have announced plans to ban the sale of ICEs in the 2030s, prompting automakers to shift toward EVs, including Volvo, which has already stopped making ICEs (Box 8.1). Virtually all major automakers are now introducing multiple all-electric models with the expectation that drivers will also make the switch—which is critically dependent on the widespread availability of EV charging infrastructure.

---

[1] Marc Monbouquette of Enel X provided useful content, references, and reviewed an earlier draft of this chapter. Carolyn Sisto of CPUC and Chris King of Siemens e-Mobility also provided assistance.

[2] Anther reason is that existing ICEs, especially the more expensive SUVs, offer much higher profit margins than current EVs.

[3] Emissions from the transport sector exceed those from the power sector in many parts of the world including California, hence the keen interest to electrify transport.

Variable Generation, Flexible Demand
ISBN 978-0-12-823810-3
https://doi.org/10.1016/B978-0-12-823810-3.00015-7

BOX 8.1 EVs: The future of personal road transport

China, which accounts for nearly half of the world's electric car sales, had a 10% plug-in vehicle sales quota for 2019, 12% in 2020, and at least 20% by 2025. The country has an aggressive New Energy Vehicle (NEV) mandate to improve urban air quality. Moreover it has identified electrified transportation as a strategic industry for both the domestic and export markets. At the same time, China has one of the most ambitious renewable portfolio standards (RPS) for a developing economy—35% by 2030—to reduce its dependence on polluting coal and imported oil. In September 2020, China announced its intention to become carbon neutral by 2060, the first major developing economy to make such an ambitious pledge.

China may be ahead but is not alone. California, the world's fifth largest economy, has set a target to make its entire economy carbon neutral by 2045. To achieve this goal, it must electrify virtually everything while making its electric grid zero carbon by 2045.[4] The state has a target of 5 million zero emission vehicles by 2030 and 260,000 public charging ports by 2025.

Similarly, the European Union and the United Kingdom have also set carbon neutrality targets by 2050—which requires massive electrification of everything, particularly electric transportation. Germany expects to have as many as 7 million EVs by 2030. All these developments suggest that EVs will be the future of personal road transport in key markets.

[4] Greenhouse gas emissions attributed to the transport sector far exceed those of the electric sector in California, making the electrification of transport a top priority.

This chapter assumes that it is a matter of time before an increasing proportion of global car sales will be EVs.[5] With an increasing number of EVs come a number of challenges *and* opportunities:

- The main *challenge* is to build the necessary infrastructure to charge millions of EVs as they enter the market and supply them with renewable energy—little is gained if one switches from ICEs to EVs if they are charged with electricity generated from polluting coal-fired plants;[6] and
- The main *opportunity* is that EVs' large batteries offer an ideal form of flexible demand since—like electric water heaters with a storage tank—

[5] Electric trucks, buses, service and delivery vehicles are also being electrified at a fast pace.

[6] There are studies that suggest that even if EVs are charged in a state such as West Virginia, which is predominantly coal-supplied, they are still less polluting than ICEs.

they need not be charged at the same time and certainly not during peak demand hours. Ideally, they can be charged during times when electricity is plentiful and cheap—such as during sunny hours of the day in places where there is excess solar generation.

As shown in Fig. 8.1, No other electricity using device, not even air conditioning, offers such a massive load with such an enormous demand flexibility potential.

The balance of the chapter is organized as follows:

- Section 2 provides a brief history of EVs and why it has taken so long for a promising technology to reach commercial scale;
- Section 3 describes the need for smart EV charging capabilities and why it is critical to develop the technology and incentives to manage when and where they are charged; and
- Section 4 considers more advanced *functionalities* including bidirectional charging such as vehicle-to-grid (V2G) integration, where a portion of the stored energy in EV batteries may be discharged to the grid during peak demand hours; followed by the chapter's conclusions.

2. Why has it taken so long?

While it is not unusual for promising innovations to sit dormant in the laboratory before they are perfected, commercialized and applied in large scale, it is a puzzle why it has taken EVs such a long time to reach where

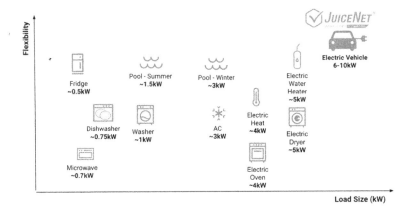

Figure 8.1 *Nothing matches the massive load* and *flexible demand of EVs.*
(Source: Enel X at https://evcharging.enelx.com/news/blog/606-why-electric-vehicle-grid-integration-matters-today.)

they are today. In strict efficiency terms—delivering energy to the wheels—they offer a far superior solution compared to ICEs.

EVs have been around at least as early as 1832 when Robert Anderson demonstrated a crude prototype. But the technology sat dormant until English inventor Thomas Parker came up with a more practical electric car in 1895 (Fig. 8.2).

Unlike ICEs, which require a combustion chamber to contain the noisy explosions and transfer power to the wheels through mechanical means, EVs deliver the stored energy in the battery to a motor, which directly turns the wheels. It is simple and elegant with few moving parts. Moreover, EVs do not need a cooling system nor various apparatus—belts, pumps, radiator, mufflers, exhaust pipes, gas tank, fuel pump, transmission and gearbox, etc. Few moving parts mean few things to repair or replace. EVs are quiet and there is no harmful emissions coming out of the tailpipe—there is no tailpipe. The main drawback has always been, and continues to be, how to store sufficient energy in the battery to match the range of typical ICEs.[7]

Figure 8.2 *Thomas Parker's electric car, 1895. (Source: Wikipedia at https://upload. wikimedia.org/wikipedia/commons/thumb/e/e7/Thomas_Parker_Electric_car.jpg/220px-Thomas_Parker_Electric_car.jpg.)*

---

[7] Gasoline packs a huge amount of energy in small space with relatively little weight—an insurmountable advantage relative to early batteries.

But for whatever reason[8] the EV technology sputtered and was sidelined for roughly a century with only a few unsuccessful attempts. In the 1960s, Henney Coachworks and the National Union Electric Company, makers of Exide batteries, produced the Henney Kilowatt, based on the French Renault Dauphine (Fig. 8.3). The 72-V version of the car had a top speed of 96 km/h (60 mph) and could travel for nearly an hour on a single charge, a major limitation. Production was halted in 1961.

A major milestone for EVs was reached in July 1971 during the Apollo 15 mission when NASA's Moon Lunar Rover became the first manned vehicle to drive on the surface of the moon (Fig. 8.4). The "Moon buggy" developed by Boeing and GM's subsidiary, Delco Electronics, featured a DC drive motor for each wheel, and a pair of 36-V nonrechargeable silver-zinc potassium hydroxide batteries. It was an elegant EV for the moon but not suitable for the mass market on earth.

In 1996, GM introduced EV1, a car with a range of 260 km (160 miles) with NiMH batteries. From the start, however, it was obvious that GM's heart was not in it. Moreover, at the time, there was no charging network to speak of, making it impractical. Production ceased in 2003.

Figure 8.3 *Henny Kilowatt, 1961. (Source: Wikipedia at https://upload.wikimedia.org/ wikipedia/commons/thumb/b/bd/Kilowatt.jpg/220px-Kilowatt.jpg.)*

---

[8] One theory is that the oil industry of the early 1990s saw ICEs as a great source of demand growth for the nascent oil business and did its best to discourage EVs to get in the way.

Figure 8.4 Lunar module, July 1971. *(Source: Wikipedia at https://upload.wikimedia.org/ wikipedia/commons/thumb/e/ed/Apollo15LunarRover.jpg/220px-Apollo15LunarRover.jpg.)*

The limited range of EVs and the lack of adequate charging infrastructure has been the proverbial *chicken-and-egg problem*: Automakers would not make EVs because of insufficient demand, and drivers would not buy them because they were not offered in sufficient quantities, styles and affordable prices—or a decent range.

One solution to the limited range was to develop *hybrids*, as with Toyota's successful Prius. As the battery technology improved, Nissan introduced the all-electric Leaf in 2010. Still, hybrids and EVs remained as a niche product appealing to a narrow market segment—primarily the environmentally or economically conscious. Nobody would buy them for style, performance, range, or roominess while the lack of an extensive network of charging stations limited the appeal of the all-electric cars.

In 2014, Elon Musk, the CEO of Tesla, seized an opportunity others had ignored up to that point: to develop a stylish high performance EV with a reasonable range sold without price haggling directly to customers—no dealers. Moreover, instead of focusing on the bottom or middle price range, Tesla initially aimed for the high end of the market offering a "sexy" product that affluent drivers would *actually* want to buy and enjoy driving.[9] The Tesla Roadster, introduced in 2008, came with lithium-ion batteries with a range of 320 km or 200 miles. It was an instant

[9] Another reason was that Tesla was not cannibalizing an existing legacy business, it was starting from scratch with nothing to lose and everything to gain.

success, selling nearly 2,500 in 30 countries by 2012—it could have sold far more only if it could have produced more. By the end of March 2020, Tesla had sold nearly a million EVs—becoming the global leader in a growing market, which is no longer considered a niche product. Tesla's market valuation exceeds those of much larger established car companies.[10]

Tesla's success forced all major automakers to seriously reconsider EVs, something they had never done before. In 2017 GM introduced the all-electric Chevy Bolt as did virtually all other major carmakers. The rest, you may say, is history. California's Governor, Gavin Newsom, signed an executive order in Sept 2020 banning ICEs starting in 2035.

While limited range and the limited charging infrastructure remains a stumbling block, several factors are pushing EVs forward including:

• Climate change and the move to decarbonize the transport sector;
• Changes in car ownership, the rise of ride sharing, advances in autonomous EVs, and the prospect of mobility-as-a-service;
• Ambitious EV targets, policy support, financial incentives, and proposed bans on ICE sales[11]; and
• Falling cost and improved performance of batteries.

By 2020 not only were EVs approaching cost parity with ICEs but increasingly recognized by both the automakers and policymakers in a number of key car markets such as China, California, the United Kingdom, and the EU, as the only viable long-term option for personal mobility. In Norway—an outlier—EVs already outsell ICEs as the country envisions a fully electrified transport sector within a decade. Today, if you are a major automaker without a range of EV models, you probably don't have a future in the business.

3. Vehicle-grid integration

While currently EVs are more expensive to buy, they have lower operating costs—that is fuel *and* maintenance costs.[12] Prior to the 2020 pandemic, the global EV sales were expected to reach 23 million by 2030 with a cumulative stock of 130 million, according to the International Energy

---

[10] By mid-2020, Tesla's market capitalization exceeded that of Toyota.
[11] The UK plans to ban the sale of ICEs as early as 2032.
[12] The relative economics varies based on fuel costs—which include taxes—versus electricity costs, tariffs applicable to EV charging, as well as other variables.

Agency (IEA).[13] The IEA projected the share of EVs in 2030 in Japan to reach 37%, over 30% in the United States and Canada, 29% in India, and 22% globally. China is expected to remain the biggest EV market, with as many as half of the global sales. By 2040, EVs may reach 60% of auto sales by some estimates. A May 2019 Bloomberg[14] report said:

> Based on analysis of the evolving economics in different vehicle segments and geographical markets, BNEF's Electric Vehicle Outlook 2019 shows electrics taking up 57% of the global passenger car sales by 2040, slightly higher than the forecast a year ago. Electric buses are set to hold 81% of municipal bus sales by the same date.

Moreover, the electrification of the transport sector extends beyond passenger cars to light-duty trucks, buses, and many utility-type vehicles with short range such as delivery vans, garbage trucks, postal delivery trucks, fire engines, bicycles, scooters, and so on.[15] At the same time, there are many uncertainties about the longer term viability of personal car ownership model, which many experts believe may be upended by car sharing and ride hailing services especially in densely populated urban areas where traffic congestion, lack of parking, and access restrictions to city centers may make private car ownership a thing of the past. Simultaneously, telecommuting and advances in autonomous EVs are likely to change personal mobility.[16] Likewise, mass transit, electric bicycles, and scooters are increasingly competing with cars in city centers where average travel distances are short. This suggests that ICEs may only remain viable in rural areas with low population density, inadequate charging infrastructure, and long travel distances.

---

[13] Refer to Global EV outlook 2019 at https://www.iea.org/reports/global-ev-outlook-2019.

[14] Refer to BNEF, May 2019 at https://about.bnef.com/blog/electric-transport-revolution-set-spread-rapidly-light-medium-commercial-vehicle-market/.

[15] The dramatic drop in oil prices following the 2020 pandemic has reduced the fuel savings advantage of EVs but not the longer-term prospects of EVs. The more serious immediate concern is that people will put off buying any cars until the crisis is over.

[16] According to a March 2020 study by Rob West of Thunder Said Energy, telecommuting may have a bigger impact than EVs, at least through 2030.

Regardless of how the future unfolds, it is fair to say that a vast network of charging stations will be needed to support the increasing numbers of EVs over time.[17]

EVs can be charged where they are parked[18]—at home, at work, at shopping centers, restaurants, at filling stations, and at rest areas on freeways. Eventually, they will be as ubiquitous as today's petrol stations. In fact, many existing stations are likely to be converted to EV charging stations. The specifics vary but:

- EV owners with a private garage typically install a low to medium voltage charger for overnight charging;
- Apartment dwellers are likely to plug into a dedicated charger in the facility;
- In city centers, public chargers are appearing on the streets, in parking garages, shopping malls, offices, etc.; and
- Fast chargers are appearing in rest areas on roads and where petrol stations are currently found.

Regardless of where the chargers are located, infrastructure is needed to allow millions of EVs to "fill up" quickly and conveniently as ICE drivers currently do.[19]

The enormity of the task becomes obvious considering where we are relative to where we have to be. In March 2020,[20] there were some 78,500 charging outlets and 25,000 charging stations for plug-in EVs in the United States—numbers that need to rapidly rise. China, the biggest EV market, has at least 3 times as many as the United States.

California, the most populous state, currently accounts for roughly half of the EVs in the United States—around 700,000—has 6,835 fast-charging stations and a total of 28,545 charge ports. As a point of reference there are currently some 168,000 gas stations with multiple pumps at

---

[17] Refer to Chargers Are the Final Roadblock to America's Electric Car Future, 1 June 2020 at https://www.bloomberg.com/news/features/2020-06-01/electric-car-chargers-will-determine-america-s-green-future.

[18] There is interest in EVs with solar PV panels built into their exterior, which can charge the battery while parked or moving, mostly to extend the range.

[19] For further details visit Vehicle Grid Integration Council (VGIC) website at https://www.vgicouncil.org/about/mission.

[20] Refer to https://www.statista.com/statistics/416750/number-of-electric-vehicle-charging-stations-outlets-united-states/.

each in the United States.[21] A similar number of charging points for EVs will probably be needed.

Once installed, every charger must be connected to the electricity distribution network.[22] Moreover, the chargers must be enabled to recognize who is receiving the charge, who is delivering the juice, and settle for billing purposes.[23] The price may vary by time, the location, and the voltage level—and each utility, distribution company, or private party offering service must develop the back-office settlement and billing system[24] (Fig. 8.5).

"Filling up" an EV battery, however, is more complicated than that of filling an empty gas tank. To start with, not all EVs manufactured by different automakers are necessarily compatible with all existing charging stations.[25] Developing standards that eventually allow all EVs to charge from all chargers, something that we take for granted for ICEs, is critical.[26]

[21] Refer to https://www.fueleconomy.gov/feg/quizzes/answerquiz16.shtml.
[22] Refer to Why Electric Vehicle-Grid Integration Matters Today, 21 Nov 2019 at https://evcharging.enelx.com/news/blog/606-why-electric-vehicle-grid-integration-matters-today.
[23] Major US EV charging operators such as EVgo, ChargePoint, and Electrify America currently have protocols to allow EVs to use multiple systems and settle internally.
[24] California has mandated that publicly accessible chargers offer energy based pricing in $/kW; other states may decide on time-based pricing or other schemes; for further details on CA refer to https://www.cdfa.ca.gov/dms/pdfs/regulations/EVSE-FinalText.pdf.
[25] The standards vary from state to state in the United States and from one country to another. All public chargers in California, for example, have to comply with the two main open standards, with Tesla as an exception.
[26] Tesla, for example, decided from the start to build its own network of fast charging stations, initially focusing on its key US market, realizing that it couldn't otherwise sell many cars. Moreover, Tesla's charging stations are exclusive to Tesla owners; other EVs cannot use this private network while Tesla drivers are given adopters that they can charge at other charging systems. Since Tesla wants to sell more EVs—selling electricity at charging stations is a mere vehicle to that end—presently it has more charging outlets in the United States than all other fast-charging networks combined. In the state of Wyoming, for example, Tesla has 10 charging stations versus a single fast-charging plug suitable for others; in WV, the respective numbers are 8 and 2 according to Bloomberg at https://www.bloomberg.com/news/features/2020-06-01/electric-car-chargers-will-determine-america-s-green-future.

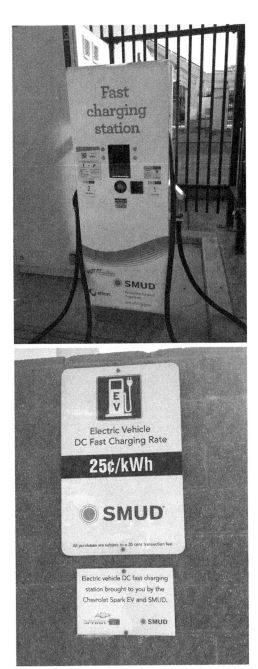

Figure 8.5 *Public fast EV charging station offering electricity at 25 cents/ kWh. (Source: Photos taken by the author at SMUD head office charging station in Sacramento, CA.)*

A second challenge is that storing electricity is more complicated than gasoline or diesel. Until EV charging stations install on-site storage, they draw the energy, kWhs, and capacity, kWs, from the distribution network at the time of charging. For this reason, *when* and *where* the charging takes place and at what voltage level makes a difference as explained in Box 8.2.

Aside from *when*, *where*, and at what *voltage level*, *how many* EVs are simultaneously charged on the same distribution network, feeder, or

BOX 8.2 A primer on EV chargers[27]

Broadly speaking, EVs can be charged at three levels:
- Level 1 offers basic service using a standard 120V[28] household outlet. This is the cheapest but also the *slowest* option providing about four to five miles of range per hour of charging. People with short commuting distance who can leave the EV plugged overnight may find this adequate. Plug-in hybrids have smaller battery packs than all electric cars and may be better candidates for Level 1 charging.
- Level 2 offers 200 V or higher, providing 12–60 miles of range per hour, depending on how much power the charger can supply and the battery can accept.
- Level 3, commonly known as "DC fast charging" or simply "fast charging" uses direct current (DC) at much higher voltage levels, as high as 800 V. This allows for very rapid charging—typically 80% full in 20–30 min.

The question of which type of charger is the right one boils down to cost versus speed. The main drawback of the fast chargers is the cost and the high voltage level, which may not be readily available. A level 2 charger may cost $2,500 versus as much as $520,000 for a level 3 charger.[29]

Most EVs sold today come with a standard 120-V portable charger and some manufacturers supply a dual 120/240-V unit.[30]

---

[27] This box is based on Understanding different EV charging levels, 8 May 2019, Enel X at https://evcharging.enelx.com/news/blog/550-different-ev-charging-levels.

[28] The 120 V is standard voltage in the Americas. Europe and much of the rest of the world uses a 220V electric supply.

[29] Refer to https://www.bloomberg.com/news/features/2020-06-01/electric-car-chargers-will-determine-america-s-green-future.

[30] A number of companies are engaged in developing the necessary software that would manage the charging of EVs including Enel X's JuiceNet and JuiceBox.

service transformer may be a limiting factor. This explains why Vehicle-Grid Integration (VGI) speaks to the intelligent management of EV charging in a way that optimally and cost-effectively integrates new load from EV charging while avoiding putting additional strain on the grid, necessitating upgrading of the distribution network or additional generation capacity.

It is easy to see that the new load expected from the proliferation of EV charging stations necessitate *intelligent* management. In this context, EVs can be a blessing or a nightmare depending on how the charging is *managed*. In the case of the state of California, for example, one projection puts the incremental EV load as high as 4.9 GW by 2030. Clearly, if this load is *unmanaged*, many drivers may decide to plug in and charge their EVs as soon as they return home after work, say between 6 and 7 p.m. This would make the daily peak of California's "Duck Curve" even more challenging than it already is.

Given the challenges represented by a large influx of EV load, a number of leading EV Service Providers (EVSPs) and automotive original equipment manufacturers (OEMs) have joined to establish the Vehicle Grid Integration Council[31] (VGIC) to promote the intelligent management of EV charging. The Council's mission is:

To support the transition to a decarbonized transportation and electric sector by ensuring the value from EV deployments and flexible EV charging and discharging is recognized and compensated in support of achieving a more reliable, affordable, and efficient electric grid.

VGI is defined as:

.... the capability of an EV or EV charging station to alter the time, charging level, or location at which a grid-connected EV charges or discharges, in a manner that optimizes EV charging with the electric grid and provides net benefits to ratepayers.

The top challenges of VGI (Box 8.3) include encouraging drivers to:
- Delay evening charging to low off-peak or overnight period;
- Charge during the sunny hours of the day at work; and
- Avoiding charging during high demand periods.

---

[31] Refer to https://www.vgicouncil.org.

BOX 8.3 Managing EV charging[32]

Currently, most EV owners plug in at home and/or at work usually over extended periods at low voltage levels. In many cases, the EV remains plugged far longer—up to 90% longer—than is necessary to fully charge the battery. Moreover, the drivers may be oblivious to the time-varying cost of electricity.

The critical variables that must be considered in optimizing a VGI strategy include:

- Battery's energy (kWh) and capacity rating (kW);
- Charger's power rating (kW);
- Time of charging and its coincidence with distribution network and grid conditions;
- Amount of time a vehicle is plugged in; and
- EV driver's mobility requirements, primarily the range.

For example, a Chevy Bolt has a 60 kWh battery and comes with a 7.5 kW Level 2 charger. In an *unmanaged* scenario, the typical driver may plug in when arriving home around 7 p.m. with a 75% charge. The aim is to be fully charged by 8 a.m. the next morning.[33] In reality, this EV needs to be plugged in for only 2 h,[34] which can be scheduled within a 13-h window. This explains the extreme flexibility of EV charging load.

For this scenario, the optimum VGI strategy would be to schedule 2 h of charging within the 13-h window to minimize the costs and the stress on the grid by avoiding distribution system overloads, avoiding high wholesale prices while absorbing as much wind or solar to reduce renewable energy curtailments, issues covered in Chapter 1.

Each kind of charging requires its own VGI solution to minimize costs and stress on the grid. For example,

- Workplace or shopping charging should focus on absorbing low- or negatively priced midday solar energy while reducing solar curtailments;
- Public fast charging should avoid or minimize peak times, which enhances the resource adequacy of the grid while avoiding dispatching polluting and expensive natural gas peaking plants;
- Fleets or hubs for e-buses or e-taxis should schedule charging to avoid peak hours, keep charging load under a given limit to minimize demand charges, avoid distribution upgrades, or buy from a retailer or distributor who offers the lowest prices; and
- DC fast charging facilities, which impose a big impact on the network, should similarly aim to curtail aggregation of chargers to provide resource adequacy, contingency reserves, or ancillary services in the event of an emergency on the grid.[35]

---

[32] This box is based on Why Electric Vehicle-Grid Integration Matters Today, 21 Nov 2019 available at https://evcharging.enelx.com/news/blog/606-why-electric-vehicle-grid-integration-matters-today this article was first published February 11, 2019 in Natural Gas & Electricity Journal.

[33] Of course, for a short commuting distance, there is no reason to fully charge the EV.

[34] 15 kWh/7.5 kW = 2 h.

[35] Such facilities increasingly invest in on-site storage—the electric equivalent of the large underground tanks holding thousands of gallons of gasoline at current filling stations.

Unsurprisingly, encouraging charging when energy prices are low provides benefits to EV owners[36] as well as non-EV owners by spreading out the fixed costs of utility infrastructure over more kWh sales and avoiding costly grid upgrades that can result from unmanaged charging.

For example, starting in 2017, Enel X employs a fleet of thousands of its *managed* charging stations to create a *virtual battery* with 30 MW/70 MWh of capacity, which participates in California Independent System Operator's (CAISO) day-ahead and real-time energy markets.[37] The virtual battery dynamically manages EV charging loads in response to CAISO's commands and provides benefits to the grid as well as EV owners. The successful experiment demonstrates that participants across a state—or country—can be aggregated to deliver energy services, primarily demand response, by curtailing charging when directed. Enel X provides participating customers with a share of the wholesale energy savings in exchange for control of their EV charging schedules.

Critically, the capital outlays for the charging stations that provide these flexibility services are generally borne by EV owners, not utility ratepayers in contrast with standalone energy storage systems. Utilizing EV batteries for grid services through VGI entails a *de minimus* incremental cost above and beyond the primary purpose of EVs, which is to provide mobility.

VGI schemes such as Enel X's demonstrate that an aggregated and managed number of EVs can provide a range of valuable services by participating in different markets such as:

Energy
- Day-ahead market;
- Real-time market; and
- Load shift/building.

Capacity
- Forward markets, such as resource adequacy;
- Day-ahead markets;
- Avoided capacity charges; and
- Distribution upgrade deferral

Ancillary services
- Spinning reserve;

---

[36] Time of Use (TOU) rates with a significant differential between on- and off-peak periods incentivize charging when electricity costs are relatively low and allow EVs to be fueled considerably cheaper per mile than ICEs.

[37] Refer tohttps://www.utilitydive.com/news/emotorwerks-provides-caiso-with-30-mw-of-dr-through-smart-ev-charging/532110/.

- Nonspinning reserve; and
- Frequency regulation.

Other studies (Box 8.4) have shown that if a significant number of EVs charge during the existing peak periods, this will

- Make the "Duck Curve" even more challenging for the CAISO;
- Require additional investment in upgrading the distribution infrastructure to be able to handle increased demand; and

BOX 8.4 Managed EV charging can be win-win-win[38]

Minimum load is among the many challenges facing CAISO and the "duck curve." The California system is currently estimated to have 13–15 GW of nondispatchable generation that is difficult or highly undesirable to ramp *down*. When CAISO's net load falls below 15 GW, there is problem.

According to the Lawrence Berkeley National Laboratory (LBL), in 2014 CAISO's *net load* fell below 15 GW for only 0.03% of the hours. By 2025, however, net load is forecast to fall below 15 GW for 14.2% of the hours—and even higher in 2030 and beyond as the state marches toward a decarbonized electricity mix.

Similarly, by 2025, CAISO's daily morning ramp *down* and evening ramp *up* rates will be twice as large as they were in 2014, reaching −8 GW/h in the morning and +11 GW/h in the evening, respectively. The LBL researchers reported that:

The additional 7 GW/h ramp up need by 2025 is equivalent to requiring an additional 35 natural gas turbines each with 600 MW of capacity to operate over a 3 h period to ramp from 0% to 100% output.

The storage capacity of the EVs, which are currently charged in an *uncontrolled* manner, can come handy. Today, most EVs "begin charging when plugged in, and do not stop until the vehicle's battery is full, or until the driver unplugs for their next trip." Consequently, "EVs with uncontrolled charging will do little to mitigate the daytime over-generation problems."

The obvious conclusion: "… there is a substantial lost opportunity if vehicles are not integrated with the grid."

Grid integrated vehicles can strategically time their charging to alleviate the daytime valley in the 'duck curve' by delaying charging to the time when net load is the lowest (i.e. when solar generation is highest).

[38] This box based on "Clean vehicles as an enabler for a clean electricity grid" by Jonathan Coignard et al., 16 May 2018 at Environmental Research Letters, Volume 13, Number 5 https://iopscience.iop.org/article/10.1088/1748-9326/aabe97/meta.

Continued

BOX 8.4 Managed EV charging can be win-win-win[38]—cont'd

According to the LBL research, "... nearly 2 GW of renewables curtailment can be avoided with V1G only vehicles, and nearly 5 GW in the case with a mix of V1G and V2G[39] vehicles; substantially more than with the (state's) 1.3 GW Storage Mandate."

Later in the day, for the case with a mix of V1G and V2G vehicles, the evening peaks can be reduced by more than 5 GW by 2025 compared with the un-controlled scenario.

Clearly, "The real strength of grid-integrated vehicles in mitigating the 'duck curve' is in avoiding large system-wide ramping."

In the V1G-only case, down- and up-ramping are both mitigated by more than 2 GW/h by 2025. In the case with a mix of V1G and V2G vehicles, however, substantially larger gains are seen.

Importantly, it is seen that EVs can maintain ramping rates at or below today's levels even with substantial renewables deployment on California's grid.

But "Ensuring that EV owners participate in controlled charging could be a challenge." There are, however, many ways to incentivize when EVs are charged,

... for example, EV owners could be paid (e.g. by accessing lower electricity rates when charging at home, or reduced parking fees when utilizing public charging) when their vehicles are grid-connected and participating in controlled charging, with possibly higher incentives for and/or sparing utilization of V2G to avoid excessive battery degradation.

The LBL researchers note that, "For valley-filling and ramp-up mitigation (i.e., the most severe of the duck curve problems), V1G-only vehicles fulfill 1 GW of storage-equivalent, a large fraction of the 1.3 GW California Storage Mandate, and provide equivalent capability as $1.45−1.75 billion of stationary storage investment. This represents a substantial cost savings for renewables integration, as V1G is readily achievable at less than $150 million over uncontrolled charging with today's technology."

In a scenario with a mix of V1G- and some V2G-capable vehicles ... EVs provide equivalent services of 5 GW of stationary storage for valley-filling and ramp-up mitigation, the equivalent of $12.8−15.4 billion in stationary storage investment.

In other words, EVs employed in the V1G mode "can fulfill the vast majority of California's Storage Mandate[40] at much lower costs." With some EVs having V2G capability, their potential contribution is further boosted, allowing more variable renewables to be integrated into the grid.

[39] V1G refers to grid-to-vehicle charging and V2G refers to the reverse, namely vehicle-to-grid discharging.
[40] In 2013, California mandated the procurement of 1.3 GW of stationary storage by 2025.

- Increase—rather than reduce—emissions from the electric sector since more polluting natural gas—fired peakers will most likely be required to meet the evening's peak demand.

Conversely, if the EV load is properly *managed*, the exact opposite can happen.

Additionally, EVs can provide two coveted ancillary services, frequency regulation and load following, much better than anything else as illustrated in Fig. 8.6.

Others who have examined the potential benefits of *managed* EV integration have also documented significant potential savings. One report by the Rocky Mountain Institute[41] shows how flexible EV demand can modify the shape of the "California Duck" curve in simulated studies as shown in Fig. 8.7.

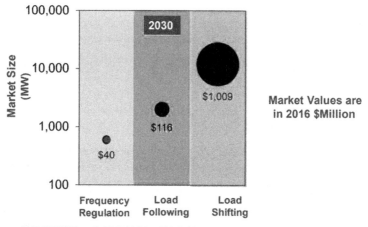

2018 42MMT Case in 2016 $/kW-yr. (CA 2018-2030 levelized value)

Figure 8.6 *Estimated market value of EVs' flexibility in 2030.* *(Source: GridWorks VGI initiative framing document at https://gridworks.org/wp-content/uploads/2019/08/Gridworks-VGI-Initiative-Framing-Document.pdf.)*

[41] Refer to https://rmi.org/wp-content/uploads/2017/04/RMI_Electric_Vehicles_as_DERs_Final_V2.pdf and others including eMotorWerks, Honda, & Southern California Edison Offer Nation's 1st Smart Charging Program" CleanTechnica. February 2018 at https://cleantechnica.com/2018/08/02/emotorwerks-honda-southern-california-edison-offer-nations-first-s mart-charging-program/.

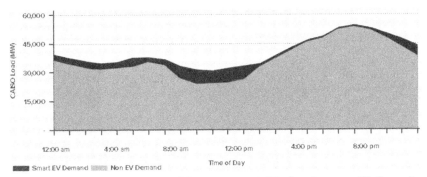

Figure 8.7 *Illustrative example of how managed EV charging can fill the valley without adding to the peak of the California "Duck Curve." (Source: Electric vehicles as DER, Rocky Mountain Institute at https://rmi.org/wp-content/uploads/2017/04/RMI_ Electric_Vehicles_as_DERs_Final_V2.pdf.)*

Despite the tremendous potential, the RMI says there are a number of technical, financial, and regulatory issues including the question of who should invest in VGI infrastructure. The RMI report identifies several business models, which the "utilities" and regulators must resolve including:

- Utility as Facilitator, where the utility treats EV charging like other loads, providing nondiscriminatory service at regulated rates, but does *not* engage in EV charging business itself;
- Utility as Manager, where in addition to delivering service to the chargers, the utility *manages* the charging operation and integrates it with the capabilities and limitations of the grid;
- Utility as Provider, where the utility provides service to the chargers, engages in charging business, and receives a cost-based payment for electricity; and
- Utility as Exclusive Provider, where vendors other than the regulated utility are prohibited from providing charging service.[42]

How these issues are resolved will vary from one country or state to another.[43]

---

[42] In many states, existing laws precludes the resale of electricity to third parties.

[43] This is among a number of thorny issues facing regulators. Under prevailing rate of return regulation, utilities, would welcome the opportunity to invest in VGI as this offers a new sources of revenues, which they can collect from all customers. But not all customers are currently EV owners, leading to major fairness and equity issues.

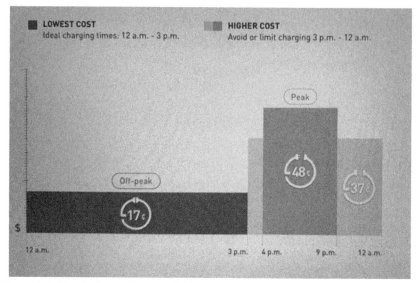

Figure 8.8 *Simple time-of-use tariffs for EV charging.* Existing EV charging tariffs for PG&E customers incentivizes charging during off-peak hours while avoiding the belly of the "duck curve" from 4 to 9 p.m. *(Source: PG&E.)*

Another important issue is what is the best means of providing the correct price signals to EV owners on when—and possibly where—to charge their vehicles. Because of their voracious demand for energy and capacity, special care must be taken to *strongly* discourage charging at times of scarcity while encouraging it during times of abundance. Obvious options include:

- Simple time-of-use pricing (TOU)—typically with high and low price periods (Fig. 8.8);
- More sophisticated pricing schemes such as super-off-peak, shoulder and super-on-peak pricing; with significant price differential between them; and
- Dynamic pricing—where prices may vary in real time based on the actual supply-and-demand conditions.[44]

---

[44] Prices can also have a locational dimension for energy via the locational marginal price (LMP) as well as a location-specific distribution or delivery component based on circuit loading.

Simple TOU tariffs such as the one in Fig. 8.8 are the easiest to administer and for EV drivers to understand while the latter two deliver more accurate price signals.[45] As noted by the RMI report,

Charging loads could be controlled directly—by grid operators, utilities or aggregators of charging infrastructure—within parameters set by the user. This approach will likely yield the best results of all: a true, real- time implementation could control the charging of individual vehicles on a distribution circuit to avoid overloading it, and could optimize all assets on the grid under a dynamic pricing regime. However, it will not be a reality until utilities, vehicles, and charging systems implement bidirectional communications systems to support it.

Various studies suggest that EV drivers *do* respond to price signals—depending on the strength of the signal and other variables.[46] Based on pilot projects the RMI found that,

The more money drivers could save by switching from undesirable to desirable charging times, the more drivers took advantage of it.

For example, if super-off-peak prices are offered during the sunny hours of the day when there is plenty of solar generation, some drivers may be motivated to charge at work during the day.

Another possibility is that many EV drivers may be willing to allow a third party—a smart aggregator—to manage the batteries' state of charge—making sure their batteries always have sufficient charge to get them to their destination while minimizing the electricity bill. Since most EVs sit idle most of the time and are normally used for limited range, this should be a relatively easy problem to solve.[47] The optimal solution varies for EVs that are extensively used such as e-taxis, versus privately owned EVs.

---

[45] There are other ways to manage charging of EVs, including treating it as a form of direct load control (DLC), an approach used to manage the highly cyclical air conditioning loads.

[46] Refer to https://sepapower.org/resource/residential-electric-vehicle-time-varying-rates-that-work-attributes-that-increase-enrollment/.

[47] Technically speaking, EVSP aggregators (unless Tesla were to get into the VGI aggregation game) do not have access to the car's SOC. This would have to be gleaned through an agreement with the OEM to use the vehicle's telematics data.

The RMI report summarized its finding as below:

Maximizing the benefits of using EV charging as a distributed energy resource will require the active support of a wide and unusually diverse range of stakeholders: regulators, transmission system operators, distribution system operators, utilities, customers, aggregators, vehicle manufacturers, commercial building owners, elected officials, and others. And realizing those benefits will be a complex task, because doing so may require policies and mechanisms that cut across conventional boundaries, such as the ones between wholesale and retail markets, or between customers and generators. This underscores the importance of including vehicle electrification in integrated resource plans.

As already noted, the electrification of the transport sector will encompass all types of road transport, each with certain level of demand flexibility. In the case of municipal busses,[48] for example, a promising option may be to rely on fast charging supercapacitors, which can be quickly recharged at frequent stops along a preset route instead of carrying heavy batteries around.[49] Such buses as well as other service vehicles following preset routes can be fast charged using overhead chargers or chargers embedded in the road where the vehicles normally stop (Fig. 8.9).

Eventually, many electrified vehicles will become autonomous, and there is speculation that the current private car-ownership model, where a single family may own multiple cars sitting idle most of the time, will be replaced with increased car sharing and mobility-as-a-service (MaaS)

Figure 8.9 *E-buses can rely on supercapacitors with overhead fast charging.* *(Source: Siemens e-Mobility.)*

---

[48] According to the Union of Concerned Scientists, an estimated 28 million trucks and buses in the United States in 2019 accounted for 28% of the emissions in the transportation sector. This explains why the California Air Resources Board has ambitious goals to electrify this heavily polluting fleet by 2045.

[49] Siemens and others have already operated successful trials using such e-buses.

options. This will have significant implications for charging, for example, resulting in far fewer cars used far more extensively[50].

Companies running such services will have strong financial incentives to charge their fleets using the most efficient technologies and at times and locations that minimize the electricity costs including incentives to invest in storage at their service hubs or depots to reduce demand charges and avoid the expensive peak demand hours.

4. Vehicle-to-grid (V2G) and more advanced schemes[51]

Once we manage to incentivize large numbers of drivers to switch to EVs—which will only happen once we have a vast and convenient network of private and public charging stations—and manage to charge them intelligently, then we can envision even more exciting opportunities including V2G functionality to truly unlock the potential of EVs, not only as a flexible load but also as a balancing resource[52] (Box 8.5).

The potential of V2G becomes intuitively obvious once we have large numbers of EVs charging when generation is plentiful and prices are low while allowing some portion of that stored energy to be fed back into the grid at times when the supplies are scarce and prices are high. Imagine what can be done in 2030 when California expects to have over 5 million EVs:

- If most of the EVs are charged during times of plentiful supply/low prices, a significant amount of the "excess" variable renewable generation can be stored in the batteries helping reduce the required ramping in the morning; and
- If some of the fully charged EVs can discharge a portion of their stored energy back to the grid during the neck of the curve, we can reduce the critical afternoon/early evening ramping.

The benefits of such a scheme, further explained in Box 8.6, are so compelling that it is hard to imagine that the V2G, in one form or another, will not emerge as a possibility, sooner or later.

---

[50] Estimates suggest that one shared car replaces as many as seven private cars, but driven far more miles over extended hours.

[51] Refer to Final Report of the Vehicle to Grid Interconnection Subgroup. California Public Utilities Commission, December 2019 at https://static1.squarespace.com/static/5b96538250a54f9cd7751faa/t/5df293184c2e38505cebb5bd/1576178475052/2019-12-11+V2G+AC+Interconnection+Subgroup+Report+in+R.18-12-006+-+FINAL.pdf.

[52] In the V2G field, there is interest in concepts such as V2H (vehicle-to-home) and V2V (vehicle-to vehicle, for use in an emergency to fill up an empty battery).

BOX 8.5 California is planning ahead, not just for VGI but V2G

Given the significance of EVs, the state of California is putting considerable effort to advance the development and deployment of VGI. In 2019, Governor Newsom signed SB 676[53] (Bradford) into law, signaling the state's intent to catalyze the marketplace for VGI and utilize it to maximize the cost-effectiveness of the state's transition to an electrified transportation sector. According to SB 676,

'Electric vehicle grid integration' means any method of altering the time, charging level, or location at which grid-connected electric vehicles charge or discharge, in a manner that optimizes plug-in electric vehicle interaction with the electrical grid and provides net benefits to ratepayers by doing any of the following: (A) Increasing electrical grid asset utilization; (B) Avoiding otherwise necessary distribution infrastructure upgrades; (C) Integrating renewable energy resources; (D) Reducing the cost of electricity supply; (E) Offering reliability services consistent with Section 380 or the Independent System Operator tariff.

The bill requires the regulator, California Public Utilities Commission (CPUC), to establish VGI strategies and quantifiable metrics to maximize the use of feasible and cost-effective VGI by January 2030[54] including rate design and the adoption of technology and customer services that provide smart EV charging, or V1G, which manages the time or rate at which an EV charges, as well as *bidirectional* EV charging, or V2G, wherein the vehicle battery discharges to provide electricity to the facility or grid.

The bill also requires investor-owned utilities to submit an annual report to demonstrate their progress in furthering the VGI strategies adopted by the CPUC and to incorporate them in their Integrated Resource Plans (IRPs).

Moreover, SB 676 gives load serving entities (LSEs) the flexibility to pursue the highest net value VGI strategies that make the most sense for their service area and customers. Similarly, technology providers have broad latitude to offer

[53] Refer to Enel X, 23 Jul 2019 Letter on SB 676 (Bradford) on Transportation electrification: electric vehicles: grid integration.

[54] Refer to the CPUC's transportation electrification rulemaking (R.18-12-006) with the Assigned Commissioner's Scoping Memo and Ruling, R.18-12-006, May 2, 2019, in Section 3. Clarifying Guidance, subsection 8. VGI Working Group at http://docs.cpuc.ca.gov/PublishedDocs/Efile/G000/M285/K712/285712622.pdf.

Continued

BOX 8.5 California is planning ahead, not just for VGI but V2G—cont'd

services that make good business sense. The intent of the law is to ensure that the increased electricity demand from EV charging will lead not to increased electricity costs, but instead to increased revenue that will *decrease* electricity rates. Such an outcome is feasible if the fixed costs of the system are spread across a greater volume of sales.[55]

[55] Refer to Synapse study at https://www.synapse-energy.com/sites/default/files/EVs -Driving-Rates-Down-8-122.pdf and E3 study at https://www.ethree.com/tools/electric-vehicle-grid-impacts-model-2/and California Transportation Electrification Assessment, Phase 2: Grid Impacts, October 23, 2014, at http://www.caletc.com/wp-content/uploads/2016/08/CalETC_TEA_Phase_2_ Final_10-23-14.pdf and https://www.mjbradley.com/reports/mjba-analyzes-state-wide-costs-and-benefits-plug-vehiclesfivenortheast-and-mid-atlantic and IRP 2017 Proposed Reference System Plan, Attachment C, "Summary of RESOLVE Inputs and Outputs," September 19, 2017, page 48, "Selected Resource Summary—Battery Storage" Row, "42 MMT_Reference" column (1992 MW) and page 51, "Selected Resource Summary—Battery Storage" Row, "42 MMT_Flexible EVs" column (1452 MW) at http://docs.cpuc.ca.gov/PublishedDocs/Efile/G000/M195/K910/195910922.pdf and IRP 2017 Proposed Reference System Plan, September 19, 2017, slides 139, 187, 203. http://cpuc.ca. gov/uploadedFiles/CPUCWebsite/Content/UtilitiesIndustries/Energy/EnergyPrograms/Elect PowerProcurementGeneration/irp/AttachmentA.CPUC_IRP_Proposed_Ref_System_Plan_2017_09_ 18.pdf and https://iopscience.iop.org/article/10.1088/1748-9326/aabe97/meta.

BOX 8.6 Smart bidirectional EV integration: The ultimate nirvana

Adding millions of EVs will add a major load to the electricity grids and a welcome source of increased revenue. More important, if properly managed, this could lead to *lower* electricity prices.

To illustrate how this could conceivably happen consider a regulated "poles and wires" company that has annual sales of 10 million kWhs and costs $1 million to operate. In this example, customers pay 10 cents/kWh to keep it whole.

Now, assume that the sales increase to 12 million kWh due to the increased EV load. Since the cost of maintaining a distribution network is mostly fixed, *increased* sales results in *lower* rates, in this example a drop to 8.3 cents/kWh, which accrues to both EV owners and nonowners.

Continued

BOX 8.6 Smart bidirectional EV integration: The ultimate nirvana—cont'd

Similar results have been documented in several studies,[56] for instance by Energy and Environmental Economics (E3) who estimates significant ratepayer and societal net benefits in a number of states. Another study by Synapse Energy Economics also concluded a downward trend on electric rates in California with managed EV charging.

Such results have been documented by others including MJ Bradley and Associates,[57] who found that for the five northeastern states, each EV added delivered an annual *net present value* (NPV) ranging from $107–265 in 2030, with the benefits rising to $349–520 per EV by 2050.

Moreover, the CPUC's [58]IRP found that flexible EV charging can displace the need for 540 MW of stationary energy storage, reduce renewable energy curtailment, and provide grid integration benefits—which become more valuable with higher renewable penetration levels.

Finally, Lawrence Berkeley National Laboratory documented the benefits of VGI in enabling California's renewable energy targets over stationary energy storage as described in Box 8.4. Specifically, the LBL study compared the potential value of California's 1.3 GW of stationary battery storage mandate with smart charging and V2G enabled-EVs. It found that smart charging with 1.5 million EVs expected by 2025 could provide the equivalent renewable integration services of $1.3–1.6 billion worth of stationary storage. These cost savings would increase to $12.65–15.25 billion of stationary energy storage if the same 1.5 million EVs had bidirectional or V2G functionality. It is safe to assume that the benefits would be even bigger by 2030 when 5 million EVs are expected in California.

---

[56] E3 study at https://www.ethree.com/tools/electric-vehicle-grid-impacts-model-2/; Synapse study at https://www.synapse-energy.com/sites/default/files/EVs-Driving-Rates-Down-8-122.pdf; California Transportation Electrification Assessment, Phase 2: Grid Impacts, October 23, 2014 at http://www.caletc.com/wp-content/uploads/2016/08/CalETC_TEA_Phase_2_Final_10-23-14.pdf.

[57] MJ Bradley study at https://www.mjbradley.com/reports/mjba-analyzes-state-wide-costs-and-benefits-plug-vehiclesfivenortheast-and-mid-atlantic, 09_18.pdf and https://iopscience.iop.org/article/10.1088/1748-9326/aabe97/meta.

[58] Refer to IRP 2017 Proposed Reference System Plan, Attachment C, "Summary of RESOLVE Inputs and Outputs," September 19, 2017. See page 48, "Selected Resource Summary—Battery Storage" Row, "42 MMT_Reference" column (1992 MW) and page 51, "Selected Resource Summary—Battery Storage" Row, "42 MMT_Flexible EVs" column (1452 MW) at http://docs.cpuc.ca.gov/PublishedDocs/Efile/G000/M195/K910/195910922.pdf and IRP 2017 Proposed Reference System Plan, September 19, 2017, slides 139, 187, 203 at http://cpuc.ca.gov/uploadedFiles/CPUCWebsite/Content/UtilitiesIndustries/Energy/EnergyPrograms/Elect PowerProcurementGeneration/irp/AttachmentA.CPUC_IRP_Proposed_Ref_System_Plan_2017_.

The potential benefits outlined in Box 8.6 are reflected in the website of the VGIC which says[59]:

As the US electric grid evolves to a distributed, bi-directional network with increasing penetrations of renewable energy, electric vehicles can serve as critical resources to support and balance the grid. In California alone, VGI system-balancing capabilities could be valued at up to $15 billion by 2030. VGI can place downward pressure on electricity rates: as consumption of electricity increases to fuel EVs, VGI can limit the grid costs ultimately allocated to electric customers.

This is illustrated in Fig. 8.10, where the VGI spectrum envisions a future where EVs can potentially provide bidirectional services.

This exciting future, of course, requires further research, testing, demonstration and pilot projects, regulatory support, and policy clarity. In its report, the RMI[60] notes that,

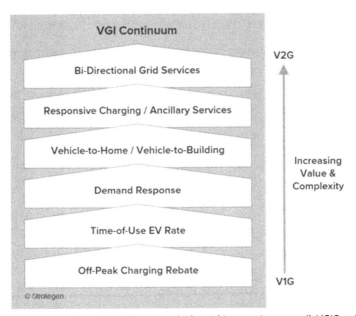

Figure 8.10 *From V1G to V2G. (Source: Vehicle grid integration council, VGIC at https://www.vgicouncil.org/why-vgi.)*

---

[59] Refer to https://www.vgicouncil.org/why-vgi.
[60] Electric Vehicles as Distributed Energy Resources. Rocky Mountain Institute e-Lab at https://rmi.org/wp-content/uploads/2017/04/RMI_Electric_Vehicles_as_DERs_Final_V2.pdf.

The auto industry needs to build V2G features into its vehicles. Currently, most manufacturers are not including onboard V2G capability in their vehicles (except for a few pilot programs and the newer Nissan Leaf models), and even where it is built-in, using it for V2G would void the vehicle warranty.

Most current EVs specifically forbid *discharging* the battery for any reason, as this reduces the number of useful charge—discharge cycles that batteries can handle, and notes that,

It's a classic chicken-and-egg problem: Manufacturers aren't including V2G features because there isn't a market, and there isn't a market because there aren't enough vehicles with those features.

Manufacturers need to allow V2G use under their warranties. Currently, using an EV battery as grid supply would void warranties.

Moreover,

Essential hardware and software infrastructure that would be needed from end to end to enable V2G is generally lacking, including real-time data exchange, advanced metering, cybersecurity layers, and standard interfaces between vehicle and grid.

Clearly, these are formidable challenges—but the potential gains are overwhelming—suggesting that some form of bidirectional charging/discharging is inevitable.

5. Conclusions

The preceding discussion highlighted why EVs—and more broadly electrified transportation—offer such a tremendous potential with their inherent flexible demand. Once large numbers of EVs are in use and their charging is intelligently managed, they can become the biggest source of demand flexibility. This would be a blessing for networks with large penetration of variable renewable generation, providing added flexibility.

As explained, for this vision to become reality, we must

- Develop a vast charging infrastructure that is convenient and accessible;
- Develop an intelligent EV charging network that encourages charging at times—and locations—where supplies are abundant and prices are low while discouraging the opposite; and
- Get the regulatory, pricing and policy support and coordination needed from all stakeholders to manage the operation and control of the entire

value chain from the generation source to the chargers without overwhelming the distribution network.

This process is already underway. The next stage with bidirectional functionality offers even greater opportunities. To achieve this, however, requires an even higher level of coordination, collaboration, advanced technology, software, pricing, and regulatory and policy clarity and incentives.

The bottom line, on which there is universal agreement, is that electrified transportation is coming and its inherent demand flexibility is simply too valuable to be squandered.

CHAPTER 9

Load flexibility: Market potential and opportunities in the United States

Ryan Hledik, Tony Lee[1]
The Brattle Group, San Francisco, CA, United States

1. Introduction

The term "distributed energy resources (DERs)" typically brings to mind emerging energy technologies like electric vehicles, rooftop solar, or battery storage. Yet, the installed capacity of another DER—demand response (DR)—exceeds that of all of those resources combined. DR has been a valuable tool in the US portfolio of energy resources for decades and currently amounts to roughly 59 GW of capacity, or the equivalent of nearly 250 medium-sized peaking units, as illustrated in Fig. 9.1.

Historically, steady load growth meant that DR could provide significant value to the electricity system simply by reducing peak demand during a handful of hours per year, thus avoiding the construction and operation of expensive peaking generation. As a result, most traditional DR programs are designed to curtail load in a small number of hour per year that are likely to coincide with the system peak. Examples of these programs include interruptible tariffs for commercial and industrial customers and direct load control of residential air conditioners.

Recently, however, several factors have limited the value of traditional peak-focused DR in many parts of the United States. There has been a reduced need for new peaking capacity due to slowing electricity demand, an accumulation of excess capacity from generation buildout predicated on higher loads, and the rise in adoption of behind-the-meter rooftop solar

---

[1] The authors would like to thank Brattle colleagues Ahmad Faruqui, Pearl Donohoo-Vallett, and John Higham for insightful contributions throughout this assessment. All perspectives and opinions are those of the authors and do not necessarily reflect those of The Brattle Group's clients or other consultants. Elements of this chapter are derived from a study by The Brattle Group titled, "The National Potential for Load Flexibility: Value and Market Potential Through 2030," June 2019.

Variable Generation, Flexible Demand
ISBN 978-0-12-823810-3
https://doi.org/10.1016/B978-0-12-823810-3.00001-7

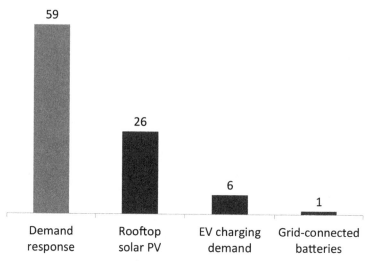

Figure 9.1 *Total US DER Capacity, in GW.* Notes: EV charging demand assumes 6 kW charging demand per EV does not account for coincidence of charging patterns. The rooftop solar PV estimate is installed capacity, does not account for derated availability during peak. Existing DR is the sum of retail DR from 2017 EIA-861 and wholesale DR from 2018 FERC Assessment of Demand Response and Advanced Metering; values are not modified to account for possible double-counting between wholesale and retail DR.

which reduces day-time peak loads. Additionally, the energy value of traditional DR has decreased, due to declining energy prices resulting from reduced natural gas prices and substantial additions of low marginal cost renewable generation. And, in some wholesale markets, increasingly stringent participation rules have reduced the ability to monetize the value of DR resources.[2] After years of growth in DR deployment, these factors have contributed to a stagnation in DR additions, as illustrated in Fig. 9.2.

Despite the recent DR deployment slowdown, there are new opportunities for continued growth if DR can be reimagined as "load flexibility" to address evolving system needs. Three industry "megatrends" present exciting opportunities for load flexibility:

- Renewables growth: Shifting consumption to low net load hours (i.e., load building) and using fast-responding DR to provide ancillary services can avoid renewables curtailment and help to maintain the delicate balance between electricity supply and demand.

[2] For example, seasonal capacity performance requirements in PJM and ISO-NE.

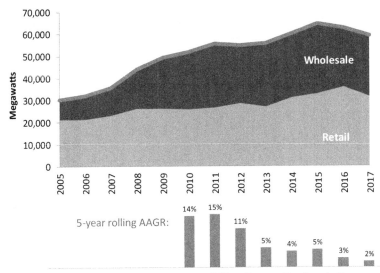

Figure 9.2 *Total US DR Peak Reduction Capability.* *(Source: The Brattle Group analysis of FERC Assessment of Demand Response and Advanced Metering reports (2006–18) and EIA-861 Data (2005–2017). Post-2012 values are not modified to account for possible double-counting between wholesale and retail DR across data sources.)*

- Grid modernization: Geographically targeted DR on distribution feeders that are approaching their capacity limits can help to defer capacity upgrades and improve system resilience.
- Electrification: Controlling new sources of load—in particular, electric vehicles and electric heating—can mitigate rising system costs while maintaining customer satisfaction and creating new value streams for smart home devices.

Ultimately, new consumer-oriented energy technologies will enable the transition from traditional DR to load flexibility. Several industry sources forecast rapid consumer adoption of behind-the-meter "smart" devices, as illustrated in Fig. 9.3. The rapid uptake of behind-the-meter (BTM) storage (82% annual growth), electric vehicles (24% annual growth), and smart appliances (53% annual growth), combined with the continued rollout of advanced metering infrastructure (22% annual growth) will lay the foundation for growth in load flexibility.

The purpose of this chapter is to provide a view on the potential for load flexibility in the United States over the coming decade. Given this book's focus on load flexibility challenges and opportunities, this chapter

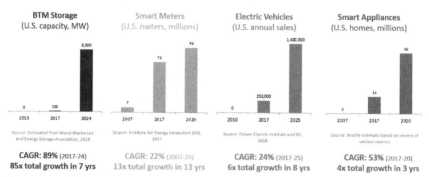

Figure 9.3 *Projected growth of Behind-the-Meter (BTM) Energy Devices in the United States. (Source: Brattle analysis of data from GTM Research (2018), Institute for Energy Innovation (IEI) (2017), Edison Electric Institute and IEI (2018), Parks Associates and Greentech Media (2018).)*

provides a sense of the magnitude of the opportunity that exists if those challenges can be overcome.

The balance of the chapter is organized as follows:

- Section 2 describes new analytical methods to simulate and value load flexibility.
- Section 3 provides an estimate of economic potential for load flexibility in the United States and identifies key drivers of future DR growth
- Section 4 concludes the chapter with three predictions regarding the future of load flexibility and provides recommendations to accelerate to deployment of load flexibility in regulatory and planning processes.

2. Quantifying load flexibility potential

The DR market potential studies that many utilities and state regulators have commissioned in the past tend to take a narrow view of DR capabilities. They typically focus on DR programs where demand reductions occur during a limited peak window and are constrained to a small number of hours per year.

However, load flexibility programs can do much more than just shave the peak. DR market potential studies will need to be expanded to accurately capture the full value of load flexibility. Beyond peak shaving, load flexibility can capture opportunities to shift and build load, provide ancillary services, and target high-value locations for T&D deferral (Dyson et al., 2015; Goldenberg et al., 2018). These additional services require accounting for value streams not previously provided by demand-side options. This requires enhanced analytical tools to simulate program operations, as these technologies could be dispatched much more frequently than traditional DR. They may also be subject to different operational constraints, based on the technology limits, consumer tolerance, and the need to recover interrupted load. Some programs can even provide multiple services from the same resource (accessing "stacked value"), but other programs may need to optimize across several competing value streams.[3]

To address these key challenges, the authors have developed a new analytical framework for assessing load flexibility potential, referred to as the "LoadFlex" model. The LoadFlex modeling framework builds on the standard approach to quantifying DR potential that has been used in prior studies[4] around the United States and internationally, but incorporates a number of differentiating features which allow for a more robust evaluation of DR programs:

- **Economically optimized enrollment**: Assumed participation in DR programs is tailored to the incentive payment levels that are cost-effective for the DR program. If only a modest incentive payment can be justified in order to maintain a benefit—cost ratio of 1.0, then the participation rate is calibrated to be lower than if a more lucrative incentive payment were offered. Prior approaches to quantifying DR potential ignore this relationship between incentive payment level and participation, which tends to understate the potential and, in some cases, incorrectly concludes that a DR program would not pass the cost-effectiveness screen.

- **Utility-calibrated load impacts**: Load impacts are calibrated to the characteristics of the utility's customer base. In the residential sector, this includes accounting for the market saturation of various end-use appliances (e.g., central air-conditioning, electric water heating). In the commercial and industrial (C&I) sector, this includes accounting for customer segmentation based on size (i.e., the customer's maximum demand) and industry (e.g., hospital, university). Load curtailment

---

[3] "Stacked value" refers to the concept of capturing multiple sources of value from a single resource. For instance, a smart thermostat program could be utilized to reduce the need for peaking generation, avoid running expensive, inefficient peaking units, and also be dispatched to relieve distribution-level capacity constraints.

[4] Examples of recent DR market potential studies include Applied Energy Group (2017), Demand Side Analytics (2018), Faruqui et al. (2014).

capability is further calibrated to the utility's experience with DR programs (e.g., impacts from existing DLC programs or dynamic pricing pilots).

- **Sophisticated DR program dispatch**: DR program dispatch is optimized subject to detailed accounting for the operational constraints of the program. In addition to tariff-related program limitations (e.g., how often the program can be called, hours of the day when it can be called), LoadFlex includes an hourly profile of load interruption capability for each program. For instance, for an EV home charging load control program, the model accounts for home charging patterns, which would provide greater average load reduction opportunities during evening hours (when EV owners have returned home from work) than in the middle of the day.

- **Realistic accounting for "value stacking"**: DR programs have the potential to simultaneously provide multiple benefits. For instance, a DR program that is dispatched to reduce the system peak and therefore avoid generation capacity costs could also be dispatched to address local transmission or distribution system constraints. However, tradeoffs must be made in pursuing these value streams—curtailing load during certain hours of the day may prohibit that same load from being curtailed again later in the day for a different purpose. LoadFlex accounts for these tradeoffs in its DR dispatch algorithm. DR program operations are simulated to maximize total benefits across multiple value streams, while recognizing the operational constraints of the program. Prior studies of load flexibility value have often assigned multiple benefits to DR programs without accounting for these tradeoffs, thus double-counting benefits.

- **Industry-validated program costs**: DR program costs are based on a detailed review of the utility's current DR offerings. For new programs, costs are based on a review of experience and studies in other jurisdictions and conversations with vendors.[5] Program costs are differentiated by type (e.g., equipment/installation, administrative) and structure (e.g., one-time investment, ongoing annual fee, per-kilowatt fee) to facilitate integration into utility resource planning models.

---

[5] The analysis relied on data from Potter and Cappers (2017) and program data from load flexibility pilots listed in Hledik et al. (2019a,b).

In evaluating the economics of load flexibility, three value streams commonly included in assessments of DR potential are accounted for:

- **Avoided generation capacity costs**: the need for new peaking capacity can be reduced by lowering system peak demand.
- **Reduced peak energy costs**: Reducing load during high priced hours leads to a reduction in energy costs. This analysis estimates net avoided energy costs, accounting for costs associated with the increase in energy consumption during lower cost hours due to "load building." The energy benefit accounts for avoided average line losses.[6] This analysis does not include the effect of any potential change in energy market prices that may result from changes in load patterns (sometimes referred to as the "demand response induced price effect," or DRIPE).
- **System-wide deferral of transmission and distribution (T&D) capacity costs**. System-wide reductions in peak demand can, on average, contribute to the reduced need for peak-driven upgrades in T&D capacity.

In addition, the analysis accounts for three value streams that can be captured by more advanced load flexibility programs:

- **Geo-targeted distribution capacity investment deferral**: DR participants may be recruited in locations on the distribution system where load reductions would defer the need for capacity upgrades.
- **Ancillary services**: The load of some end-uses can be increased or decreased in real time to mitigate system imbalances. The ability of qualifying DR programs to provide frequency regulation was modeled, as this is the highest-value ancillary service, albeit with limited system need.
- **Load building/valley filling**: Load can be shifted to off-peak hours to reduce wind curtailments or take advantage of low or negatively priced hours. DR was dispatched against hourly energy price series to capture the economic incentive that energy prices provide for this service.

Beyond accounting for new value streams, it also is important to take an expanded view of the types of DR programs considered in a market potential analysis. Whereas traditional market potential studies have focused on conventional options like direct load control and interruptible tariffs, emerging opportunities like grid-interactive water heating, EV managed

---

[6] The analysis likely includes a conservative estimate of this value, as peak line losses are greater than off-peak line losses.

charging, and behavioral DR could potentially add value in new ways. In this study, 12 DR and load flexibility programs are evaluated.[7]

- Direct load control (DLC): Participant's central air conditioner is remotely cycled using a switch on the compressor.
- Smart thermostats: An alternative to conventional DLC, smart thermostats allow the temperature setpoint to be remotely controlled to reduce A/C usage during peak times. Customers could provide their own thermostat, or purchase one from the utility.
- Interruptible rates: Participants agree to reduce demand to a prespecified level and receive an incentive payment in the form of a discounted rate.
- Demand bidding: Participants submit hourly curtailment schedules on a daily basis and, if the bids are accepted, must curtail the bid load amount to receive the bid incentive payment or may be subject to a noncompliance penalty.
- Time-of-use (TOU) rate: Static price signal with higher price during peak hours (assumed 5-h period aligned with system peak) on nonholiday weekdays. Modeled for all customers as well as for EV charging.
- Critical peak pricing (CPP) rate: Provides customers with a discounted rate during most hours of the year, and a much higher rate (typically between 50 cents/kWh and $1/kWh) during peak hours on 10–15 days per year.
- Behavioral DR: Customers are informed of the need for load reductions during peak times without being provided an accompanying financial incentive. Customers are typically informed of the need for load reductions on a day-ahead basis and events are called somewhat sparingly throughout the year. Behavioral DR programs have been piloted by several utilities, including Consumers Energy, Green Mountain Power, the City of Glendale, Baltimore Gas and Electric, and four Minnesota cooperatives.[8]
- EV managed charging: Using communications-enabled smart chargers allows the utility to shift charging load of individual EVs plugged-in from on-peak to off-peak hours. Customers who do not opt out of

---

[7] Note that these load flexibility programs focus on managing customer load. Behind-the-meter generation (e.g., from backup generators, rooftop solar, or batteries) was not included in the scope of this analysis but would be a valuable extension of the research.

[8] Generally, the studies have found that customers provide a statistically significant reduction in peak demand, though a reduction that is lower than if the information were accompanied by a financial incentive to reduce load.

an event receive a financial incentive. Chapter 8 of this book addresses EV managed charging in more detail.

- Timed water heating: The heating element of electric resistance water heaters can be set to heat water during off-peak hours of the day. The thermal storage capabilities of the water tank provide sufficient hot water during peak hours without needing to activate the heating element.

- Smart water heating: Offers improved flexibility and functionality in the control of the heating element in the water heater. The thermostat can be modulated across a range of temperatures. Multiple load control strategies are possible, such as peak shaving, energy price arbitrage through day/night thermal storage, or the provision of ancillary services such as frequency regulation. They are modeled for electric resistance water heaters, as these represent the vast majority of electric water heaters and are currently the most attractive candidates for a range of advanced load control strategies.[9]

- Ice-based thermal storage: Commercial customers shift peak cooling demand to off-peak hours using ice-based storage systems. The thermal storage unit acts as a battery for the customer's A/C unit, charging at night (freezing water) and discharging (allowing ice to thaw to provide cooling) during the day.

- C&I Auto-DR: Auto-DR technology automates the control of various C&I end-uses. Features of the technology allow for deep curtailment during peak events, moderate load shifting on a daily basis, and load increases and decreases to provide ancillary services. Modeled end-uses include HVAC and lighting (both luminaire and zonal lighting options).

To illustrate the national potential for load flexibility, a utility system with characteristics that are roughly representative of national average conditions was defined. Those utility-level results were then scaled up to the national level.

3. The potential for load flexibility in the United States

The load flexibility market potential analysis indicates that a portfolio of load flexibility programs could triple existing US DR capacity, increasing

---

[9] Chapter 7 of this book discusses the role of aggregators in collectively bidding a large number of individual load flexibility resources, such as water heaters, into markets at the same scale as a larger generation resource.

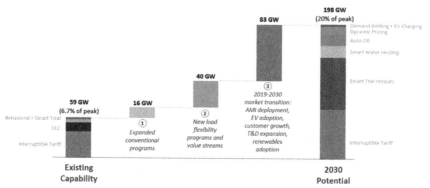

Figure 9.4 *US Cost-Effective Load Flexibility Potential, in GW of peak reduction. (Source: The Brattle Group analysis.)*

total DR from 59 GW (6.7% of peak) today to 198 GW (20% of peak) by 2030. Fig. 9.4 illustrates the three drivers that drive that potential, as well as the breakdown of potential across programs.

There are three factors that explain the difference between existing DR capability and the 200 GW of potential identified in this study: untapped opportunities for expanding conventional programs, potential in new programs and value streams under current market conditions, and expected changes in market conditions (the "market transition") over the next decade.

First, existing conventional programs often have untapped potential that can be harnessed through revamped customer marketing and outreach, modified program rules, and redesigned incentive structures. These programs mostly contribute peak capacity value, but often can do so cost-effectively by leveraging existing program infrastructure. Interruptible tariff programs for commercial and industrial customers have the greatest potential for program expansion. According to this study, modernizing expanding conventional programs is estimated to increase existing DR capability by 16 GW, or 27%.

Second, under current market conditions, new load flexibility programs and the use of load flexibility to capture additional value streams will drive significant growth. Under current national average market conditions, the most significant cost-effective potential is in residential smart thermostat programs and dynamic pricing for all customer segments. The introduction of new programs is estimated to increase existing DR capability by 40 GW, or 67%.

Third, over the coming decade, the rapid growth in adoption of AMI, EVs, smart thermostats, and other smart appliances will enable expanded participation in load flexibility programs and more advanced forms of load control. Higher levels of renewable generation deployment will introduce more energy price variability and a greater need for ancillary services, increasing the value of load flexibility programs that can shift load and have fast-response capability. Additionally, continued expansion and modernization of the T&D system introduces a growing opportunity for nonwires alternatives and T&D deferral, increasing the value of programs that are deployed on a geographically targeted basis. These market developments justify greater incentive payments for customer participation in load flexibility programs and also justify the cost-effective introduction of robust smart water heating and Auto-DR programs. These changes in market conditions are estimated to increase DR capability by 83 GW, or 140% relative to existing capability.

What is this potential worth? In 2030, the cost-effective DR portfolio modeled in this study provides over $16 billion in gross annual system benefits. Fig. 9.5 illustrates the composition of benefits. Note that these results are based on national average conditions and that there is significant regional variation to the findings of this study.

Capacity remains the dominant source of load flexibility value through 2030, accounting for over half of the total value of the load flexibility portfolio. In many (but not all) parts of the United States, there will

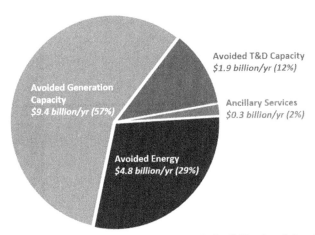

Figure 9.5 *2030 Annual Benefits of National Load Flexibility Portfolio.* (Source: The Brattle Group analysis.)

continue to be opportunities to avoid the addition of gas-fired combustion turbine units. However, capacity value will vary significantly by region; load flexibility is poised to provide most capacity value in regions with pending capacity retirements, supply needs in transmission-constrained locations, or unexpected supply shortages.

The "energy value" of load flexibility is the reduced resource cost, mostly reduced fuel costs, associated with shifting load to hours with a lower cost to serve. The estimate of energy value is based on production cost simulations for a utility with a supply mix that is over 40% zero-carbon (29% wind and solar, 12% nuclear, 2% hydro).[10] Energy value is best captured through programs that provide daily flexibility year-round, such as Auto-DR for C&I customers, TOU rates, EV charging load control, and smart water heating.

Avoided T&D value includes system-wide benefits of peak demand reduction (assumed at $10/kW-year, plus added benefit of geographically targeted T&D investment deferral (assumed at $45/kW-year but limited to 29 GW of system need).[11] Geo-targeted T&D deferral opportunities are typically high value but limited in quantity of near-term need. The size of this need may grow if utility T&D data collection and planning processes improve and make it easier to identify deferral opportunities.

The provision of ancillary services remains a relatively niche opportunity for load flexibility, primarily due to ancillary services being limited in quantity of need, and facing increasing competition from other flexible resources such as energy storage. In this study, ancillary services value is represented only by frequency regulation and assumes a need equal to 0.5% of system peak demand, which is roughly the middle of the range of requirements in wholesale markets around the United States. Additional value may exist if considering other ancillary services products. Frequency

[10] These assumptions are based on the "Accelerated Fleet Change" scenario of a MISO transmission planning study conducted in the Midwestern United States.

[11] T&D assumptions were based on a review of assumptions and distribution system data from various US utility planning studies.

regulation provides very high value to a small amount of capacity; in this analysis, the full need for frequency regulation can be served through a robust smart water heating program.[12]

It is important to note that these findings could vary significantly from one utility, state, or country to the next. The results of this study are based on US national average conditions, in order to provide an order-of-magnitude estimate of the national potential in load flexibility programs. Some individual utilities will vary significantly from these conditions. For instance, the value and mix of load flexibility opportunities for winter peaking utilities will look different than those of summer peaking utilities. Similarly, utilities vary significantly across important fundamental drivers of the potential estimate, such as customer mix, market saturation of various end-uses, existing and future generation mix, load growth, plans for unit retirements, and other factors. These differences exist across the United States as well as internationally.

To illustrate this point, consider the difference between two states where the potential for load flexibility was recently analyzed: Minnesota and California. In Minnesota, resource planning decisions were focused on replacing the retirement of 1400 MW of coal generation. The primary source of renewables additions in the Upper Midwest are wind generation, though gas units are forecasted to remain on the margin in most hours, with limited wind curtailments due to significant transmission connections. As a result, a study on the potential for load flexibility in Minnesota found that capacity and energy were the primary drivers of load flexibility value (Hledik et al., 2019a,b).

By contrast, in California, there is not a projected need for conventional peaking capacity. Resource planning decisions are driven more by considerations around integrating large amounts of solar generation and addressing local constraints on the distribution system. In this case, an assessment of load flexibility potential found that the most significant opportunities were in managing load to address evening generation ramping constraints, and a need for real-time grid balancing (Alstone et al., 2017).

---

[12] Frequency regulation is a service procured by utilities and system operators to ensure that the frequency of the system does not get too high or too low. Only certain types of fast-responding resources can provide this service. Typically, the quantity of the service that system operators need to acquire to keep the system balanced is less than 1% of peak demand. Grid interactive water heating programs provide this service by ramping water heating load up and down in response to a real-time signal from the system operator.

State	Primary drivers of need for load flexibility	Primary source of renewable generation additions	System value: Generation capacity	System Value: Energy (load shifting)	System Value: Ancillary services	System Value: T&D deferral	Load Flexibility Study
Minnesota	Pending retirement of 1,400 MW of coal generation	Wind	◕	◑	◔	◔	The Brattle Group, "The Potential for Load Flexibility in Xcel Energy's NSP Service Territory," June 2019
California	Renewables integration, local capacity constraints	Solar PV	◔	◑	◐	Not Quantified	LBNL, "2015 California Demand Response Potential Study," November 2016

◕ = *Primary source of value*   ◑ = *Moderate source of value*   ◔ = *Modest source of value*

Figure 9.6 *2030 Annual Benefits of National Load Flexibility Portfolio.* (Source: The Brattle Group analysis.)

Fig. 9.6 illustrates the contrast between California and Minnesota with respect to the value of load flexibility in those markets.

4. Conclusions

The results presented here are specific to the United States. Results could vary significantly if the same methodology were applied to other countries, due to differences in key drivers such as end-use mix and load patterns. Regardless, the authors' experience conducting demand response analysis in a variety of international jurisdictions suggests that there is significant load flexibility potential in most markets around the globe.

Based on the findings of this study, the following are three predictions for how load flexibility will evolve in the coming decade.

Prediction #1: Utility load flexibility programs will get smarter before they get bigger. Many existing programs have been underutilized for decades. There is low-hanging fruit in simply modernizing these programs to serve the growing need for system flexibility. For example, transitioning compressor switch-based DLC programs to smart thermostat-based programs, or updating the rules, incentives, and operation of interruptible tariffs. Programs could be redesigned with more flexibility around curtailment windows and offer higher incentives for more frequent interruptions or load shifting.

Prediction #2: Residential load flexibility additions will exceed those of C&I. For reasons entirely unrelated to demand response, customers are increasingly adopting technologies with load flexibility capabilities (e.g., electric vehicles, smart thermostats). At the same time, mass market smart metering deployments continue and customers are gradually being introduced to time-varying rates. While the C&I sector has provided 70% of US retail DR up to this point, these factors will combine to (finally) capitalize on the untapped potential of the residential sector

Prediction #3: New regulatory incentives will drive growth in load flexibility. There is a renewed industry-wide interest in regulatory models that provide utilities with incentives to pursue demand-side options rather than infrastructure investments. Experience with these incentive mechanisms will ultimately instill industry stakeholders with the confidence that both consumers and utilities can benefit from load flexibility.

What should happen next? A number of barriers to load flexibility deployment and adoption must be addressed in the areas of market design, technology development, regulation, utility resource planning and implementation, and program design. And there are feasible options for overcoming these barriers, such as the introduction of new regulatory and financial incentive models, technology codes and standards, and innovative retail rate and program designs. First, however, given the nascent state of load flexibility as a widely deployed resource, detailed and utility-specific load flexibility market assessments will establish a roadmap for pursuing the most cost-effective and impactful options. Such market assessments can inform integrated resource plans, renewables integration studies, and the setting of DR policy targets.

Bibliography

Alstone, P., Potter, J., Piette, M.A., Schwartz, P., Berger, M.A., Dunn, L.N., et al., 2017. 2025 California Demand Response Potential Study   Charting California's Demand Response Future: Final Report on Phase 2 Results.

Applied Energy Group, 2017. State of Michigan Demand Response Potential Study — Technical Assessment. Prepared for the State of Michigan.

Demand Side Analytics, LLC, 2018. Potential for Peak Demand Reduction in Indiana. Rocky Mountain Institute.

Dyson, M., Mandel, J., Bronski, P., Lehrman, M., Morris, J., Palazzi, T., et al., 2015. The Economics of Demand Flexibility: How "Flexiwatts" Create Quantifiable Value for Customers and the Grid. Rocky Mountain Institute.

Faruqui, A., Hledik, R., Lineweber, D., 2014. Demand Response Market Potential in Xcel Energy's Northern States Power Service Territory. Prepared for Xcel Energy.

Goldenberg, C., Dyson, M., Masters, H., 2018. Demand Flexibility — the Key to Enabling a Low-Cost, Low-Carbon Grid. Rocky Mountain Institute.

Hledik, R., Faruqui, A., Donohoo-Vallett, P., Lee, T., 2019a. The potential for load flexibility. In: Xcel Energy's NSP Service Territory. Prepared for Xcel Energy.

Hledik, R., Faruqui, A., Lee, T., Higham, J., 2019b. The National Potential for Load Flexibility: Value Market Potential Through 2030. The Brattle Group.

Potter, J., Cappers, P., 2017. Demand Response Advanced Controls Framework and Assessment of Enabling Technology Costs. Prepared for the Office of Energy Efficiency and Renewable Energy, U.S. Department of Energy.

CHAPTER 10

Demand response in the US wholesale markets: Recent trends, new models, and forecasts

Udi Helman
Helman Analytics, San Francisco, CA, United States

1. Introduction

This chapter examines how demand response (DR), defined as a change in retail customer load from expected levels in response to wholesale market prices and/or retail rate incentives, has participated in the organized US wholesale electric power markets operated by independent system operators (ISOs) in recent years.[1] There is also some data and results provided on DR in utility operations and planning. These historical trends are important to understand as a prelude to assessment of any further expansion or alignment of DR to meet the operational and reliability challenges of high penetration by variable energy resources and decarbonization of electric power systems, as examined in other chapters in this volume. As such, it is primarily a historical reference chapter, and also provides information on key data sources and how readers can continue to track trends in DR over time. Chapters 9 and 11 cover some similar ground, with the latter entirely focused on the PJM wholesale markets.

This chapter does not evaluate DR technical or market potential or speculate extensively about specific future applications or transitions more broadly to other methods for eliciting responsive load (e.g., under types of dynamic retail rates) as technology evolves and decarbonization proceeds. This topic is addressed in other chapters in this volume (and several prior volumes in this series, such as Sioshansi, 2019). As those chapters suggest, there is enormous potential for new types of responsive demand to evolve

---

[1] The term ISO is used generically here to describe all regional market operators, including those which are formally designated in the United States as Independent System Operators (ISOs) and Regional Transmission Operators (RTOs).

Variable Generation, Flexible Demand
ISBN 978-0-12-823810-3
https://doi.org/10.1016/B978-0-12-823810-3.00025-X

and expand through combinations of retail rate reforms and revisions to wholesale market participation models. However, the chapter does include some review of DR selection in recent long-term resource plans and load forecasts and considers the factors which may influence such selection. While the chapter is focused on the United States, similar regulatory developments and wholesale market issues are relevant to other countries, as also described in other chapters.

The remainder of the chapter is organized as follows:

- Section 2 summarizes selected aggregate data on DR in the United States;
- Sections 3–6 survey how DR participates in the US wholesale markets and also has some discussion about the transition to the most recent distributed resource models;
- Section 7 briefly discusses DR for T&D deferral and multiple uses;
- Section 8 discusses ISO preparation for aggregated DER models which can utilize DR;
- Section 9 reviews additional considerations for DR wholesale market participation;
- Section 10 examines results of DR forecasts for planning functions;
- Section 11 provides a brief note on data sources, followed by the chapter's conclusions.

The chapter, however, does not consider many other technical aspects of DR participation in the ISO markets in much detail, such as measurement and verification, communications, metering, and telemetry requirements.

2. Aggregate data on demand response

Data on DR aggregated on a utility, state, wholesale market, and national basis across the United States, including both retail and wholesale DR, are gathered by several federal government agencies, including annually in US Energy Information Agency (EIA) Form EIA-861 and by the Federal Energy Regulatory Commission (FERC) as well as by the North American Electric Reliability Corporation (NERC). Some state energy agencies also consistently gather data, and there are other regional sources which are issued on a regular basis, such as ISO reports and utility integrated resource plans (some discussed further below).

With respect to wholesale markets, since 2006, FERC has issued an annual assessment of demand response and advanced metering, which gathers data on DR from the EIA data, ISO/RTO markets, and jurisdictional utilities, and calculates some metrics (FERC, 2020b).

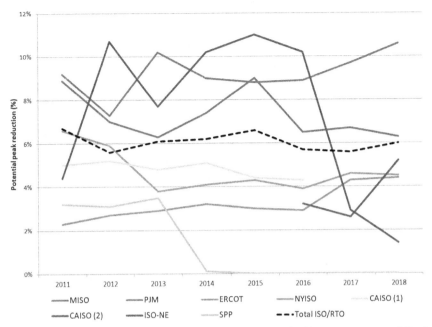

Figure 10.1 ISO and RTO Demand Response capacity (MW) as percentage (%) of potential peak reduction, 2011–18. *(Source: FERC 2020b, editions from 2012–19.)*

Since most DR currently is utilized for peak load reduction, Fig. 10.1 shows the FERC estimates of the potential contribution as a percentage of peak load of DR resources from 2011–18 in each wholesale market region (FERC 2020b).[2] As discussed further below, these consist primarily of reliability DR, with lesser participation by economic DR. Major trend lines in the past few years include the significant decline of DR in ISO–New England and SPP, smaller declines in PJM, and Fig. 10.8 clarifies that there is also a decline in California utility DR. At the same time, there has been some DR increase in ERCOT and in the MISO region. The FERC reports also show these same statistics for the nonmarket regions, which are not shown here. The variations from year to year result from a number of causes, including changes in market rules or operational

---

[2] Since the FERC estimates have sometimes utilized different measurements from year to year, in the case of CAISO, the figure shows two lines: CAISO (1) is a broader measure of California utility DR, whereas CAISO (2) is a narrower measure of DR which has qualified for market services and resource adequacy requirements. There is further discussion of this below.

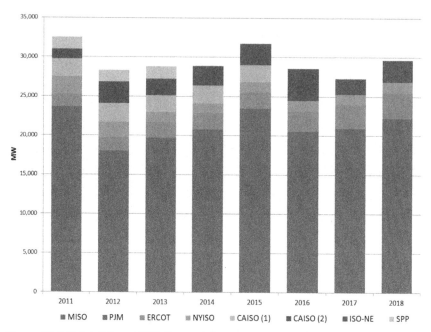

Figure 10.2 Total ISO and RTO Demand Response capacity (MW), 2011–18. *(Source: FERC 2020b, editions from 2012–19.)*

methods, and changes in market value which lead to entry or exit of DR. The FERC reports (FERC, 2020b) provide brief explanations of some of these changes, and selected examples are mentioned in the subsequent sections of this chapter and some other chapters (notably Chapter 11 on PJM).

Fig. 10.2 shows the total quantity of DR capacity in the ISOs from year to year measured in MW (rather than as percentage). The order of the bars is roughly from the ISOs with larger DR quantities on the bottom to the ones with lower quantities on the top.

Fig. 10.3, derived from the FERC annual assessment for 2017 (FERC, 2018), helps to visualize the distribution of current DR sources across the entire country, organized by reliability region. This indicates that the potential for industrial/manufacturing, commercial, and residential DR varies by region, but on average across the country, about 50% is from industrial sources, with around 30% from residential and 20% from commercial. A few ISOs provide technology data on the sources of DR in

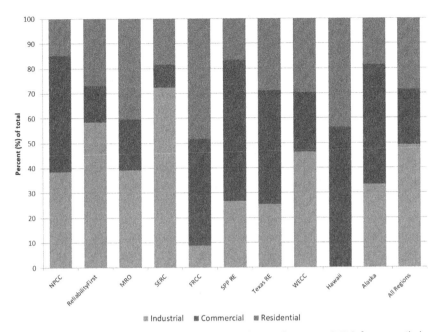

Figure 10.3 Percentage (%) of potential peak demand savings (MW) from retail demand response programs, by customer class and region in 2017. *(Source: Adapted from Table 3.2, Federal Energy Regulatory Commission (FERC), 2020b, 2019 version of report.)*

aggregate and when providing different wholesale products; some of these are discussed below.[3]

3. Demand response in the US wholesale markets

Turning from the aggregated data above, the next several sections examine additional data on DR participation in the organized wholesale markets derived from ISO and RTO reports. This section provides an overview of the history of DR in the wholesale markets. It then describes the market participation models being used for DR. Finally, it explains the definition of the market products and the sequence of the US markets, before the subsequent sections examine these products in more detail.

---

[3] For example, in 2020, PJM reported that 48% of DR was from manufacturing, 25% was from HVAC, 15% from generators, 9% from lighting, and the remainder from water heaters, batteries, refrigeration, and plug-in load (PJM, 2020a).

3.1 History of DR in the wholesale markets

In the United States, the organized wholesale markets operated by ISOs were introduced beginning in the late 1990s on a single state or multistate basis, subject to regulation by the Federal Energy Regulatory Commission (FERC), with the exception of the Texas market, which is regulated by the Public Utilities Commission of Texas (PUCT). There are seven such regional markets, which cover more than 75% of the nation's electric power systems measured by demand.[4]

In most regions of the country, utility DR programs predated the organized wholesale markets by several decades, and were initially incorporated into the nascent markets on the basis of their historical terms of utilization. There are generally two types of DR—load which participates directly in the markets through economic offers, and load which is only available for reliability-based curtailment instructions (i.e., system emergencies), but which may also incorporate offer prices. After several years of regional deference on how these historical DR programs were utilized, FERC sought to facilitate increased DR participation in the wholesale markets and transmission planning through a series of major orders applicable to all its jurisdictional ISOs and planning entities, which are briefly summarized in Table 10.1.[5] Compliance with these orders has proceeded at different rates in the different regions (Chapter 11 of this volume covers in more detail how PJM complied with the FERC orders).

One of the major developments in wholesale DR over the past decade has been the refinement and generalization of market participation models to accommodate nonutility entities, such as third-party aggregators or sufficiently large retail customers directly.[6] These models, which have had

---

[4] These are, in alphabetical order, California ISO (CAISO), Electric Reliability Council of Texas (ERCOT), ISO-New England, New York ISO (NYISO), Midcontinent ISO (MISO), PJM, and Southwest Power Pool (SPP). In addition, the CAISO operates the Energy Imbalance Market (EIM) in other parts of California and several other western states.

[5] All FERC orders in this table are available organized by date on the FERC Major Orders & Regulations webpage, https://www.ferc.gov/enforcement-legal/legal/major-orders-regulations. In addition, where relevant these orders on rehearing are available on this webpage, but ISO compliance filings and resulting FERC orders are accessible through FERC's eLibrary under the relevant docket numbers.

[6] For additional discussion of PJM, see PJM (2014), PJM (2017) and Chapter 11, this volume. In addition, ISO-NE (2017) provides recent developments in ISO-New England and MISO (2020) is a detailed business practice manual. CAISO (2020c) is a detailed schematic comparison of its main DR and DER models.

Table 10.1 Major FERC orders with provisions for demand response.

FERC order	Date of issuance	DR relevant requirements
Order 1000—Transmission Planning and Cost Allocation by Transmission Owning and Operating Public Utilities	July 2011	Regional transmission planning process must consider transmission needs due to "policy-driven" requirements; requires "comparable consideration" of transmission and nontransmission alternatives
Order 745—Demand Response Compensation in Organized Wholesale Energy Markets	March 2011	Requires that DR which passes a net benefits test and is dispatched is compensated at the energy market locational marginal price (LMP).
Order 719—Wholesale Competition in Regions with Organized Electric Markets	Oct. 2008	Facilitates DR participation in ancillary service markets on a comparable basis to generators, eliminates charges to load which scheduled day-ahead but curtail during emergencies, allows aggregators of retail customers to participate in the wholesale market, implements scarcity pricing during operating reserve shortages, and identifies further barriers to DR participation
Order 890—Preventing Undue Discrimination and Preference in Transmission Service	Feb. 2007	DR and other nonwires solutions can be considered in the transmission planning process on a comparable basis to transmission solutions

some success, are discussed further below. However, in each region, there have been other continuing reforms to market rules which affect DR participation.

A next phase of DR now unfolding is the integration of some DR capability into the more generalized distributed energy resource (DER) models, allowing for more flexible aggregations utilizing different resource types at different times. Most ISOs are in the process of defining requirements for aggregated DER, including locational restrictions for individual aggregations and metering and telemetry needed for them to actively participate in the markets (e.g., CAISO, 2020c; NYISO, 2019b; FERC, 2020a). NYISO has already folded several of its DR models into its aggregated DER model (NYISO 2019b). In a related vein, there has been attention to the potential evolution of "prosumers" which utilize combinations of on-site generation, storage, and load response. In some regions, transactions involving these new types of entities may also devolve to distribution system operators and other types of new arrangements (Sioshansi, 2019). However, progress toward these new institutions has been fairly slow.

3.2 Wholesale market programs and participation models

DR programs and market participation models are the procurement mechanisms and set of rules which govern the requirements and parameters for different classes of resources to participate in the wholesale markets generally, and for each specific service. In each of the ISO markets, DR can utilize several different programs and associated market participation models, some of which are fairly recent and respond to changing conditions on the grid. Table 10.2 summarizes these models as they currently exist for DR, along with some of the other dispatchable load models, and the aggregated DER models already implemented.[7] Additional details of participation rules for DR relevant to particular wholesale products can be found in Tables 3.6 and IRC (2018) (as well as the tariffs and technical manuals available for each ISO market).

3.3 Wholesale market products and the market sequence

Each US ISO operates centralized auction markets for energy and several ancillary services, and most operate or facilitate markets for Resource Adequacy (RA) capacity. A fourth type of market product, financial transmission rights, will not be examined here. Each of the ISOs runs these wholesale markets in a sequence from years to months ahead for capacity, to the day-ahead markets the day before physical operations, and then finally to real-time markets and physical operations, from midnight to midnight of each day. DR can participate along the full range of this sequence, as explained next.

---

[7] This table was produced by reviewing ISO documents; however, a periodic useful summary comparison of these models is provided by the ISO/RTO Council (IRC, 2018). The IRC summary also lists inactive DR participation models, and some details may be out of date.

Table 10.2 Current ISO/RTO demand response and other market participation models which support responsive load.

ISO	Demand response and other responsive load models	Description and eligible market services
ISO-NE	Forward Capacity Market (FCM)—On-Peak and Seasonal Peak Resources Demand Response Resource (DRR)/Price-Responsive Demand (PRD)	Reliability DR. Eligible for the capacity market, not required to offer into energy markets. Economic or Reliability DR general model for participation in energy, operating reserve, or capacity markets with bids in markets.
	Dispatchable Asset Related Demand (DARD)	Designed for demand-side participation of dispatchable load (e.g., pumped storage plants when pumping) and BTM DER participation in the operating reserve markets.
NYISO	Installed Capacity-Special Case Resource Program (ICAP-SCR)	Reliability DR. Eligible as capacity resources.
	Distributed Energy Resource Aggregation (DERA)	Economic DR (or other DER). Eligible for all market services.
	Emergency Demand Response Program (EDRP)	Reliability DR. Eligible for the Energy markets. Strike price can be higher of $500/MWh or LMP. Response to NYISO instruction is voluntary.
PJM	Price Responsive Demand (PRD)	PRD is an adjustment to capacity requirements. PRD is not dispatchable by PJM, but has sufficiently predictable response to energy market prices that it can be submitted as demand reduction in the capacity market. Initially allowed on a restricted basis, from the 2019/2020 Delivery Year onward it is unrestricted.
	Demand-Side Resources—Emergency Demand Response	There are current three variants: (1) Emergency or Pre-Emergency Capacity Only; (2) Emergency or Pre-Emergency

Continued

Table 10.2 Current ISO/RTO demand response and other market participation models which support responsive load.—cont'd

ISO	Demand response and other responsive load models	Description and eligible market services
		Full (Capacity and Energy); and (3) Emergency Energy Only
SPP	Demand-Side Resources—Economic Demand Response	Economic DR. There are currently three variants: (1) Economic (Energy, Spinning Reserves, Day-Ahead Scheduling Reserves, Regulation); Economic (Energy Only); Economic (Regulation Only)
	Demand Response Resource (DRR)—Dispatchable Demand Response (DDR)	Economic DR. Demand reduction provided by controllable load and/or a behind the meter generator that is dispatchable on a 5-minute basis. Eligible for day-ahead market, RUC and real-time energy, regulation, spinning reserves, and supplemental reserves.
	Demand Response Resource (DRR)—Block Demand Response (BDR)	Demand reduction which is not dispatchable on a 5-minute basis and is modeled and dispatched in hourly blocks. Eligible for day-ahead Economic DR.
MISO	Demand Response Resource (DRR)—Type I and Type II	Type I provides a fixed, prespecified quantity of curtailment in Energy and Operating Reserves when instructed; Type II can be continuously dispatched across a range and can provide Energy, Operating Reserves, Regulation, and Ramping Capability.
	Load Modifying Resource (LMR)	LMRs must curtail in emergencies and are counted toward Planning Reserve Margin Requirements (PRMR)
	Emergency Demand Response Resource (EDRR)	EDRRs must curtail in emergencies but are not counted toward Planning Reserve Margin Requirements (PRMR)

Table 10.2 Current ISO/RTO demand response and other market participation models which support responsive load.—cont'd

ISO	Demand response and other responsive load models	Description and eligible market services
ERCOT	Emergency Response Service (ERS)	Reliability DR models. Load reduction service deployed in late stages of grid emergency. Ten-minute and 30-min ERS options, with both weather sensitive and nonweather sensitive variants.
	Load Resources (noncontrollable)/Controllable Load Resource (CLR)	Economic DR. LRs are eligible to provide Responsive Reserve Service (RRS)—Under Frequency Relay type (see Section 5, below). CLRs are eligible to provide economic offers into RRS, contingency reserves, and energy.
CAISO	Proxy Demand Resources (PDR)	Economic DR. PDR allows third parties to offer load curtailment into the energy, spinning and nonspinning reserve markets, as well as the reliability unit commitment (PDR does not support supply of frequency regulation, although this has been considered in the stakeholder process). PDR is also eligible for Flexible Ramping Product (FRP) payments.
	PDR—Load Shift Resource (PDR-LSR)	Economic DR. PDR "bidirectional dispatch product" which can increase consumption during negative pricing.
	Reliability Demand Response Resources (RDRR)	Reliability DR. RDRR allows reliability retail demand response programs to be utilized for load curtailments

Continued

Table 10.2 Current ISO/RTO demand response and other market participation models which support responsive load.—cont'd

ISO	Demand response and other responsive load models	Description and eligible market services
		during emergencies. It is eligible for energy market compensation.
	Non-Generating Resource (NGR)—Dispatchable Demand Response (DDR)	A participation model which allows continuous response across the full operating range of storage devices; later extended to DR which can meet the same requirements. DDR can provide energy, operating reserves, and regulation.
	Participating Load	The Participating Load model allows dispatchable loads (such as agricultural pumps and pumped storage in pumping mode) to provide energy and contingency reserves. While responsive load, it is not counted as DR by California state energy agencies as it cannot be relied on to be available during peak load hours.
	Distributed Energy Resource Provider (DERP)	DER aggregation model which can include responsive load.

4. Energy markets and ramping reserves

Energy is defined as the injection or withdrawal of real power. In the short-term markets, DR provides energy by both curtailing load (equivalent to "negative" generation, also called "shed" DR) or by shifting curtailed load to other times of day (also called "shift" DR), or by providing more rapid subhourly response.[8] The energy markets are the primary source of wholesale market value (or utility production costs), typically accounting for over 80% of the cost of wholesale procurement to buyers. However, for DR, they have historically been a small source of economic benefit, since DR's offer prices are typically well above those of other resources, and hence curtail in the energy markets only when prices reach the highest levels or as triggered by reliability conditions.

4.1 Overview of energy market design

The ISO energy markets include day-ahead markets and real-time markets. There are two general categories of design elements which have been introduced and refined over the years which can improve DR participation: locational marginal pricing (LMP) and higher offer caps combined with scarcity pricing. LMP incorporates the impact of marginal congestion and losses at hundreds or thousands of locations on the grid and hence provides the most accurate price signal for where energy is most valuable. Increasing offer caps can draw more DR into the market, and scarcity pricing is used in each ISO market to further administratively set high energy prices (and ancillary service prices) during reserve shortage intervals.[9] Table 10.3 shows the number of LMP locations, as well as some details on offer caps. In some US regions, locational pricing on the distribution network is also being examined, which would reflect the LMP plus the impact of distribution-level constraints, and potentially further improve the locational incentives for responsive load. The table also shows the different economic and reliability DR programs which can participate in the energy markets, but it should be noted that while emergency programs typically submit energy bids, they are intended to curtail only during system emergencies.

Both shed and shift DR are being considered also to provide additional operational flexibility to address increasing system ramps and surplus conditions due to the increasing production by variable energy resources. The energy markets as currently designed will generally produce higher prices due to high ramps, whether in the day-ahead or real-time markets. In

---

[8] For further discussion of the terms "shift" and "shed" DR see LBNL (2017).

[9] Scarcity pricing avoids having to rely on resource offers to set high prices, while also in principle allowing for closer alignment of prices with estimates of value of lost load.

Table 10.3 Key characteristics of the ISO energy markets and ramping products and eligible DR participation models.

	PJM	ISO-NE	NYISO	MISO	SPP	ERCOT	CAISO
Energy pricing method	Locational Marginal Price (LMP)	LMP	Location-based Marginal Price (LBMP)	LMP	LMP	LMP	LMP
DR market participation models	Economic DR, Emergency DR	DRR/PRD—day-ahead and real-time energy markets	DERA, SCR, EDRP	DRR Type I and Type II, EDR	DDR/BDR—day-ahead; DDR—real time	CLR, ERS variants	PDR, PDR-LSR, RDRR
Minimum DR eligible size (MW)	0.1 MW	0.1 MW	0.1 MW	DRR Type I and II—1 MW EDR—0.1 MW	0.1 MW	CLR—0.1 MW ERS variants—0.1/0.5 MW	PDR—0.1 MW RDRR—0.5 MW
Minimum offer size (MW)	0.1 MW	0.1 MW	0.1 MW	0.1 MW	0.1 MW	CLR—0.1 MW ERS variants—0.1/0.5 MW	PDR—0.01 MW RDRR—0.5 MW
Energy offer caps ($/MWh) (market offers/cost-justified offers)	$1000/$2000	$1000/$2000	$1000/$2000	$1000/$2000	$1000/$2000	$9000	$1000/$2000
Name of ramping reserve, if implemented				Ramp Capability Product			Flexible Ramping Product
Eligible DR model for ramping reserve				DRR Type II			PDR

Sources: Modified from tables in EPRI, 2016.

addition, two ISOs, MISO and CAISO, have recently implemented a ramping reserve in the energy markets.[10] Any dispatchable energy resource is eligible to provide these reserves, and both MISO and CAISO compensate DR which is dispatchable in the real-time energy markets. However, at least in CAISO, there is currently an issue with insufficient response to 5-minute dispatches by DR designated as ramping reserves (CAISO, 2020).

4.2 DR participation in energy markets

How DR participates in the energy markets has some similar requirements to other resources, and some differences. There are two methods generally for all resources, including DR, to participate in the energy markets. For DR, a "self-schedule" requires submitting a fixed quantity (MW) of load curtailment for specific intervals in the market, and to be a "price-taker" of the market's resulting LMP for that interval. Alternatively, the DR submits a bid ($/MWh) to curtail and lets the energy auction market optimization determine in which intervals to curtail load and at what particular quantity. DR resources with bids which are selected through the market auction or in response to emergency instructions can set LMPs.[11]

Historically the minimum size for DR to participate in the energy markets has been as little as 100 kW, and such small size requirements are now required for other distributed resources (FERC 2020b).[12] In the day-ahead markets, the minimum duration for providing energy is 1 h; in real time, the markets clear on 5-minute intervals, and in one case (CAISO), also 15 min.

Most ISOs release public bid data for the energy markets, including for DR, and a few ISOs also provide aggregated price data on the DR offer curve. These generally show that the great majority of DR offers are at or

---

[10] Ramping reserves are characterized as energy products in some ISOs, and as ancillary services in others. These reserves are intended to ensure that there is sufficient ramping range to address uncertainty over ramping requirements during real-time operations.

[11] Whether self-scheduled or bid, the ISO markets are designed such that positive deviations from day-ahead schedules are financially settled at real-time prices, while negative deviations require the entity to "buy-back" its day-ahead position at real-time prices. This applies to all resources, including DR.

[12] Under new requirements from FERC, the minimum size of energy storage, whether in front or behind the meter, has been reduced to 100 kW (FERC 2018). Historically, aggregated DER minimum sizes have been in the range of 100-500 kW (e.g., CAISO 2020c), and are required to be 100 kW under FERC (2020a).

close to the offer caps, with only a small percentage likely to get dispatched. For example, for 2018—19, ISO-New England reports that in 2018 and 2019, 72% and 75% of capacity, respectively, was priced the offer cap of $1000/MWh, while offers at $200/MWh and below averaged around 6% (ISO-NE, 2020a, pg. 109). For these reasons, DR is rarely dispatched in the energy markets. With respect to total quantities of offers, PJM (2020a) reports the total quantity of economic DR offers in the energy markets: since early 2006, these have fluctuated from just under 1000 MW (in early 2006) to 3500 MW (in early 2016), and was around 1300 MW in summer 2020.

As noted, "shift" DR may obtain further opportunities for high value participation as penetration of variable energy resources increases (LBNL, 2017, Chapter 9, this volume). However, most of the United States does not currently provide much time-shift value (whether for DR or stand-alone energy storage). Table 10.4 shows some metrics for such energy time-shift in the ISO day-ahead energy markets in 2019. The calculation is for 4 h time-shift energy savings, such that a marginal DR resource shifts consumption from the highest cost 4 h to the lowest cost 4 h.[13] Any set of time-shift hours could be evaluated; the 4 h calculation is one metric. For reference, the table shows the value using aggregated LMPs at particular load zones or trading hubs; actual DR would settle at nodal prices. In 2019, the only US market offering substantial potential benefit for shift DR (or shed DR) was ERCOT, due to a large number of hours with very high prices.[14] The CAISO energy market provided the next highest

---

[13] There are a few limitations to this calculation. First, this calculation did not consider whether the highest price and lowest price hours are contiguous, although in most cases they are. If only contiguous hours were considered, then this value would be reduced. Second, if sufficient demand conducted time-shift, then the high prices would be reduced, and the low prices would increase; this calculation is for a hypothetical marginal DR resource, and does not consider this effect (but see Chapter 9).

[14] Some further explanation of the table is as follows. The third column shows the average daily 4 h time-shift ($/MW) for the full year, and the fourth column shows the annualized value in $/kW-year. For example, in the case of ISO-New England, in the first row, this means that if a 1 MW DR resource did this operation for 4 h every day, for the full year, making an average of $78/MW per day, it would earn just under $28,500 or $28.50/kW-year. The fifth and sixth columns show the revenues from the sum of the value of the top 10 days and the top 20 days. This shows that DR in ERCOT could have earned significant value in just a few days of 2019. Finally, the last two columns show the number of days with value over ≥ $100/MW and ≥ $50/MW.

Table 10.4 Selected metrics for marginal four (4) hour energy time-shift value in the U.S. ISO day-ahead markets, 2019.

	Market zone	Average 4-h time-shift value ($/MW) per day	Annualized 4-h time-shift value ($/kW-year)	Value ($) of top day 4-h time-shift		Number of days with 4-h time-shift value exceeding $50/MW and $100/MW	
				Top 10 days	Top 20 days	≥ $100/MW	≥ $50/MW
ISO-New England (ISO-NE)	Northeast Mass. zone (Boston area)	$78	$29	$1,630	$2,880	26	154
New York ISO (NYISO)	New York City zone	$69	$25	$2,434	$3,947	48	241
PJM	RTO zone	$66	$24	$2,098	$3,516	48	224
Midcontinent ISO (MISO)	Illinois Hub	$61	$22	$1,149	$2,019	7	93
SPP	SPP South Hub	$103	$38	$3,091	$5,272	161	332
ERCOT (Texas)	Houston Hub	$420	$153	$81,513	$86,799	178	325
California ISO (CAISO)	Southern California Edison Load Aggregation Point (LAP)	$163	$60	$4,778	$8,158	322	365

potential time-shift value, due to the expanding solar production creating the diurnal price patterns discussed in the book introduction and Chapter 1.[15]

5. Ancillary services

Ancillary services are procured by ISOs through two types of mechanisms: auction markets and tariff-based formulas (and in some cases, competitive solicitations). The market-based ancillary services include frequency regulation (usually called just regulation), and contingency reserves (which include spinning reserves and nonspinning reserves). Each region also has a primary frequency responsive obligation (NERC, 2017), but several ISOs (including ERCOT and CAISO) have evaluated procuring the requirement through a separate market reserve product. If they meet eligibility requirements, DR can generally provide each of these products. The tariff-based ancillary services include voltage control/reactive power and blackstart; while DR resources are not eligible for blackstart, they can provide voltage control in some situations.

Ancillary services are a small economic component of the ISO markets (generally accounting for under 5% and as little as 0.5% of wholesale costs), and currently provide minimal revenues to DR. However, there is a lot of interest in expanding DR's role for operational flexibility, and these services may account for more market value as conventional generation retires and production shifts to variable energy resources.

5.1 Participation in ancillary services

To participate in the ancillary service markets, all resources have to demonstrate that they can meet all control and telemetry requirements required for their particular resource and service. All resources providing ancillary services must follow ISO dispatch, whether the automatic control signals for frequency regulation, or economic dispatch instructions for contingency reserves in the event of contingencies. Resources can typically either submit offers into the auction for each product, or (if allowed) "self-schedule," meaning that the resource is scheduled for use with a

---

[15] In CAISO, the years 2018 and 2017 offered higher time-shift value than 2019, but due to decreased demand, the first half of 2020 has been lower than all three prior years despite the continued solar production.

predetermined quantity and compensated as a price-taker. Unlike the energy markets, the sale of ancillary services in the day-ahead market cannot be "sold back" in the event that the resource cannot perform or has other uses in real-time operations; rather, there can be refunds of payments to sellers and possibly penalties for nonperformance.[16]

5.2 Contingency reserves and primary frequency response

Contingency reserves are held to recover from unplanned outages of the largest generation or transmission elements, while primary frequency response is an automatic response to system frequency excursions commonly provided by committed synchronous generation and DR, and now being procured as a reserve product in some regions (see, e.g., EPRI, 2019).[17] Table 10.5 provides some key characteristics of these products along with DR participation models. Blanks in the table indicate that there is no such component.

If it can meet participation requirements, DR is well suited to providing contingency reserves, as the payments for these services are steady (if typically low value) over the year (see 2019 prices in Table 10.4), but the number of actual events which require load curtailment is generally small and of limited durations.[18] Fig. 10.4 shows the percentage of spinning reserves provided by DR in PJM and NYISO. Fig. 10.5 then shows the average monthly supply (MWh) (on the right axis) and average hourly monthly revenues ($/MWh) (on the left axis) to DR for spinning reserves in PJM, from 2017 to mid-2020. This illustrates the

---

[16] For example, CAISO logs the rescission of payments to DR for nonperformance as non-spinning reserves in its annual DR reports (CAISO, 2020a).

[17] Spinning reserves are provided from headroom on eligible on-line synchronized resources, inverter-based resources, and economic or reliability DR, which can offer the range of output (or load drop) which they can achieve in 10 min and sustain from 30 min to 1 h (see Table 10.5). Nonspinning reserves are from off-line generation and both economic and reliability DR which can respond within the same time requirements (but including also longer time response reserves such as 30-minute reserves) and sustain for similar periods. A generator or storage resource responds to a contingency event by injecting energy, while DR responds by curtailing. Primary frequency response is an automated response which does not require operator dispatch in seconds and minutes.

[18] For example, from Jan. 2010 to Sept. 2019, there were 233 synchronized reserve events in PJM, with an average duration of 11.9 min (Monitoring Analytics, 2020).

Table 10.5 Key characteristics of the ISO contingency reserve markets and eligible DR participation models.

	ISO-NE	NYISO	PJM	MISO	SPP	ERCOT	CAISO
Product name—10-min spinning reserve (SR)	Ten-Minute Spinning Reserve	Spinning Reserve	Synchronized Reserve (Tier 1 and Tier 2)	Spinning Reserve	Spinning Reserve	Responsive Reserve Service	Spinning Reserve
DR market participation models	DRR	DERA	DSR	DRR Type I and Type II	DDR/BDR	LR/CLR	PDR, DDR, Participating Load
Max DR contribution (%), if specified			33%	40%		60%	
Product name—10-min nonspinning reserves (NSR); other reserves	Ten-Minute Nonspinning Reserve; Thirty-Minute Operating Reserve	Nonspinning Reserve	Nonsynchronized Reserve	Supplemental reserve	Supplemental reserve	Contingency Reserve—Nonspinning Reserve	Nonspinning Reserve
DR market participation models	DRR	DERA	DSR	DRR Type I and Type II	DDR/BDR	LR/CLR	PDR, DDR, Participating Load
Forward procurement	✓ (FRM)		✓[a]				
Day-ahead procurement	✓	✓	✓	✓	✓	✓	✓
Real-time procurement	✓	✓	✓	✓	✓	✓	✓
Real-time procurement	0.1 MW	0.1 MW	0.1 MW	1 MW	1 MW	0.1 MW	0.5 MW

Minimum DR eligible size (MW)[b]	0.1 MW	0.1 MW	0.1 MW	0.1 MW	0.1 MW	0.1 MW	0.01 MW
Minimum offer size (MW)							
Offer cap ($/MW)	FRM: $14/kW–month	$1000	SR –Tier 2: resource's O&M cost + $7.50/MWh margin	$100	$100	$7000	$250
Minimum continuous energy when dispatched	30 min	30 min	30 min	30 min	1 h	30 min	30 min
Pricing locations	3 zones	3 zones	System; subzones may be defined	7 zones	4 zones	System	2 fixed zones; up to 8 add'l subzones
Average SR/NSR hourly market price ($/MW) in 2019	~$2 (SR)[c]	$4.39 (SR)/$3.82 (NSR) (NYC zone)	$3.01 (SR)/$0.24 (NSR) (RTO zone)	$2.23 (SR)/$0.54 (NSR)	$5.17 (SR)/$1.75 (NSR)	$26.61 (RRS)/$13.44 (NSR)	$7.39 (SR)/$0.75 (NSR)

Acronyms: *CR*, Contingency Reserves (ERCOT); *FRM*, Forward Reserve Market (ISO-New England); *NSR*, nonspinning reserve; *RRS*, Responsive Reserve Service (ERCOT); *SR*, spinning reserve.
[a]Day-ahead scheduling reserves (DASR) in PJM.
[b]Can be aggregations.
[c]Real-time 10-min spinning reserve price for all hours.
Source: Based on tables in EPRI, 2016.

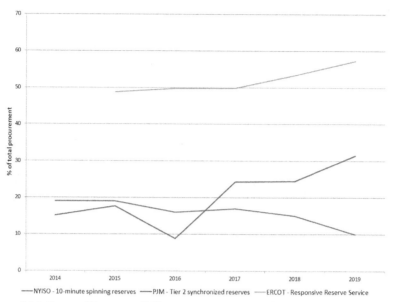

 — legend: ——NYISO - 10-minute spinning reserves ——PJM - Tier 2 synchronized reserves ——ERCOT - Responsive Reserve Service

Figure 10.4 Demand Response (DR) as a percentage (%) of total spinning reserve and responsive reserve procurement, 2014–19. *(Sources: Monitoring Analytics (2014–20); ERCOT, 2020; Potomac Economics, 2020a.)*

variability in selection of DR for this service and how revenues fluctuate over the year. In the other ISOs, DR is a smaller contributor to spinning reserves or does not currently participate (notably in the CAISO and SPP markets).

Some ISOs provide data on the sources of DR providing spinning reserves. In 2020, PJM reports that 79% of these reserves were from manufacturing, 13% from behind-the-meter generators, 5% from HVAC, with the remaining small percentages from lighting, refrigeration, and water heaters (PJM, 2020a).[19]

With respect to nonspinning reserves, these are typically lower value than spinning reserves (see 2019 prices in Table 10.5). DR provides nonspinning reserves in some ISOs, but not others. In CAISO, in 2018, DR submitted offers and self-schedules which could have provided almost 23% of the annual nonspinning reserve requirement, but awarded bids and self-schedules resulted in about 4% of the annual supply, with an average

---

[19] Data on the contribution of DERs supporting DR in this market can be found in PJM, 2020b.

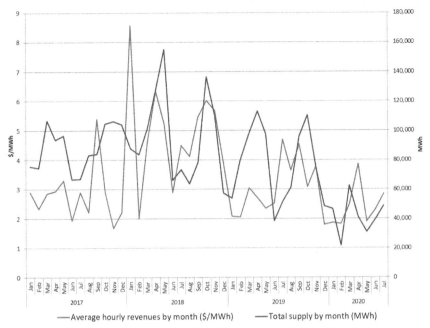

Figure 10.5 Average hourly revenue ($/MWh) per month (left axis) and total monthly supply (MWh) (right axis) of DR providing Synchronized Reserves in PJM, 2017–2020 (July). *(Sources: PJM 2020 (and earlier years).)*

hourly revenue of $6.08/MW.[20] In contrast, DR does not provide any nonsynchronized reserves in PJM (PJM, 2020a) or ERCOT (ERCOT, 2020).

In addition to DR operated from behind the meter, there can be other wholesale dispatchable loads which provide contingency reserves. These include bulk energy storage, such as pumped storage hydro, during pumping operations, and agricultural water pumps. In these cases, these resources can stop pumping on contingency reserve time frames, and then resume later. In some markets, such resources can be significant suppliers of contingency reserves at times of day.

ERCOT's Responsive Reserve Service encompasses three services: Primary Frequency Response (PFR), Fast Frequency Response, and Under Frequency Relay (UFR). Fig. 10.4 also shows the percentage of DR

---

[20] This estimate was derived from DR data, and the nonspinning reserve requirements reported in CAISO DMM (2019).

providing the UFR component of Responsive Reserve Service in ERCOT. UFR comprises 60% of the RRS total (ERCOT, 2020).[21]

5.3 Frequency regulation

Frequency regulation—or simply regulation—is energy injected (regulation up) or withdrawn (regulation down) in response to automatic control signals, typically on time-frames of 2—6 s, for purposes of maintaining system frequency in between economic dispatch (which takes place on 5-minute intervals in the ISOs). For DR, load follows these control signals in the opposite direction from generation: decreasing load to provide upwards regulation and increasing load to provide downwards regulation.

Table 10.6 provides details on several key characteristics of the regulation markets in each ISO, including identifying the relevant DR participation models, average procurement requirements and recent prices.[22] Each ISO has one or more DR participation models (although CAISO does not currently allow its PDR model, the most widely used economic DR model, to provide Regulation).

In PJM, since 2014, when data were first provided publicly, DR has provided between 0.6% (2014) and 2.4% (2019) of the Regulation

---

[21] This maximum supply from load resources is "subject to a minimum capacity of 1150 MW required from Resource providing RRS using Primary Frequency Response."

[22] As is shown, the ISOs all procure a range for providing regulation from eligible resources, which is called regulation capacity (MW), and all except for ERCOT pay for performance on the basis of actual movement (or mileage) by the selected resources. In some ISOs, the Regulation product combines upwards and downwards regulation, whereas in others, up and down are different products. The latter approach can facilitate participation by certain types of DR. For example, a hot water heater may be turned off to provide Regulation Up, but may not have as much capability to provide Regulation Down when turned back on (that is, the amount of Regulation Up may not be symmetrical with Regulation Down). In addition, there are differences in the control signals which can facilitate DR participation when utilizing behind-the-meter storage, as well as differences in procurement methods and the pricing rules which affect the value of the product (see last row of Table 10.5).For example, both PJM and CAISO have offered energy neutral control signals on cycles of 15—30 min, to facilitate participation by short-duration storage but which can be followed by DR resources as well (when eligible). Otherwise, generation resources have historically been expected to have a minimum continuous capability to follow the control signal for 30—60 min.

Table 10.6 Key characteristics of the ISO regulation markets and eligible DR participation models.

	ISO-NE	NYISO	PJM	MISO	SPP	ERCOT	CAISO
Regulation capacity name	Regulation Capacity	Regulation Capacity	Regulation Capability	Regulating Capacity	Regulation-Up Service, Regulation-Down Service	Regulation Service, Fast Responding Regulation Service	Regulation Capacity
Performance component name	Regulation service	Regulation movement	Regulation performance	Regulating mileage	Regulation-up mileage, Regulation-down mileage	N/A	Regulation mileage up Regulation mileage down
"Fast" Regulation Signal	Yes	No	Yes—RegD	No	No	Yes—FRRS	No
DR Market participation model	None	DERA	DSR	DRR Type II	DDR	LR/CLR	DDR
Minimum eligible DR resource size (MW)[a]	N/A	0.1 MW	0.1 MW	1 MW	1 MW	0.1 MW	0.5 MW
Minimum offer size (MW)	0.1 MW	0.1 MW	0.1 MW	0.1 MW	0.1 MW	0.1 MW	0.01 MW

Continued

Table 10.6 Key characteristics of the ISO regulation markets and eligible DR participation models.—cont'd

	ISO-NE	NYISO	PJM	MISO	SPP	ERCOT	CAISO
Max DR contribution (%), if specified			25%				
Avg. hourly Reg. Requirement (MW)	50–200	175–300	525–800	~400	Varies	100–700	350–500
Day-ahead procurement	✓	✓		✓	✓		✓
Real-time procurement		✓	✓	✓	✓	✓	✓
Offer cap ($/MW)	$100 for Reg. Capacity; $10 for Reg. Service	$1000	$100	$500	$500	$9000	$250
Average hourly market price ($/MW) in 2019	$21.96 (Regulation capacity)	$9.08	$16.30	$8.15	RU–$9.45 RD–$7.60	RU–$23.14 RD–$9.06	RU–$13.27 RD–$11.74

DR acronyms see Table 10.1; *RD*, Regulation Down; *RU*, Regulation Up.
[a]Can be aggregations.
Source: Modified from tables in EPRI, 2016.

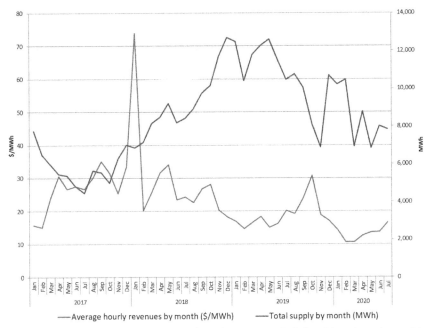

Figure 10.6 Average hourly revenue ($/MWh) per month (left axis) and total monthly supply (MWh) (right axis) of DR providing Regulation in PJM, 2017–2020 (July). *(Sources: PJM 2020 (and earlier years).)*

requirement, with small incremental increases year by year (Monitoring Analytics, 2015–20, also Chapter 11 this volume). Fig. 10.6 shows the average monthly supply (MWh) (on the right axis) and average hourly monthly revenues ($/MWh) (on the left axis) to DR for Regulation in PJM, from 2017 to mid-2020. In 2020, 84% of this supply was from residential hot water heaters and 10% from behind-the-meter batteries, with the remainder from generators and HVAC systems (PJM, 2020a). In MISO, large industrial processes have also provided regulation. In the other regions, frequency regulation has not yet attracted substantial participation by DR.[23] Despite the large quantity of behind-the-meter batteries already installed in California, as of this writing, Regulation market participation has not yet begun.

---

[23] ERCOT lists supply of Regulation in its DR reports (ERCOT, 2020), but at least in recent years, this is energy storage resources which are categorized as Controllable Load Resources (CLR) when they are charging.

The regulation markets clearly provide expansion potential for types of DR, particularly as conventional generation retires and behind-the-meter batteries and hybrid vehicles become more common. There is significant untapped capability from hot water heaters identified in technical potential studies. However, DR will also be competing with many other existing and new sources of this service, notably transmission and distribution-connected batteries, which may significantly suppress market prices by the end of the decade.[24]

6. Resource adequacy requirements and capacity markets

In the United States, each of the ISOs, with the exception of ERCOT, and most utilities outside those markets have Resource Adequacy (RA) requirements based on a forecast of future peak demand and reserves (denominated in MW). Load-serving entities (LSEs) are allocated a load-ratio share of capacity obligations, which must be fulfilled on different time-frames consistent with reliability and product rules (see Table 10.7). Depending on the region, compliance mechanisms include centralized capacity markets, bilateral contracts, and self-supply; Table 10.7 provides some basic details.[25] DR is currently participating either directly in the

---

[24] For example, transmission and distribution-connected lithium ion batteries have already accounted for over 40% of annual Regulation procurement in PJM at times over the past few years (see Monitoring Analytics, 2015—20), and also eventually there will be substantial potential supply from variable energy resources with smart inverters.

[25] Three ISOs (PJM, ISO-New England, and NYISO) have centralized capacity markets, in which all RA capacity—whether procured through the auctions, self-supplied, or bilaterally contracted—is cleared at transparent market-clearing prices. In PJM and NYISO, supply offers clear against a downward sloping demand curve. MISO has a capacity market design in which about 65% of regional capacity is cleared through a 2-month ahead capacity auction (see discussion in Potomac Economics, 2020b). Load-serving entities in SPP and CAISO primarily meet state resource adequacy requirements through self-owned or bilaterally contracted resources (CAISO also conducts some very limited residual capacity procurement through its backstop mechanism). In these bilateral markets, there is not a transparent market clearing price for capacity but there may be reporting of bilateral prices (e.g., CPUC, 2019). In California, state regulators and the CAISO have also added a "flexible capacity" requirement, which establishes that some of the procured capacity must be capable of supporting the evolving system ramps being created largely by solar energy production. As noted, ERCOT does not have a resource adequacy requirement (and is called an "energy-only" market), but conducts periodic assessments of supply adequacy which are used to set energy market bid caps, which are higher than in other markets (see Table 10.2).

Table 10.7 Key characteristics of the ISO capacity market or procurement mechanisms and eligible DR participation models.

Name	ISO-NE	NYISO	PJM	MISO	CPUC/CAISO
Name	Forward Capacity Market (FCM)	Installed Capacity (ICAP) Market	Reliability Pricing Model (RPM)	Planning Resource Auction (PRA)	CPUC Resource Adequacy program/CAISO Capacity Procurement Mechanism
Sloped demand curve	yes (only for system-wide)	yes	yes	No	no
Max. price	Net CONE	Net CONE	Net CONE	CONE	Bilateral market
Number of locations	3 zones	4 zones	9 zones	10 zones	8 local areas
Minimum offer price rule	yes	yes	yes	no	no
Must-offer requirement	DAM energy and RTM energy	DAM energy	DAM energy	DAM energy	DAM energy
Forward Period	3 years ahead	1 month prior	3 years ahead	2 months prior	3 years ahead
Commitment Period	1 year	Seasonal (6 months), monthly	1 year	1 year	1 year (local, flexible), monthly (system)
DR market participation models	FCM: On-Peak and Seasonal Peak Resources; DRR–PRD	DERA, SCR	PRD, Emergency Demand Response	LMR, DRR Type I and Type II, Emergency Demand Response	CPUC—utility programs; CAISO—PDR, RDRR
DR operational requirements	FCM On-peak: available during summer and winter daily peak hours; FCM Seasonal peak: seasonal peak hours	SCD: available all hours; triggered by operational procedures. 4 h minimum response	Availability all year, from 6 to 15 h depending on DR model and season.	LMR: Availability all hours during summer; 4 h minimum deployment.	PDR: System and local RA is available all hours, all markets, 4 h continuous dispatch; Flexible RA is available for 3 h during designated ramp hours; at least one start per day and 5 dispatches per month. RDRR: responds only during reliability conditions up to 15 events and/or 48 h per term

capacity markets, or is counted as an off-set against LSE capacity obligations. Capacity revenues provide the majority of DR market compensation or valuation as an offset. Other types of DER which may include responsive load are also now becoming eligible as capacity resources.

In the ISOs where there are centralized capacity markets with transparent prices (PJM, New York ISO, and ISO-New England), procurement of capacity can range from 10% to 20% of total wholesale market costs. Capacity prices are highest when new capacity resources need to clear the auctions, and are lowest when prices are set by existing resources (i.e., during periods of low load growth or excess supply). When energy market revenues decline, capacity market revenues in principle act to support the "going forward" costs of existing capacity resources needed for reliability, to avoid mothballing and retirement. While outside the scope of this chapter, a complicating factor now emerging across the US regions is that new RA capacity is being introduced through state policies, such as renewable portfolio standards (RPS) and more recently, storage policies.

6.1 Participation in resource adequacy requirements and capacity markets

As noted above, DR resources have been the primary type of distributed resource participating to date in the capacity markets. There are a number of emergency and economic DR models for participation as capacity resources (see Tables 10.2 and 10.7); key requirements include the number of allowed load interruptions (which can range from a small number to unlimited), the availability windows (which range from all hours to daily peak hours to seasonal peak hours), and the minimum duration of the interruption (which can range from 1 h to 10 h or more, depending on the resource type and ISO).

Fig. 10.7 shows percentage (%) of DR which was counted toward LSE capacity obligations or cleared the PJM, ISO-NE, and NYISO capacity markets in recent years (similar percentages are evident in MISO, but consistent data for multiple years was not found). The PJM and ISO-NE markets clear 3 years ahead, so the quantities are shown for the delivery year, while the NYISO market is seasonal and month-ahead (and hence the future periods shown in the figure have not yet been addressed).

For aggregated DER which includes responsive load, participation rules for capacity markets are still in development (FERC 2020a), but given that aggregations can consist of multiple types of resources, there is the opportunity to combine resources which on their own would not achieve the

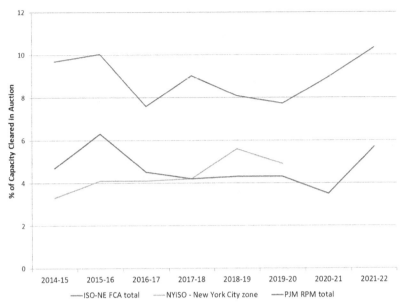

Figure 10.7 Demand Response (DR) as a percentage (%) of total capacity cleared in US central capacity market auctions, 2014/15–2021/22. *(Sources: PJM, LLC, 2020a; ISO-New England, 2020a; Potomac Economics, 2020a.)*

minimum requirements to qualify as capacity resources. The emerging practice has been to require DER aggregations to meet the same capacity requirements as conventional generation, rather than the limited DR availability obligations shown in Table 10.7 (e.g., CAISO, 2020c). ISOs also differ in the types of other distributed demand–side resources which can qualify as capacity resources. For example, in several markets (PJM, MISO and ISO-New England), energy efficiency projects have been incorporated (see also discussion in Chapter 11).

In regions with RA requirements but without centralized capacity markets, utilities, and in some cases, third-party aggregators, can still count DR toward resource adequacy obligations. For example, Fig. 10.8 shows the DR counted toward the CPUC RA requirement from the three large investor owned utilities in California—Southern California Edison (SCE), Pacific Gas and Electric (PG&E), and San Diego Gas and Electric (SDG&E)—from 2008 to 19. This is both from utility DR programs and from the Demand Response Auction Mechanism (DRAM), a CPUC pilot auction market procurement implemented by these utilities to facilitate participation by third-party aggregators, which has operated from 2016 to

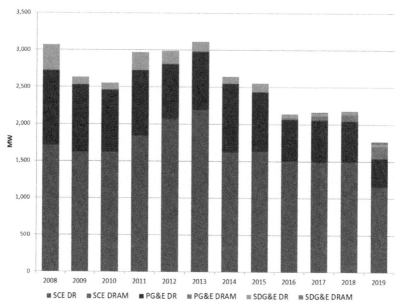

Figure 10.8 California IOU and DRAM DR counted toward CPUC Resource Adequacy requirements, 2008–19. *(Source: CPUC, 2019.)*

19. In 2018, the sum of these procurements constituted about 4.4% of total RA resources committed (CPUC, 2019).[26] Another feature of the California RA program is that it is the first to create specific rules for DR to qualify as "flexible RA," which is required to be able to support the expanding CAISO power system ramps for a minimum of 3 h (see Table 10.7 and CAISO, 2020c).

7. DR for T&D deferral and multiple uses

While this chapter is focused on wholesale markets, DR is also eligible for transmission and distribution upgrade deferral, and in principle to simultaneously provide other wholesale and retail services, which in the United States has been called "multiple uses" (FERC, 2017). As noted in

---

[26] Note that as community choice aggregators (CCAs) progressively make up much of the load in the California investor-owned utility territories, the cost of the utility procured DR is allocated to the CCAs through distribution charges, and the CCAs are also undertaking limited DR procurement (not shown in the figure). However, the figure shows an overall decline in DR contribution to RA since 2013.

Table 10.1, FERC established requirements for such consideration in transmission planning in Order 890 and Order 1000, and more recently has provided policy guidance on multiple use projects which could incorporate DR and other "nonwires alternative" resources (FERC, 2017).

While policy and market interest in nonwires alternatives to high-voltage transmission upgrades has been growing around the country, the ISOs with the most experience to date include CAISO and NYISO. It is important to note that nonwires alternatives to ISO transmission upgrades are generally not eligible as ISO transmission assets with rate recovery through transmission access charges (see, e.g., discussion in CAISO, 2020d, 2013). Hence, they may be approved by the ISOs within the transmission planning process as a least cost alternative solution to a transmission need, but will need to obtain state regulatory approval for cost recovery or obtain sufficient revenues otherwise through bilateral contracts and wholesale market value.

For DR and other DER used to avoid distribution network upgrades, in the United States, state regulators will typically have oversight over the preferred set of distributed resources which utilities or third-parties select as well as the procurement methods and rate incentives. A recent example is the large distribution deferral project by Consolidated Edison in the New York City area approved by state regulators, which included procurement of DR through auctions (NYPSC, 2014; Consolidated Edison, 2016).[27]

While multiple use nonwires alternatives, such as DR, which can also provide wholesale services, are still a very small portion of the wholesale markets (in part due to regulatory barriers), they present coordination issues which will require new market rules which can be enforced by each off-taker of the project's services. The most obvious requirement is a set of enforceable rules which establish the constraints on operations of the aggregated DER, including responsive load. For example, just like all resources interconnected to the transmission or distribution systems, if DERs offer wholesale reliability services (such as ancillary services and operating as a capacity resource) they must perform according to the market rules or face penalties (e.g., FERC, 2020a).

---

[27] Reports on this project can be obtained through the New York Public Service Commission website under Case 14-E-0302.

8. Emergence of DER models which can utilize responsive load

As discussed above, aggregated DER models are the latest wholesale (and retail) market opportunity for flexible demand typically within an aggregation of other resources, including customer-sited generation and stationary or non-stationary storage, at a single or multiple locations. Several chapters in this book examine variants on such models (see also Sioshansi, 2019; FERC 2020a). The primary differences between conventional DR and these new DER models are that the latter can inject energy back to the grid, may have more flexibility to provide multiple services, and may be subject to other locational requirements. In addition, when delivering wholesale services, the aggregated DER provider may in the future interface directly with the ISO or via the distribution system operator platforms now being evaluated in some regions (e.g., New York Department of Public Service DPS, 2020). In the United States, these new aggregated DER models are most advanced in California and New York, but are being widely tested and are now subject to the additional federal requirements noted earlier (FERC 2020a).

9. Other considerations for DR wholesale market participation

This section examines further factors which affect the DR sector over time, and are considered by regulators and ISOs when designing DR programs and market participation models.

9.1 State of competition in DR

A large number of different types of entities are participating in the DR market, including most obviously retail load-serving entities and the newer entrants, such as third-party aggregators. While in principle, where it is permitted, the third-party DR sector should have low barriers to entry, in practice, it has become fairly concentrated in most regions (although this does not imply that DR aggregators can exercise market power, since they are typically small percentages of the wholesale markets). This is for a number of reasons. There is a financial cost to different types of wholesale market participation (or in utility retail customer programs), such as provision of ancillary services which requires metering and telemetry, which may inhibit smaller market participants, or limit them to services with the lowest participation costs, but which may in turn provide the least

Table 10.8 Demand Response market competition metrics and other data.

ISO	Report	Metrics and other data
PJM	Monitoring Analytics (2020)	- DR product Herfindahl–Hirschman Index (HHI); by capacity and by hourly MWh - Percentage share of 4 largest firms - Number of CSPs and participants in each CSP
ISO–New England	ISO-NE (2020a)	- number of entities with capacity supply obligations and share of total
New York ISO	NYISO (2019a,c)	- participation by DR agent type and program
California ISO	CAISO (2020a)	- Names of DR providers and number of resource IDs (sometimes redacted in public version)
	CAISO DMM (2020b)	- total number of DR resources

operational flexibility. For DR aggregators, customer acquisition and maintenance can be costly, and difficult to maintain during periods of lower market prices. These factors affect trends in market concentration in the DR sector.

Several ISOs and some state regulators (e.g., CPUC) provide data on the state of competition in the DR sector, some including the names of the companies. Table 10.8 provides selected examples.

9.2 Market design risk

ISO market rules and operations are complicated, can vary in key details between ISOs, and at times can change dramatically in general or specific details. These include changes to market design for particular products, software upgrades, and changes to the DR rules themselves (see also Chapter 11). In some cases, changes in rules have increased the benefits to DR, and in other cases, they have resulted in declines. FERC's annual assessments provide brief observations on these trends, as do some of the ISO reports. In some cases, the correlation between DR and particular changes to market rules is clear. For example, in NYISO, changes to auditing and baseline methodologies in 2010 led to a 50% reduction over a few years in reliability DR eligible as capacity using the SCR model, and the quantity has not recovered since, in part also due to low capacity prices (Potomac Economics, 2020a). Similarly, in ISO-NE, changes to capacity resource performance requirements applicable to all resources are thought to have influenced a 48% decrease in DR participation from 2017 to 2018 (FERC, 2020b).

9.3 Market price risk

Market prices for energy, ancillary services, and capacity change over time, sometimes sharply from year to year, creating revenue uncertainty for wholesale DR which are not otherwise receiving payments through utility programs.

Since most DR derives the majority of its value from the capacity markets, it will be particularly sensitive to prices in those markets. Most US regions have had large planning reserve margins and low load growth over the past decade, which have kept capacity prices down, although these margins have been declining. With respect to energy prices, the trend in recent years has been toward lower energy value due primarily to low natural gas prices and in some regions, the increased penetration of variable energy resources. These trends in market prices influence the fluctuations in DR shown in the data above.

In California, the impact of solar energy on energy prices is commonly understood to provide opportunities for operationally flexible resources (e.g., see Chapter 1 this volume); however, the average energy price in the past few years has continued to decline and the net load prices during the ramp periods, with a few exceptions, are not particularly higher than the highest hourly peak load prices in prior years. Hence, there are as yet few additional high prices to trigger DR in the ramp periods. At the same time, price differentials between the solar production hours and the early evening hours are generally growing (although not linearly), possibly creating opportunities for DR which can time-shift, as shown in Table 10.4.

9.4 DR performance

DR resources have experienced a range of performance quality, from high to low, which varies by program incentives, market requirements, and the type of market service. A detailed assessment of this performance is beyond the scope of this chapter. Some ISOs and their market monitors have provided public quantitative analysis of performance. Many of the problematic issues, such as inadequate measurement and verification, and insufficient penalties for nonperformance, have been addressed through market rule revisions, and some continue to be evaluated (e.g., PG&E, 2020). Chapter 11 provides a review of such performance issues in PJM. Elsewhere, public data may be less accessible, but, for example, see the

CAISO hourly data released annually on DR not responding to dispatch instructions (see, e.g., CAISO, 2020a).

9.5 Impacts of state policies on DR

In the United States, due to the lack of federal clean energy policy in recent years (with the primary exception of federal tax incentives), state policy-makers have provided most of the impetus for the clean energy transition through a variety of mechanisms, including different types of clean energy procurement standards (notably renewable portfolio standards), additional financial incentives, and resource planning requirements.

State policies have both direct and indirect impacts on DR. Notably, state regulators create the requirements and provide oversight of utility DR programs, which still comprise the majority of DR capability in most regions of the country, and hence regulatory review is often integral to how those programs evolve and are compensated, and whether they are migrated into wholesale markets. In some states, notably California, state regulators have directed large quantities of financial incentives to behind-the-meter resources, including energy storage which can support DR, and are also attempting to foster additional third-party DR participation in the wholesale markets by providing enhanced procurement mechanisms. Several studies forecast that the combination of technology evolution, state regulatory innovation, and modifications to wholesale market design will facilitate greater DR participation as decarbonization increases (e.g., LBNL, 2017; Chapter 9, this volume).

At the same time, there is an indirect effect on DR in that state policies are both simultaneously creating the conditions for new DR opportunities (such as needs for additional operational flexibility) but also providing the incentives or mandates for resources which compete with DR as clean energy peaking resources (such as both in-front-of-the-meter and behind-the-meter solar PV and more recently energy storage) or flexibility resources (such as energy storage). These state policies vary widely, and hence interactions with existing or potential DR will be different in each case (an example of a research study which examines these interactions is Mills and Wiser (2014)).

10. DR forecasting for planning functions

For researchers and commercial entities interested in forecasts of DR in the United States, many of these can be found in resource planning scenarios undertaken by different entities in the US power sector. These include utility and state regulator long-term resource plans, ISO and other regional planner load and resource forecasts, and commercial trade forecasts. These results can be compared to the findings from research studies of the technical and economic potential for DR such as those reviewed in this book. While this chapter will not survey this broad area, a few examples are described below and summarized in Table 10.9.

10.1 DR in recent integrated resource plans

With respect to utility resource planning, integrated resource planning (IRP) began to be utilized extensively in the 1980s by state regulators to improve utility planning and in particular to integrate demand-side management into planning. With the expansion of competitive wholesale markets in beginning in the late 1990s, such planning was initially terminated in the market regions. However, with the growth in state clean energy policies in the 2000s, variants on resource planning and forecasting across all electric power domains are again being utilized extensively across the country, both outside the wholesale market regions, but also within them, including by state regulators in some restructured states (such as California and Connecticut).

A sampling of recent IRPs which evaluate scenarios with high renewable penetration and decarbonization reveals that there is as yet, little evidence of expectations of a significant increase in DR, even in of higher renewable penetration. Rather, these planning studies currently foresee a very large increase in wind and solar power along with energy storage, mostly in the form of lithium-ion batteries. The solar power provides much of the peak-shaving during daylight hours in the scenarios, while the energy storage is used to address the "net load" peaks. This simply leaves little opportunity for DR, and it is reflected in the results. However, readers should note that this survey is indicative, and not comprehensive.

Examples of these types of results can be found in the ongoing IRPs in California. The California legislature reestablished IRPs on a state-wide basis in 2015, and currently has the most extensive and advanced clean energy policies of any large US region. The CPUC conducts an IRP proceeding on a 2-year cycle for its jurisdictional LSEs (the first two cycles are 2018 and 2020), while the CEC provides oversight over the publicly owned utilities, which conduct their IRPs on a 1—5 year cycle (readers can follow this process over time at the referenced websites).[28] The remainder of this discussion will focus on the CPUC process. In each cycle, the CPUC identifies a Reference System Plan (RSP) on a 10-year horizon for LSEs to use as the basis for formulating their procurement plans. The agency utilizes a publicly available capacity expansion model,[29] which includes a range of other scenarios in addition to the RSP, including several which extend to full decarbonization in 2045. In its current iteration, the model includes "shed" DR as an endogenous selection, while "shift" DR is an exogenous addition.[30] For the 2020 cycle, the CPUC RSP scenario has a carbon emissions target of 46 million metric tons per year by 2030, which requires an effective 60% RPS. The RSP resource mix resulting from the capacity expansion model, and approved by the CPUC, selects only 222 MW of additional "shed" DR, in addition to the 2195 MW of existing shed DR in the baseline set of resources, for a total of 2481 MW of DR available over the next decade (CPUC, 2020). In contrast, the model selects almost 11 GW of new storage by 2030 such that between already approved battery storage and existing pumped storage, the region would have around 14.7 GW of energy storage, along with 46 GW of solar generation.[31] While the CPUC allows LSEs to make their own determinations about the resources to select, the many complex issues associated with DR are noted in some of the resulting IRPs (see, e.g., PG&E, 2020). Since IRP results typically change from year to year, these results are presented as illustrative of a case where additional DR is apparently being largely displaced as a summer peaking resource by solar and as a flexibility resource by energy storage.

10.2 DR in ISO load and resource forecasts

In the United States, ISOs and RTOs are not resource planners, but they may develop or modify forecasts of load and future resources for

---

[28] For more details, see the CPUC IRP webpage at https://www.cpuc.ca.gov/irp/and the CEC IRP webpage at https://www.energy.ca.gov/rules-and-regulations/energy-suppliers-reporting/clean-energy-and-pollution-reduction-act-sb-350-1. All the submitted IRPs in the state will be found on these regulatory agency websites.

[29] This model can be accessed at https://www.cpuc.ca.gov/General.aspx?id=6442459770.

[30] Note that additional shed DR is expected to cost from \$75-\$200/kW-year.

[31] Moreover, although much more speculative, the same scenario extended to 2045 under an effective 97% RPS also selects no more DR, but has almost 60 GW of energy storage in the final year.

transmission planning, operational planning and seasonal reliability assessments. Hence, they are another source of information on DR forecasts.

With respect to transmission planning, ISO and RTOs use load forecasts and resource scenarios as inputs into the planning process, in particular to evaluate the need for transmission expansion to meet public policy goals and otherwise planned new resources. In some cases, these scenarios are derived directly from state planning assumptions (e.g., the CAISO uses the CPUC's IRP scenarios described above in its transmission planning process, and NYISO uses state policy scenarios to evaluate the need for public policy-driven transmission projects), whereas in others, the ISO builds its own forecasts and may use its own capacity expansion modeling (e.g., MISO). DR forecasts are generally part of long-term load forecasting for reliability purposes (e.g., references to PJM, ISO-NE, NYISO), and may enter into transmission planning processes. Table 10.6 shows that in the selected ISO DR forecasts reviewed for this chapter, there is not a significant forecast of increased DR.

Of the ISOs reviewed, only MISO had developed a range of DR forecasts reflecting alternative scenarios. MISO's Transmission Expansion Planning (MTEP) process uses scenario-based resource forecasts based on capacity expansion modeling and stakeholder inputs. In the 2019 approved plan (MISO, 2019), there are four 15 year scenarios, ranging from 15% to 35% renewable penetration (see Table 10.9). Depending on the scenario, DR selection in the plan increases slightly over current capacity, ranging from 6 to 8 GW by 2033.

The conclusion of this brief survey of resource forecasts is that there is still a gap between the expectations of state regulators, ISOs and utilities regarding the potential for a major expansion of DR, and the results of research studies on DR technical potential. This may be the result of one or more factors, potentially in combination, such as limitations of the modeling methods of resource planning and conservatism on the part of planners with respect to assumptions of further DR expansion, even as the resource mix is otherwise substantially transformed to achieve decarbonization.

---

[32] Energy Efficiency (EE) is cited when converted into peak reductions (MW). DER capacity forecasts may not be cited in particular IRPs, but may be factored into the load forecast. For EE and DG with asterisks (*), the measure is based on contributions to peak reductions.

Table 10.9 Selected results on DR in recent US resource planning and load forecasting.

Study author and function	Scenario name	Starting year/Ending year	Renewable energy penetration in ending year (%)	DR in starting year (GW)	DR in ending year (GW)	Other demand-side solutions and energy storage in ending year (GW)[32]
California Public Utilities Commission (CPUC), 2020 IRP; see also PG&E (2020)	2020 Reference System Plan (RSP)	2020/2030	60	2.2	2.4	ES—14
Duke Energy Carolinas (2019) IRP	DSM summer peak reductions	2020/2034	13	1.1	1.15	EE—0.775 ES—0.2
Portland General Electric (2019) IRP	DR summer	2020/2025	Not specified	0.032	0.108 −0.333	EE—0.157—0.167
MISO (2019) MISO Transmission Expansion Plan (MTEP19)	Limited Fleet Change	2019/2033	15	6	6	DER—6
	Continued Fleet Change		20	6	7	EE—6 DER—7
	Accelerated Fleet Change		35	6	7	EE—7 DER—7
	Distributed and Emerging Technologies		25	6	8	EE—7 DER—10 ES—2
PJM—2019 Regional Transmission Expansion Planning (RTEP)/Load Forecast Report	Load Forecast Report Demand Resources	2019/2029	Not specified—based on queue	8.1	9.4	Not specified

Continued

Table 10.9 Selected results on DR in recent US resource planning and load forecasting.—cont'd

Study author and function	Scenario name	Starting year/ Ending year	Renewable energy penetration in ending year (%)	DR in starting year (GW)	DR in ending year (GW)	Other demand-side solutions and energy storage in ending year (GW)[32]
ISO-NE (2020b) CELT report	Load Forecast, Demand Response with forward capacity obligations	2020/ 2029	Not specified	0.494	0.592	EE—3* DG—0.265*
NYISO, 2019 Load and Capacity Data	Summer Capability Period—Planning Assumptions	2019/ 2029	6.5	1.3	1.3	EE—4.3 DER—1.3 ES—3.8

11. Note on data sources

The data in this chapter have been compiled from a variety of sources, many of which track the evolution of DR in the United States on a consistent basis. The US Energy Information Agency (U.S. EIA, 2020) gathered utility demand-side management program data from 2001 to 12, after which it transitioned to category of Demand Response in its Annual Electric Power Industry Report, Form 861. All these data are archived on the EIA website (U.S. EIA, multiple years). FERC's annual assessments (e.g., FERC, 2020b) utilize the EIA data as well as many other sources.

For the ISOs, the details of DR market participation rules, and the associated wholesale market product designs, are found in the ISO tariffs and business practice manuals. Generally, data on DR market participation and performance can be tracked in the state of the market reports issued by the internal and external ISO market monitors as well as in other data on DR released by the ISOs. Despite an attempt at common market metrics by FERC, the data on DR released by the ISOs are not standardized, and there can be inconsistencies between the same report and between market monitoring reports and the data shown in other ISO DR reports. Of the ISOs, PJM provides the most data consistently on DR in its market monitoring reports (Monitoring Analytics, multiple years). Some market monitors also provide data on the structure of the DR providers, such as concentration of the firms. Monthly and/or annual DR program reports are also released by PJM (2020), ISO-New England (2020c), ERCOT (2020) and CAISO (2019). These differ widely in the type of data provided, which generally includes data on the type of services provided but may also include detail on sources of DR and its distribution around the region.

For insight into the actual impact of analysis of DR technical potential, the many utility and state resource planning reports are also a useful reference; these are typically updated on an annual or semiannual basis, depending on the state and utility.

12. Conclusions

This chapter has provided historical perspective on DR in the United States over the past decade or more, with an emphasis on wholesale market participation and some review of results in recent long-term load and resource planning. The primary objective of the chapter has been to organize data and sources, such that readers can continue to compile information on the further evolution of DR during the coming period of rapid electric power system transformation.

The general finding is that in several US wholesale markets, DR has already proven its capability to provide a large percentage of peaking capacity resources and operating reserves. However, these results are uneven across the country, and fluctuate, sometimes substantially, from year to year. The reasons relate to changes in wholesale market rules, market prices, and the impact of state policies. Moreover, while some ISOs and regions continue to experience increases in DR penetration, others have been experiencing declines, for a variety of reasons. One reason is that energy, ancillary service, and capacity prices have generally been low in recent years, due to low natural gas prices and the increasing penetration of gas-fired resources and renewables in some regions. At the same time, in much of the country, reserve margins have been high until recently.

Of particular interest, in California, operating conditions have emerged which could provide opportunities for the types of flexible DR discussed in this book (and evaluated in technical DR potential studies of California and elsewhere). However, recent California resource plans suggest that the commercial potential for flexible DR could be adversely affected by the combination of other resources, such as energy storage, which are entering the market at decreasing costs and with financial incentives. Many research studies have begun to examine these interactions between alternative types of clean energy resources, and more research is needed.

Similarly, in other US regions, with a few exceptions, planners and load forecasters are generally not identifying major increases in DR over the coming 10—15 years even as renewable generation greatly increases. These forecasts reflect a current snapshot, which may change over time as more new types of resources are utilized, and hence do not necessarily refute the projections in other chapters of this book for much greater utilization of flexible demand as policies and markets evolve. But they do suggest the need to more fully evaluate how other new clean energy resources affect the potential for expanded DR.

References

California Independent System Operator (CAISO), Annual Report Evaluating Demand Response Participation in the CAISO, Under Docket No. ER06-615-000, Accessed 2020a, released annually since 2007, [Online] Available: www.ferc.gov. [*Note: most market data redacted until 2016 report*].

California Independent System Operator (CAISO). 2019 Annual Report on Market Issues and Performance. 2020b CAISO Department of Market Monitoring.

California Independent System Operator (CAISO), PDR-DERP-NGR Summary Comparison Matrix, No Date, Accessed 2020c [Online] Available: http://www.caiso.com/Documents/ParticipationComparison-ProxyDemand-DistributedEnergy-Storage.pdf.

California Independent System Operator (CAISO), 2019—2020 Transmission Plan, March 25, 2020d [Online] Available: http://www.caiso.com/planning/Pages/TransmissionPlanning/2019-2020TransmissionPlanningProcess.aspx.

California Independent System Operator (CAISO), "Consideration of Alternatives to Transmission or Conventional Generation to Address Local Needs in the Transmission Planning Process," September 4, 2013, [Online] Available: http://www.caiso.com/Documents/Paper-Non-ConventionalAlternatives-2013-2014TransmissionPlanningProcess.pdf.

California Public Utilities Commission (CPUC), 2019—2020 Electric Resource Portfolios to Inform Integrated Resource Plans and Transmission Planning, Rulemaking 16-02-007, April 6, 2020 [Online] Available: https://docs.cpuc.ca.gov/PublishedDocs/Published/G000/M331/K772/331772681.PDF.

California Public Utilities Commission (CPUC), 2018 Resource Adequacy Report, August 2019 [Online] Available: https://www.cpuc.ca.gov/RA/.

Consolidated Edison, Brooklyn Queens Demand Management Demand Response Program Guidelines, June 28, 2016 [Online] Available: https://conedbqdmauctiondotcom.files.wordpress.com/2016/03/bqdm-dr-program-overview-6-28-161.pdf.

Duke Energy Carolinas, 2019 Integrated Resource Plan Update, September 2019 [Online] Available: https://starw1.ncuc.net/NCUC/ViewFile.aspx?Id=40bbb323-936d-4f06-b0ba-7b7683a136de.

Electric Power Research Institute (EPRI), Ancillary Services in the United States: Technical Requirements, Market Designs and Price Trends. EPRI, Palo Alto, CA: 2019. 3002015670. [Online] Available: www.epri.com.

Electric Power Research Institute (EPRI), Wholesale Electricity Market Design Initiatives in the United States: Survey and Research Needs. EPRI, Palo Alto, CA: 2016. 3002009273. [Online] Available: www.epri.com.

ERCOT, Annual Report of Demand Response in the ERCOT Region, Multiple Years, Accessed 2020, typically published March of subsequent year [Online] Available: http://www.ercot.com/services/programs/load.

Federal Energy Regulatory Commission (FERC), 2017. Utilization of Electric Storage Resources for Multiple Services When Receiving Cost-Based Rate Recovery. Policy Statement.

Federal Energy Regulatory Commission (FERC), 2018. Order 841, Electric Storage Participation in Markets Operated by Regional Transmission Organizations and Independent System Operators. 162 FERC, 61,127, 18 CFR Part 35, 2018 [Docket Nos. RM16-23-000; AD16-20-000].

Federal Energy Regulatory Commission (FERC), 2020a. Order 2222, Participation of Distributed Energy Resource Aggregations in Markets Operated by Regional Transmission Organizations and Independent System Operators. 172 FERC, 61, 247, 18 CFR Part 35, September 17, 2020 [Docket No. RM18-9-000].

Federal Energy Regulatory Commission (FERC), 2020b. Assessments of Demand Response and Advanced Metering, Accessed in 2020, published 2006—2019 [Online] Available: https://www.ferc.gov/industries/electric/indus-act/demand-response/dem-res-adv-metering.asp.

ISO New England and New England Power Pool, Revisions to Implement Full Integration of Demand Response, July 27, 2017 [Online] Available: https://www.iso-ne.com/static-assets/documents/2017/07/prd_implement_full_integration.pdf.

ISO-New England, 2019 Annual Markets Report, ISO New England Inc. Internal Market Monitor, May 26, 2020a.

ISO-New England, 2020 CELT Report: 2020—2029 Forecast Report of Capacity, Energy, Loads and Transmission, May 2020b [Online] Available: https://www.iso-ne.com/system-planning/system-plans-studies/celt.

ISO New England, Demand Resources Working Group, Monthly Statistics Report, Demand Resource Strategy Department, Accessed 2020c [note: statistics published periodically since 2004, monthly report structure in recent years] [Online] Available: https://www.iso-ne.com/markets-operations/markets/demand-resources.

ISO/RTO Council (IRC), North American Wholesale Electricity Demand Response Program Comparison, 2018 ed., Updated November 2018, [Online] Available: https://isorto.org/reports-and-filings/.

Lawrence Berkeley National Laboratory (LBNL), 2025 California Demand Response Potential Study, Final Report on Phase 2 Results, 2017 [Online] Available: https://www.cpuc.ca.gov/General.aspx?id=10622.

Midcontinent ISO (MISO), 2019 MISO Transmission Expansion Planning (MTEP), December 2019 [Online] Available: https://www.misoenergy.org/planning/planning/mtep-2019-/.

Midcontinent ISO (MISO), Demand Response Business Practices Manual, BPM-026-r5, Effective Date: July 01, 2020 [Online] Available: https://www.misoenergy.org/legal/business-practice-manuals/.

Mills, A., and R. Wiser, Strategies for Mitigating the Reduction in Economic Value of Variable Generation with Increasing Penetration Levels. Lawrence Berkeley National Laboratory, 2014 [Online]. Available: https://emp.lbl.gov/sites/all/files/lbnl-6590e.pdf.

Monitoring Analytics, LLC, State of the Market Reports for PJM, Accessed 2020 for multiple years [Online] Available: http://www.monitoringanalytics.com/reports/PJM_State_of_the_Market/.

New York Public Service Commission (NYPSC), Order Establishing Brooklyn/Queens Demand Management Program, Case 14-E-0302, Issued December 12, 2014.

New York Department of Public Service (DPS), 2020. Webpage on "REV — Demonstration Projects", Accessed July 2020 [Online] Available: http://www3.dps.ny.gov/W/PSCWeb.nsf/All/B2D9D834B0D307C685257F3F006FF1D9?OpenDocument.

New York Independent System Operator (NYISO), Inc. NYISO 2019 Annual Report on Demand Response Programs (See Also Prior Years), 2019a [Online] Available: filed with FERC.

New York Independent System Operator (NYISO), Inc., 2019b. Proposed Tariff Revisions Regarding Establishment of Participation Model for Aggregations of Resources, Including Distributed Energy Resources, and Proposed Effective Dates; Docket No. ER19-2276-000, filed June 27, 2019.

New York Independent System Operator (NYISO), Inc. Semi-annual Reports on New Generation Projects and Demand Response Programs, Docket Nos. ER03-647-000 and ER01-3001-000, May 30, 2019c.

North American Electric Reliability Corporation (NERC). BAL-003-1 Frequency Response Obligation Allocation for Operation Year 2017, North American Electric Reliability Corporation (NERC), 2017.

Pacific Gas & Electric (PG&E), Integrated Resource Plan 2020, Prepared for the California Public Utilities Commission, September 2020.

PJM, LLC. 2020 Demand Response Operations Markets Activity Report: August 2020, James McAnany, PJM Demand Side Response Operations, August 7, 2020a (annual report) [Online] Available: https://www.pjm.com/~/media/markets-ops/dsr/2020-demand-response-activity-report.ashx.

PJM, L.L.C., February 2020b. 2019 Distributed Energy Resources (DER) that Participate in PJM Markets as Demand Response, PJM Demand Side Response Operations (annual report) [Online] Available: https://www.pjm.com/~/media/markets-ops/demand-response/2019-der-annual-report.ashx.

PJM, LLC. Demand Response Strategy, PJM Interconnection, June 28, 2017 [Online] Available: https://www.pjm.com/~/media/library/reports-notices/demand-response/20170628-pjm-demand-response-strategy.ashx.

PJM, LLC. The Evolution of Demand Response in the PJM Wholesale Market, PJM Interconnection, October 6, 2014 [Online] Available: https://www.pjm.com/~/media/library/reports-notices/demand-response/20141007-pjm-whitepaper-on-the-evolution-of-demand-response-in-the-pjm-wholesale-market.ashx.

Portland General Electric, 2019 Integrated Resource Plan, July 2019 [Online] Available: https://www.portlandgeneral.com/our-company/energy-strategy/resource-planning/integrated-resourceplanning.

Potomac Economics, 2019 State of the Market Report for the MISO Electricity Markets, June 2020b [Online] Available: https://www.potomaceconomics.com/markets-monitored/miso/.

Potomac Economics, 2019 State of the Market Report for the New York ISO Electricity Markets, May 2020a [Online] Available: https://www.potomaceconomics.com/markets-monitored/new-york-iso/.

Sioshansi, F. (Ed.), 2019. Consumer, Prosumer, Prosumager: How Service Innovations Will Disrupt the Utility Business Model. Elsevier.

U.S. Energy Information Agency (EIA), Annual Electric Power Industry Report, Form EIA-861 Detailed Data Files, Accessed 2020 [Online] Available: https://www.eia.gov/electricity/data/eia861/.

CHAPTER 11

What is limiting flexible demand from playing a bigger role in the US organized markets? The PJM experience

Joseph E. Bowring
Monitoring Analytics, LLC, Eagleville, PA, United States

1. Introduction

Well-functioning wholesale power markets require active participation by both supply and demand. For a number of reasons, however, the development of the demand side of wholesale power markets has lagged behind the development of the supply side. The demand side of the wholesale power markets has significant potential but the current market designs are a barrier to the realization of that potential, topics covered in Chapters 9 and 10. In PJM, the design of wholesale market demand-side programs has been modeled, to a significant extent, on the premarket regulated utility curtailment tariff design. As a result, in PJM the design of wholesale market demand-side programs has largely been a failure. That approach should be ended. But, an effective demand side is essential to the competitive functioning of wholesale power markets. In PJM, and wholesale power markets generally, a new, market-based approach is required in order to maximize the opportunities for the economic participation of demand-side resources. An improved market design will provide the opportunity for load to benefit from the option to flexibly respond to both energy and capacity prices and to receive the benefits of that response.

The market approach should include:
- Treatment of demand as demand;
- Actual real-time locational marginal pricing;
- Improved assignment of capacity costs;
- Better load forecasting by the RTO;
- Provision of real-time and forecast information by the RTO to customers; and

Variable Generation, Flexible Demand
ISBN 978-0-12-823810-3
https://doi.org/10.1016/B978-0-12-823810-3.00007-8

259

- Better coordination with retail tariffs.

This chapter, which is focused on PJM, is organized as follows:

- Section 2 provides background on PJM;
- Section 3 reviews salient points in the development of demand-side programs in PJM;
- Section 4 addresses the current status of demand-side resources in PJM and the impacts of those resources on PJM markets;
- Section 5 reviews the significant issues with demand-side programs in PJM;
- Section 6 describes a market-based approach to an effective demand side in PJM; followed by the chapter's conclusions.

2. PJM basics

The PJM Interconnection, L.L.C. (PJM) operates a centrally dispatched, competitive wholesale electric power market that, as of March 31, 2020, had installed generating capacity of 185,189 megawatts (MW) and 1049 members including market buyers, sellers, and traders of electricity in a region including more than 65 million people in all or parts of 13 states (Delaware, Illinois, Indiana, Kentucky, Maryland, Michigan, New Jersey, North Carolina, Ohio, Pennsylvania, Tennessee, Virginia, West Virginia) and the District of Columbia (Fig. 11.1).

PJM introduced energy pricing with cost-based offers and market-clearing nodal prices on April 1, 1998, and market-clearing nodal prices with market-based offers on April 1, 1999. PJM operates the Day-Ahead Energy Market, the Real-Time Energy Market, the Capacity Market, the Regulation Market, the Synchronized Reserve Market, the Day-Ahead Scheduling Reserve (DASR) Market, and the Financial Transmission Rights (FTR) Market.[1]

Table 11.1 shows selected 2018 and 2019 summary statistics for PJM markets.[2]

---

[1] More details are available in the *2019 State of the Market Report for PJM* (March 12, 2020) http://www.monitoringanalytics.com/reports/PJM_State_of_the_Market/2019. shtml.

[2] See the *2019 State of the Market Report for PJM* (March 12, 2020) http://www. monitoringanalytics.com/reports/PJM_State_of_the_Market/2019.shtml.

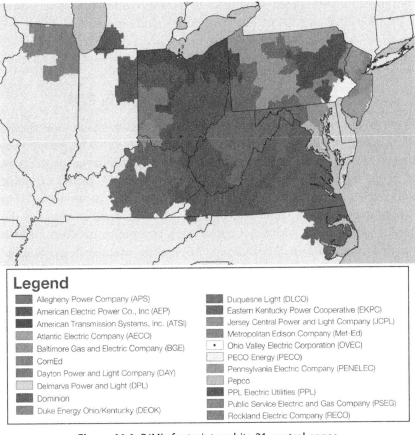

Legend

Allegheny Power Company (APS)
American Electric Power Co., Inc (AEP)
American Transmission Systems, Inc. (ATSi)
Atlantic Electric Company (AECO)
Baltimore Gas and Electric Company (BGE)
ComEd
Dayton Power and Light Company (DAY)
Delmarva Power and Light (DPL)
Dominion
Duke Energy Ohio/Kentucky (DEOK)

Duquesne Light (DLCO)
Eastern Kentucky Power Cooperative (EKPC)
Jersey Central Power and Light Company (JCPL)
Metropolitan Edison Company (Met-Ed)
Ohio Valley Electric Corporation (OVEC)
PECO Energy (PECO)
Pennsylvania Electric Company (PENELEC)
Pepco
PPL Electric Utilities (PPL)
Public Service Electric and Gas Company (PSEG)
Rockland Electric Company (RECO)

Figure 11.1 PJM's footprint and its 21 control zones.

Table 11.1 PJM market summary statistics.

	2018	2019	Percent change
Average hourly load (MW)	90,308	88,120	(2.4%)
Average hourly generation (MW)	94,236	93,434	(0.9%)
Peak load (MW)	147,042	148,228	(0.8%)
Installed capacity at December 31 (MW)	185,952	184,744	(0.6%)
Load weighted average real time LMP ($/MWh)	$38.24	$27.32	(28.6%)
Total congestion costs ($ million)	$1309.9	$583.3	(55.5%)
Total uplift credits ($ million)	$198.1	$88.6	(55.3%)
Total PJM billing ($ billion)	$49.79	$39.20	(21.3%)

Table 11.2 shows the components of the total cost of wholesale power in PJM for 2018 and 2019.[3] Energy, capacity, and transmission charges are the three largest components of the total price per MWh of wholesale power, comprising 97.3% of the total price per MWh in 2019.

3. Development of demand side programs in PJM

The demand-side programs in the PJM market have their roots in the utility curtailable load programs that preceded organized wholesale power markets. Those curtailable load programs were largely a device used in cost of service studies in regulated utility rate cases to reduce the allocation of both generation and transmission capacity costs to large customers and thus to reduce their rates. Those cost of service studies typically allocated the costs of generating capacity and transmission facilities on the basis of load during a small number of peak hours. The studies assumed that curtailable customers would be offline during those hours with the result that the fixed costs allocated to those customers were reduced significantly as were the corresponding revenue requirements. Curtailable customers received significant rate reductions as a result. In concept, curtailable customers could be called to reduce load during an emergency. But curtailable customers were seldom, if ever, actually called on to reduce load under the utility curtailment programs.

The early demand-side programs in PJM mimicked the utility curtailable load programs. Many of the issues in the design of PJM demand-side programs, and in the design of RTO/ISO demand-side programs more generally, are a result of simply retaining the core concepts of the prior design rather than recognizing that the introduction of markets required a rethinking of the design. Reliance on the assumption that only a very small number of hours determined investment in generating units and the design of the program as an emergency only program were two persistent features. Some large customers benefited from the rate design associated with the utility programs and continued to benefit from the repetition of that design in the PJM demand-side programs.

The first PJM demand-side program was the Active Load Management (ALM) program under the PJM Capacity Credit Market (CCM) capacity

---

[3] See the *2019 State of the Market Report for PJM* (March 12, 2020). http://www.monitoringanalytics.com/reports/PJM_State_of_the_Market/2019.shtml.

Table 11.2 Total PJM price per MWh by category: 2018 and 2019.

Category	2018 $/MWh	2018 ($ million)	2018 Percent of total	2019 $/MWh	2019 ($ millions)	2019 Percent of total	Percent change
Load-weighted energy	$38.24	$30,253	61.4%	$27.32	$21,088	54.3%	(28.6%)
Capacity	$13.02	$10,298	20.9%	$11.27	$8700	22.4%	(13.4%)
Capacity	$12.97	$10,260	20.8%	$11.25	$8686	22.4%	(13.2%)
Capacity (FRR)	$0.00	$0	0.0%	$0.00	$0	0.0%	(0.0%)
Capacity (RMR)	$0.05	$38	0.1%	$0.02	$14	0.0%	(62.1%)
Transmission	$9.47	$7494	15.2%	$10.39	$8019	20.6%	(9.7%)
Transmission service charges	$8.81	$6966	14.1%	$9.75	$7524	19.4%	(10.7%)
Transmission enhancement cost recovery	$0.57	$454	0.9%	$0.55	$427	1.1%	(3.6%)
Transmission owner (schedule 1A)	$0.09	$74	0.2%	$0.09	$69	0.2%	(5.0%)
Transmission Seams Elimination Cost Assignment (SECA)	$0.00	$0	0.0%	$0.00	$0	0.0%	(0.0%)
Transmission facility charges	$0.00	$0	0.0%	$0.00	$0	0.0%	(0.0%)
Ancillary	$0.80	$632	1.3%	$0.72	$557	1.4%	(9.6%)
Reactive	$0.41	$321	0.7%	$0.44	$339	0.9%	(8.2%)
Regulation	$0.18	$145	0.3%	$0.12	$90	0.2%	(36.3%)
Black start	$0.08	$65	0.1%	$0.08	$65	0.2%	(2.1%)
Synchronized reserves	$0.06	$50	0.1%	$0.04	$34	0.1%	(29.5%)
Nonsynchronized reserves	$0.02	$14	0.0%	$0.02	$12	0.0%	(11.6%)
Day-ahead Scheduling Reserve (DASR)	$0.05	$37	0.1%	$0.02	$18	0.0%	(51.2%)
Administration	$0.50	$399	0.8%	$0.51	$394	1.0%	1.2%

Continued

Table 11.2 Total PJM price per MWh by category: 2018 and 2019.—cont'd

Category	2018 $/MWh	2018 ($ million)	2018 Percent of total	2019 $/MWh	2019 ($ millions)	2019 Percent of total	Percent change
PJM administrative fees	$0.47	$371	0.8%	$0.47	$365	0.9%	(0.7%)
NERC/RFC	$0.03	$25	0.1%	$0.03	$27	0.1%	(9.1%)
RTO startup and expansion	$0.00	$2	0.0%	$0.00	$2	0.0%	(3.3%)
Energy uplift (operating reserves)	$0.23	$185	0.4%	$0.11	$88	0.2%	(51.2%)
Demand response	$0.01	$5	0.0%	$0.00	$4	0.0%	(26.6%)
Load response	$0.01	$5	0.0%	$0.00	$3	0.0%	(48.4%)
Emergency load response	$0.00	$0	0.0%	$0.00	$1	0.0%	(0.0%)
Emergency energy	$0.00	$0	0.0%	$0.00	$0	0.0%	(0.0%)
Total price	**$62.27**	**$49,265**	**100.0%**	**$50.33**	**$38,850**	**100.0%**	**(19.2%)**
Total load (GWh)	**791,094**			**771,929**			**(2.4%)**
Total billing ($ billions)	**$49.27**			**$38.85**			**(21.1%)**

market design.[4] Under the ALM program, load serving entities (LSEs) could reduce their peak loads based on contracts with individual customers to reduce specified amounts of load when PJM declared an emergency. ALM credits reduced LSEs' capacity obligations and their capacity costs and reduced the total PJM capacity obligation correspondingly. While ALM continued the structure of the utility curtailable programs, ALM was on the demand side of the market and was an actual demand-side program.

Effective June 1, 2007, the RPM capacity market design was implemented in PJM, replacing the CCM capacity market design that had been in place since 1999.[5] The RPM was a forward-looking, annual, locational market, with a must offer requirement for capacity, an associated obligation to offer energy in the market every day, and mandatory participation by load. CCM was a daily, single-price, voluntary balancing market that included less than 10% of total PJM capacity. Under RPM, there was an administratively determined demand curve that, with the supply curve based on capacity market offers, determined market prices. Under CCM the demand for capacity was defined by participant buy bids.

The beginning of the treatment of demand as supply in PJM began with the RPM capacity market design. The PJM ALM program was replaced by the PJM load management (LM) program. Under ALM, providers had received an MW credit which offset their capacity obligation. The LM program introduced two RPM-related products. Demand response (DR) resources were load resources that offered into capacity auctions as capacity supply and received the relevant locational clearing price. Interruptible Load for Reliability (ILR) resources were load resource that were not offered into the capacity auctions, but received the final, zonal ILR price determined after the close of the second incremental auction. Under the

[4] See the *2007 State of the Market Report*, Volume 2: Section 5, "Capacity Market." http://www.monitoringanalytics.com/reports/PJM_State_of_the_Market/2007.shtml. J. Bowring, "Capacity Markets in PJM," *Economics of Energy & Environmental Policy*, Vol. 2, No. 2. Pages 47–64. IAEE 2013. J. E. Bowring, "The Evolution of the PJM Capacity Market: Does It Address the Revenue Sufficiency Problem?" in *Evolution of Global Electricity Markets: New Paradigms, New Challenges, New Approaches,* Ed. F. P. Sioshansi, Oxford, UK: Elsevier Ltd., 2013. J. E. Bowring, "The Evolution of PJM's Capacity Market," in Competitive Electricity Markets: Design Implementation, Performance. Ed. F.P. Sioshansi, Oxford, UK: Elsevier Ltd., 2008.
[5] See the *2007 State of the Market Report*, Volume 2: Section 5, "Capacity Market." http://www.monitoringanalytics.com/reports/PJM_State_of_the_Market/2007.shtml.

ALM program in the daily capacity market, resources could be nominated a day in advance of the capacity market day. Under RPM, DR resources had to be offered into the auction 3 years in advance of the delivery year, while ILR resources had to be certified only 3 months prior to the delivery year.

The performance obligation of DR was extremely limited compared to the 8760 h per year performance obligation of all other capacity resources. DR was required to respond only during the hours of 12:00 p.m. to 8:00 p.m. on nonholiday weekdays during the months of June through September; only for a maximum of 10 times during that period each year; and only for a maximum of 6 h when called.

ILR was a flawed concept. Allowing participants to choose whether to offer an unlimited amount of ILR in the market after the price was known was not consistent with basic market principles. The ILR product was a free option to sell an unlimited quantity of DR at the market clearing price which was known 3 years in advance of deciding whether to sell, just prior to the delivery year and with only a limited obligation to perform.

PJM made ongoing ad hoc changes to the capacity market design in an effort to accommodate the inefficient rules governing the role of demand side resources and to offset the negative impacts of demand side resources on the capacity market. But the ad hoc changes created separate, distortionary effects on the capacity market.[6]

In an attempt to account for the potential impact of ILR on the capacity market, PJM made an arbitrary reduction of 1.2% in the administrative demand curve for capacity in the 2011/2012 capacity auction run in May 2008. The ILR product was eliminated on March 26, 2009, for the 2012/2013 capacity market auction run in May 2009.[7] When ILR was eliminated, PJM made another arbitrary reduction in the administrative demand

---

[6] While PJM plays a large role in market design, the final market design is the result of PJM positions within the complex PJM stakeholder decision-making process that affects market design choices based on the narrow self-interest of groups of participants, and ultimately FERC decisions. FERC's policy preferences play a significant role in market design. FERC has been a strong supporter of the existing demand-side market design. The Independent Market Monitor's (IMM) role includes making market design recommendations to improve efficiency and competitiveness both in the PJM stakeholder process and at FERC.

[7] 126 FERC ¶ 61,275 (2009).

curve of 2.5% in each BRA. For the 2012/2013 BRA, the 2.5% reduction resulted in the removal of 3343.3 MW from the demand curve. The stated rationale for this reduction in demand was to permit procurement of DR, asserted to be a short lead time resource, in later Incremental Auctions (IA) for the same delivery year. This reduction was referred to as the Short-Term Resource Procurement Target. PJM eliminated the 2.5% offset in 2016 for the 2019/2020 and subsequent BRAs. Removing a portion of demand affects prices at the margin, which is where the critical signal to the market is determined. The reduction in demand suppressed capacity prices below the competitive level for multiple delivery years.[8]

Energy efficiency (EE) investments, if effective, reduce energy and capacity usage and the benefits of that reduced energy usage automatically flow to the customer through the normal functioning of markets. The inclusion of EE in the capacity market as a supply side resource was an ad hoc change to the capacity market design intended to reflect the inadequacies in PJM's load forecasting method, particularly the reliance on load in prior years as a predictor of expected load. Rather than directly addressing the multiple issues with the load forecasting method, PJM included EE as a supply side capacity resource. The concept was that the impact of the EE would not be fully reflected in the PJM load forecast for 4 years and that therefore the customer's peak load used to assign capacity costs would not be fully reduced for 4 years. Any specific EE resource was eligible to offer in the PJM Capacity Market for only four consecutive years. EE resources became eligible to offer in the PJM capacity market auctions as a supply side resource effective with the capacity market run in 2009 for the delivery year 2012/2013.[9]

The inclusion of EE in the capacity market as supply required another ad hoc modification to the capacity market design. PJM increased the demand for capacity in the capacity auction (BRA) to offset the impact of EE resources. This EE add back adjustment to the capacity market demand curve was intended to avoid double counting, as EE reductions for the delivery year are reflected in the final load forecast model for that delivery

---

[8] "Analysis of the 2012/2013 RPM Base Residual Auction," (August 6, 2009). http://www.monitoringanalytics.com/reports/Reports/2009/Analysis_of_2012_2013_RPM_Base_Residual_Auction_20090806.pdf.

[9] *PJM Interconnection, L.L.C.,* Tariff Amendments, Docket No. ER09-412-000 (December 12, 2008).

year. But PJM's actual implementation of the add back does affect capacity market prices because it is not implemented correctly.[10]

The result was the creation of an inferior capacity market product, demand-side resources, that had only extremely limited performance obligations but were paid the same price as generation capacity resources that were required to be available in every hour of the year, and that displaced generation resources in the capacity market. The result was also the creation of another unnecessary and inferior capacity market product, EE. EE had no performance obligations and no meaningful measurement and verification but nonetheless affected capacity market results.[11]

Not surprisingly, the role of demand-side resources in the capacity market increased dramatically as a result of the favorable rules. Demand-side participation was 962.9 MW for delivery year 2010/2011, increasing to 1826.6 MW for delivery year 2011/2012, to 8740.9 MW for delivery year 2012/2013 and to 10,779.6 MW for delivery year 2013/2014, an increase of 1019% in 3 years (See Table 11.3). Demand-side resources increased from 0.7 of cleared capacity resources to 6.7% of cleared capacity resources over this period.

In 2010, after the capacity market cleared for 2013/2014, PJM recognized that the flawed, existing demand-side products had significant, negative market impacts. The limited obligation of DR to respond in real time resulted in DR crowding out resources without such limitations and potentially putting reliability at risk. PJM recognized the negative impact on reliability of the inferior demand-side product. PJM filed at FERC on December 2, 2010, to create two new demand-side products, in recognition of the fact that "the current product definition is no longer adequate to ensure that reliability requirements are met."[12] FERC approved the filing.[13]

---

[10] "Analysis of the 2021/2022 Base Residual Auction," (December 17, 2017) http://www.monitoringanalytics.com/reports/Reports/2017/IMM_Analysis_of_the_20202021_RPM_BRA_20171117.pdf.

[11] The IMM has provided detailed analysis of each base auction in the capacity market and made associated recommendations. For a complete list of related reports, see the 2019 State of the Market Report for PJM (March 12, 2020), page 261. http://www.monitoringanalytics.com/reports/PJM_State_of_the_Market/2019.shtml.

[12] PJM Interconnection, L.L.C., Docket No. ER11-2288-00 (December 2, 2010).

[13] 134 FERC ¶ 61,066 (2011).

Table 11.3 Demand-side MW in the PJM capacity market.

	UCAP (MW)		
	DR RPM	Total RPM	DR percent
	Cleared	Cleared	Cleared
2007/2008	127.6	129,409.2	0.1%
2008/2009	559.4	130,629.8	0.4%
2009/2010	892.9	134,030.2	0.7%
2010/2011	962.9	134,036.2	0.7%
2011/2012	1826.6	134,182.6	1.4%
2012/2013	8740.9	141,283.9	6.2%
2013/2014	10,779.6	159,844.5	6.7%
2014/2015	14,943.0	161,205.0	9.3%
2015/2016	15,453.7	173,519.4	8.9%
2016/2017	13,265.3	179,749.0	7.4%
2017/2018	11,870.7	180,590.5	6.6%
2018/2019	11,435.4	175,996.0	6.5%
2019/2020	10,703.1	177,064.2	6.0%
2020/2021	9445.7	173,688.5	5.4%
2021/2022	11,415.5	165,770.5	6.9%

But PJM left the existing limited product in place, created a complex new design to mitigate the limitations of the existing product, and created a new product with limitations that also had to be mitigated, rather than addressing the fundamental issues associated with the role of demand side in the PJM markets. The fundamental issues included that the demand-side product: was treated as a supply side resource; was required to respond for only extremely limited times; did not have a must offer obligation in either the energy market or the capacity market; and was permitted to offer marketing plans into capacity market auctions rather than actual physical resources as required for all other capacity resources.

As a result, effective for the 2014/2015 delivery year, the PJM capacity market design incorporated annual and extended summer DR product types, in addition to the existing limited DR product type.[14,15] Each DR product type was subject to a defined period of availability, a maximum number of interruptions, and a maximum duration of interruptions. The rule changes also included setting a maximum level of Limited DR and a

---

[14] 134 FERC ¶ 61,066 (2011).
[15] *PJM Interconnection, L.L.C.,* Docket No. ER11-2288-00 (December 2, 2010).

maximum level of Extended Summer DR cleared in the auction. The new rules defined a Minimum Annual Resource Requirement and a Minimum Extended Summer Resource Requirement for PJM and for local deliverability areas (LDAs) with a separate demand (VRR) curve.[16] Annual Resources include generation resources and Annual DR. The Minimum Resource Requirements are targets established by PJM to ensure that enough Annual Resources were cleared in order to address the reliability concerns created by the Extended Summer and Limited DR products. The targets were also intended to ensure that enough Annual Resources plus Extended Summer Resources were cleared to address the reliability concerns created by the Limited DR product. The reliability risk of relying on either the Extended Summer or Limited DR products was a result of the fact that inferior resources not required to respond at all times of the year when needed for reliability were substituting directly for more reliable generating resources. The new rules established a maximum level of Limited DR and a maximum level of Extended Summer DR because additional purchases of these products were not consistent with reliability.

But PJM's implementation of these new products was done incorrectly.[17] The quantity of Limited DR and Extended Summer DR was not capped, as intended, at a fixed MW level. As implemented, if the Minimum Annual Resource Requirement constraint were binding, rather than being capped, the Extended Summer and Limited DR products would fill in the balance of capacity needed to meet the VRR curve. As a fix, effective with the 2017/2018 Delivery Year, the Minimum Annual and Extended Summer Resource Requirements were replaced by Limited and Sub-Annual Resource Constraints.[18] The Limited Resource Constraint limits the quantity of Limited DR that can be procured, and the Sub-Annual Constraint limits the quantity of Limited DR and Extended Summer DR that can be procured. The modifications to the rules for the 2017/2018 Delivery Year reduced the impact of Limited and Extended Summer DR on market outcomes compared to what the impact would have been without the rule changes.

---

[16] The LDAs for which Minimum Resource Requirements are established were subsequently revised. See 135 FERC ¶ 61,102 (2011).
[17] "Analysis of the 2017/2018 Base Residual Auction," (October 14, 2014) Pg 16. http://www.monitoringanalytics.com/reports/Reports/2014/IMM_Analysis_of_the_2017_2018_RPM_Base_Residual_Auction_20141006.pdf.
[18] 146 FERC ¶ 61,052 (2014).

Another significant issue was identified in the design of the PJM demand-side programs in 2011. On February 4, 2011, PJM and PJM's Independent Market Monitor (IMM) issued a Joint Statement on the double counting issue related to the measurement and verification of compliance by demand-side resources. The statement made clear that PJM and the IMM believed that this was an issue of market manipulation that resulted in overpayments to CSPs and could have a negative effect on reliability. CSPs are aggregators, termed curtailment service providers.

Double counting exists when a customer pays for capacity equal to its defined peak load (PLC), receives credit when called on for its reduction from its PLC to its reduced load, but the CSP which enrolls it in the PJM DR program receives credit for the same reduction between the actual load and PLC. It is double counting because the customer pays for capacity only equal to its PLC, but the CSP is paid a capacity payment for the customer reducing its load from actual load to PLC. In other words, the customer, through the CSP, is selling back to PJM capacity that it did not purchase.

The issue was resolved by an FERC order on November 4, 2011, that accepted PJM's proposed solution, which eliminated double counting and provided for a transition.[19]

The PJM capacity market was significantly redesigned again in 2017. The Capacity Performance design changes to the capacity market were partially effective beginning with the 2018/2019 Delivery Year and fully implemented in the 2020/2021 Delivery Year. As a result, there was a complex transition in the 2018/2019 and 2019/2020 Delivery Years in which there were multiple demand-side products, some following the old rules and some following the new rules. Effective for the 2018/2019 and the 2019/2020 Delivery Years, there were two types of demand resource and energy efficiency resource products included in the RPM market design: Base Capacity Demand Resources and Base Capacity Energy Efficiency Resources; and Capacity Performance Resource Annual Demand Resources and Capacity Performance Annual Energy Efficiency Resources.[20,21] Effective with the 2020/2021 Delivery Year, the Capacity Performance product is the only capacity product type, but includes two

---

[19] 137 FERC ¶ 61,108.
[20] 151 FERC ¶ 61,208.
[21] "Reliability Assurance Agreement Among Load Serving Entities in the PJM Region," Article 1.

season types, annual and summer demand resources and annual and summer energy efficiency resources.

4. Current status and impacts of demand-side resources in PJM markets

The primary source of revenue for demand-side resources in PJM markets is and has been the capacity market. More than 95% of market revenue for demand-side resources has come from the PJM capacity markets. Fig. 11.2 shows all revenue from PJM demand response programs by market for 2008 through 2019.[22] In 2019, demand resource revenue, which includes capacity and emergency energy revenue, accounted for 98.8% of all revenue received by demand response providers, the economic program for 0.2%, synchronized reserve for 0.6%, and the regulation market for 0.5%. The changes in cleared quantities reflected the underlying changes in the rules governing the potential revenues and risks associated with selling demand-side resources in the capacity market.

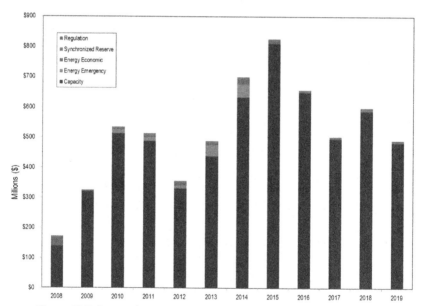

Figure 11.2 Demand response revenue by market: 2008 through 2019.

[22] See the *2019 State of the Market Report for PJM*, (March 12, 2020). http://www.monitoringanalytics.com/reports/PJM_State_of_the_Market/2019.shtml.

The inclusion of demand resources and energy efficiency resources has had a significant impact on the PJM capacity market by suppressing prices and by displacing resources with performance obligations for every hour of the year. For example, based on actual auction clearing prices and quantities and make whole MW in the last BRA, the inclusion of demand resources and energy efficiency resources resulted in a $1,729,462,670 reduction, 15.7%, in capacity market revenues for the 2021/2022 BRA compared to what capacity market revenues would have been without any demand resources or energy efficiency resources, from $11,030,339,776 to $9,300,877,106.[23]

The 2021/2022 BRA was the third BRA held using the EE add back mechanism. The EE add back mechanism has had a significant impact on the PJM Capacity Market. For example, based on actual auction clearing prices and quantities and make whole MW, the inclusion of energy efficiency resource offers and the EE add back MW, resulted in a 10.1% increase in revenues for the 2021/2022 BRA compared to what RPM revenues would have been if energy efficiency projects were reflected in the demand and EE resources did not participate on the supply side. The add back mechanism, as implemented, overcompensated for the EE reductions even though it was intended to be neutral. Correctly implemented, the add back mechanism would have meant that EE did not affect the capacity market prices although EE would displace other capacity resources.

The accuracy of the peak load forecast has also had a significant impact on the PJM capacity market. The load forecasts used to define the demand for capacity have been overstated and the result has been that capacity market prices have been too high. An analysis of the auctions for the 2014/2015 through 2018/2019 delivery years shows that the peak load forecasts for the Third Incremental Auctions, the final forecast before the actual delivery year, have been on average 5.8% lower than the peak load forecast used to define the demand for capacity in the corresponding BRA. Using PJM's peak load forecast for the 2021/2022 Base Residual Auction resulted in a 42.9% increase in RPM revenues for the 2021/2022 RPM Base Residual Auction compared to what revenues would have been using a load forecast that is 5.8% below the PJM peak load forecast. As a related

---

[23] "Analysis of the 2021/2022 Base Residual Auction," (December 17, 2017) http://www.monitoringanalytics.com/reports/Reports/2017/IMM_Analysis_of_the_20202021_RPM_BRA_20171117.pdf.

result, PJM has systematically procured substantially more capacity than required for reliability. PJM should improve its load forecasting.

5. PJM demand side programs: A failed design

Demand-side programs in the PJM capacity market have been characterized by a series of fundamental market design failures. The core market design failure was to treat demand-side resources as if they were supply-side resources rather than as demand side. A related market design failure was to treat demand-side resources in the capacity market design as full substitutes for supply-side resources when they were defined to not be full substitutes. Demand-side resources are paid the same as cleared supply-side resources. But demand-side resources are not required to actually be substitutes for the capacity resources they displace in the market clearing. Demand-side resources had only a very limited obligation to respond during the summer only and then only for limited hours, although there are now both annual and summer products. Demand-side resources are treated as emergency only resources, rather than economic resources, despite being treated in the market as substitutes for economic resources.

The potential hours of obligation have been expanded, but defining demand-side resources as emergency only, rather than economic resources, means that they are seldom called on. Economic resources have a daily must offer requirement in the energy market and are used in merit order. Instead, demand-side resources are used only as emergency resources and are paid very high strike prices regardless of actual market prices when they are called on. Demand-side resources are permitted to enter a strike price which they are guaranteed to be paid when called. These strike prices may be as high as $1800 per MWh despite the fact that offer caps for all other resources are $1000 per MWh. There is no requirement to support or explain strike prices.

Demand-side resources, unlike all other capacity resources, are not required to offer actual physical assets into capacity auctions.[24] Demand-side resources are required only to have a marketing plan at the time of offers in the capacity market. As a result, sellers of demand-side resources have had the unique ability to treat the sale of capacity as an option and

---

[24] See the "Analysis of Replacement Capacity for RPM Commitments: June 1, 2007 to June 1, 2017," http://www.monitoringanalytics.com/reports/Reports/2017/IMM_Report_on_Capacity_Replacement_Activity_4_20171214.pdf (December 14, 2017).

demand-side resources have disproportionately replaced their obligation to sell capacity by buying replacement capacity in incremental auctions at steep discounts to the base auction sale prices, profiting from the price difference while not actually providing demand-side resources. The behavior of sellers of demand-side resources has suppressed the price of capacity in the BRA compared to the competitive result in part because sellers shift demand for physical capacity resources from the relatively high price BRA to the much lower price IAs.

Despite the fact that PJM is an LMP market, demand-side resources, unlike all other capacity resources, are not required to be locational for dispatch purposes. The dispatch of demand-side resources by PJM does not match the resources to the locational need and often dispatches demand-side resources where they are not needed and where they may be counterproductive, for example when reduced load on one side of a constrained transmission line results in increased flow on that line and an increased, rather than a decreased, need for constraint relief.

Performance and testing metrics for demand-side resources systematically overstate the capability and performance of demand-side and energy efficiency resources. Measurement and verification of demand-side resources has been based on counterfactual assumptions.[25] Testing is scheduled by demand-side resources rather than PJM and does not replicate actual market conditions.

Actually and accurately determining the level of demand-side activity by individual customers is almost impossible. The counterfactual measurement and verification methods typically used in these programs are based on unverifiable assumptions that generally result in overestimates of actual reductions. The current rules for limited, extended summer, and annual demand response use the average reduction for the duration of an event. The average duration across multiple hours does not provide an accurate metric for each 5 min interval of the event and is inconsistent with the measurement of generation resources and 5 min locational pricing. The calculation methods of event and test compliance do not provide reliable results. Settlement locations with a negative load reduction value (load increase) are not netted by PJM within registrations or within demand response portfolios. PJM limits compliance shortfall values during a demand response event to 0 MW.

---

[25] See the *2019 State of the Market Report for PJM*, Volume 2, Section 6: Demand Response (March 12, 2020).

Energy efficiency measurement issues have been egregious even by the standards of the demand side-programs. Energy efficiency resources selling in the capacity market have been largely lighting programs with measurement and verification plans based solely on unverified assumptions. Measurement and verification plans do not require a demonstration that the payments by energy efficiency CSPs have changed the behavior of customers. The energy efficiency programs have taken advantage of the PJM rules but failed to add any demonstrable energy efficiency capability. Energy efficiency is a valuable resource, but the treatment of energy efficiency in the PJM capacity market has created inappropriate incentives that do not result in an increase in actual, incremental energy efficiency. Most EE in the PJM capacity market does not change activity at the margin and is based on highly speculative and inaccurate measurement and verification methods. Effective energy efficiency measures reduce energy usage and capacity usage directly. The reduced market payments that result directly and automatically from using less energy are the appropriate compensation.

While demand-side programs have been less significant in the PJM energy market, there have also been market design issues in the PJM energy market. On April 1, 2012, FERC Order No. 745 was implemented in the PJM economic program, requiring payment of full LMP for dispatched demand resources when a net benefits test (NBT) price threshold is exceeded.[26] This approach replaced the payment of LMP minus the charges for wholesale power and transmission included in customers' tariff rates. PJM calculates the NBT price threshold by first taking the generation offers from the same month of the previous year. For example, the NBT price calculation for February 2017 was calculated using generation offers from February 2016. PJM then adjusts these offers to account for changes in fuel prices and uses these adjusted offers to create an average monthly supply curve. PJM estimates a function that best fits this supply curve and then finds the point on this curve where the elasticity is equal to one.[27] The price at this point is the NBT threshold price.

When the zonal LMP is above the NBT threshold price, economic demand response resources that reduce their power consumption are paid the full zonal LMP. When the zonal LMP is below the NBT threshold price, economic demand response resources are not paid for any load reductions.

[26] 134 FERC ¶ 61,187 (2011).
[27] "PJM Manual 11: Energy & Ancillary Services Market Operations," §10.3.1, Rev. 108 (Dec. 3, 2019).

The NBT test is a crude tool that is not based in market logic. The NBT threshold price is a monthly estimate calculated from a monthly supply curve that does not incorporate real-time or day-ahead prices. In addition, it is a single threshold price used to trigger payments to economic demand response resources throughout the entire RTO, regardless of their location and regardless of locational prices.

The asserted necessity for the NBT test is an illustration of the illogical approach to demand-side compensation embodied in paying full LMP to demand resources. The benefit of demand-side resources is not that they suppress market prices, as assumed in the NBT approach, but that customers can choose not to consume at the current price of power, that individual customers benefit from their choices and that the choices of all customers are reflected in market prices. If customers, or their designated intermediaries, face the market price, customers have the ability to not purchase power and the market impact of that choice does not require a test for appropriateness.

Even for customers prepared to be price responsive, customers do not pay a price based on the marginal cost of supplying power. In PJM, customers do not pay the marginal cost of the next increment of demand as they would in an efficient market design. Customers do not pay the locational marginal price for wholesale power. Retail rate designs generally do not incorporate locational or real-time prices.

The demand side of a market is simply customers deciding whether or not to purchase another unit of a product. The price of the product is a key variable in that decision, as is the relative size of spending on that product in customers' overall spending. In PJM, despite the use of locational marginal pricing since the start of the markets on April 1, 1999, load does not actually pay the nodal or locational price. While generators are paid the locational marginal price, load pays the average zonal price or a price based on an aggregation of nodes by a competitive load serving entity (LSE). A small number of very large customers have become their own LSEs and have paid nodal prices. This failure to fully implement locational pricing for both the supply and the demand sides of the market means that loads do not receive an accurate price signal from the wholesale power market about the value of the power at their location. Load that increases flow on a constrained transmission line pays the same price as load that helps relieve that constraint. Under locational pricing for load, the customer that increases flow on a constrained transmission line would pay a higher price, reflecting the locational marginal cost of meeting their load. The incentive to interrupt to help the constraint would correspond to the locational marginal cost of that power.

For electricity customers, the final price of electricity is defined by the local retail tariff. The retail tariff is designed to ensure that customers pay for the costs of the local distribution system, the transmission system, and the generation of wholesale power. Retail tariffs are generally the complex result of years of cost of service ratemaking including the allocation of costs using cost of service studies and rate design analyses. The retail tariffs may include rates that are flat, or that vary by consumption levels or by season or time of day, or a combination, or that are a function of demand at a specific time or times. Seasonal and time of day tariffs were a crude attempt to reflect actual temporal differentials in the cost of generating power but are generally correct only by accident. The actual cost of power varies by day and by time of day in ways that cannot be captured in a static tariff design because the variation is a result of ongoing and unpredictable changes in the factors that affect power prices from the weather to the cost of fuels to unit outages to transmission outages to changes in behavior by generators and load. If the goal is to reflect the cost of wholesale energy, rates need to be dynamic. The only solution is to have a direct pass through of the relevant locational marginal price to customers. But the cost of wholesale power is only about a third of the retail price of electricity. As a result, the impact of variations in the cost of wholesale power is attenuated. The costs of the distribution system and the costs of the transmission system do not vary in the same way as the costs of wholesale power.

In PJM, about 20% of the cost of wholesale power is the cost of capacity (Table 11.2). The cost of capacity is assigned to load serving entities on the basis of demand during the single system coincident peak hour during the year. The cost of that capacity is generally assigned to individual customers on the basis of a customer's demand during the five system coincident peak hours during the year, the customer's peak load contribution (PLC).

A customer's demand during the five coincident peak demand hours of the year can thus have a significant impact on the customer's bill and this impact is amplified by the fact that the retail rate design frequently also assigns transmission system costs and sometimes distribution system costs on the basis of the same 5 h. In PJM, large customers have managed their coincident peak demands, beginning under utility regulation, in order to manage their power bills, based on this retail rate design, even without formal demand-side programs.

It is clear that the incentive to build or maintain a thermal power plant in PJM markets is not a function of demand and prices during 1 h per year or 5 h per year. If the costs of capacity were spread over more hours, the ability to avoid capacity costs by interrupting for just a few hours per year would be appropriately limited. The incentives to build and maintain power plants would be better aligned with the incentives to avoid paying for capacity. In an energy only wholesale power market, customers would have an incentive to avoid using power when energy prices were relatively high. High prices are generally not limited to a few hours per year and the high price hours can occur in the winter, in the summer and during shoulder periods. Matching the allocation of the capacity costs to high demand hours would better align the capacity market price signals with the market and provide a better price signal to demand-side participants.

The current demand-side market design should be ended, including DR and EE and the inclusion of demand-side response as supply. The goal should be to replace the current demand-side design with a design that permits more flexible participation by all customers or their designated intermediaries, participation that does not depend on the pretense that demand is supply with all the complex associated rules, and participation that is expected to result in savings to customers based on their behavior, which could easily exceed current DR-related payments to customers.

6. The path forward: flexible demand response in wholesale power markets

Demand-side flexibility means that the demand for power will react to the price of power in real time. There is no ideal or target level of demand-side activity. The goal should be to make customers' decisions as simple as possible. As in any market, optimal demand-side participation would reflect the price of power, including energy and capacity, and the costs of interruption by customers. Intermediaries can play a key role. Aggregators, termed curtailment service providers (CSPs) in PJM, can aggregate customers, take on the transaction costs, and share savings with end-use customers. The basic incentives are the same except that end-use customers assign demand responsiveness to intermediaries. All customers, no matter how small, should have the opportunity to manage their demand in response to prices.

In a well-functioning market, customers or their designated intermediaries would pay a price for wholesale power that corresponds to the real-time, locational marginal cost of producing that power. Customers would have the incentive to increase or decrease usage based on the price. That approach applies whether the wholesale power market is an energy-only market or includes both an energy and capacity market. The price of power has both locational and temporal aspects. The accurate locational price of wholesale power varies on both the locational and temporal dimensions. It is no more accurate to pay a locational price averaged over a day or a week or a month than it is to pay a 5-min price aggregated over a zone. The locational marginal cost of power can change every 5 min in PJM. The price signal to customers should reflect both the spatial (nodal location) and temporal (5 min) granularity because that price signal reflects the marginal cost of meeting the load at that location at that time.

A fully functional demand side of the electricity market would mean that customers or their designated intermediaries can see real-time energy price signals in real time, can react to real-time prices in real time, and can receive the direct benefits or pay the costs of changes in real-time energy use. In PJM, where there is a capacity market, this would also mean that customers can see current capacity prices, can react to capacity prices, and can receive the direct benefits or pay the costs of changes in the demand for capacity. A functional demand side of these markets would mean that customers have the ability to make decisions about levels of power consumption based both on how customers value the power and on the actual cost of that power.

In a market with such significant renewables participation that the price of energy is zero or negative at times, the value of power is zero or negative, and the price signal to load should reflect that value. Demand-side participation has a zero or negative value at that time. But when the price of energy increases, the value of demand-side participation tracks that price. At the margin, if power prices are too low to induce customers to reduce demand and so low as to provide incentives for supply to retire, the new equilibrium will define the balance between economic supply and economic demand-side resources.

Relevant demand-side activity can be short term or long term. Demand-side flexibility is a measure of short-term demand-side potential. Demand-side flexibility illustrates the ability of load to react in the short run to energy market price signals. But expectations of price levels can also provide an incentive to modify customers' capital investment including everything from smart meters, to energy management systems, to better structure design, to more efficient devices, to smarter devices, and to more automated systems. Some customer capital investments improve the ability to react to prices. Some customer capital investments reduce demand regardless of the real-time price, but nonetheless have a significant impact on the market and on market prices. Energy efficiency investments are such investments. More efficient and smarter appliances and buildings are examples of energy efficiency investments that are more likely and more effectively to result from improved standards than special market programs. The approach to demand-side flexibility in wholesale power markets should break from the history of utility curtailment programs but also break from the representation of the demand side of the market as if it were the supply side of the market. The demand side is not the supply side. The demand side of the market should be treated as the demand side. Although treating the demand side as if it were the supply side was intended to encourage the growth of demand-side resources, the result has been a series of unintended consequences and ad hoc patches to the capacity market rules with unintended but pervasive negative consequences. The PJM market design, and wholesale market design generally, has not facilitated the active participation of the demand side of the market. Pricing for load is not locational. The allocation of capacity costs is not based on the actual reasons that capacity is added and maintained. Retail tariffs are a patchwork of ineffective and inaccurate approaches to real-time pricing.

The PJM markets have paid a lot for a high level of demand resources in the capacity market. But the payments and MW are not evidence of success in including an active role for the demand side of the market. The level of response is evidence that market participants respond to market incentives. The level of response has not provided significant demand side flexibility and responsiveness to prices. The demand-side programs in PJM have paid for resources that are relatively inflexible and that are rarely called on.

In a market with a well-designed demand side that provides appropriate and transparent price signals to participants, it would be expected that participation would be diverse and unconcentrated. But the PJM market design, including the complex PJM capacity market rules, have been in part responsible for the fact that the ownership of both energy and capacity demand resources is highly concentrated in PJM. The HHI for economic resources was 8261 in 2019. The HHI for emergency demand response committed MW was 1838 for the 2019/2020 Delivery Year.[28] In the

---

[28] The Herfindahl—Hirschman Index (HHI) concentration ratio is calculated by summing the squares of the market shares of all firms in a market. The "Merger Policy Statement" of FERC states that a market can be broadly characterized as unconcentrated when HHI is below 1000; moderately concentrated when HHI is between 1000 and 1800; and highly concentrated when HHI greater than 1800. See *Inquiry Concerning the Commission's Merger Policy under the Federal Power Act: Policy Statement,* 77 FERC ¶ 61,263 *mimeo* at 80 (1996).

2019/2020 Delivery Year, the four largest companies owned 78.8% of all committed demand response UCAP MW.[29]

Markets do not work without clear rules. If the active participation of the demand side is to be encouraged and facilitated, the goal should be to ensure that the market signals reflect the actual marginal costs of producing energy, whether that includes scarcity pricing or a combination of scarcity pricing and a capacity market. Those market signals require a good wholesale power market design including good locational pricing and a good retail pricing design. There is no reason to pay above market prices or provide other special incentives for demand-side participation.

There is a potential way forward to incorporate demand side in the power markets and to facilitate a flexible demand side that has the tools to respond to market prices but is not constrained by the excess complexity of trying to fit demand-side response into PJM's capacity market rules.

As a preferred alternative to being a substitute for generation in the capacity market, demand response resources should be on the demand side of the capacity market rather than on the supply side. Rather than detailed demand response programs with complex and difficult to administer rules that are subject to gaming, customers would be able to avoid capacity and energy charges by not using capacity and energy at their discretion. The level of energy and capacity paid for would be defined by metered usage rather than complex and inaccurate counterfactual measurement protocols.

The RTO could facilitate demand response in both the energy and the capacity markets by providing transparent real-time information including ongoing, updated, real-time forecasts of locational load and prices to customers. The RTO could facilitate rational prices in the capacity market by ensuring that load forecasts used in the capacity market reflect actual and expected customer behavior. Participating load or its intermediary would inform PJM prior to a capacity auction of the MW participating, the months and hours of participation, the temperature humidity index (THI) threshold at which load would be reduced, and/or the locational price at which load would be reduced. PJM would reduce the load forecast used in the RPM auction based on the designated reductions. Load would agree to curtail demand to at or below a defined level, less than the customer's measured peak demand (based on a reasonably defined capacity market

---

[29] See the *2019 State of the Market Report for PJM,* Volume 2, Section 6: Demand Response (March 12, 2020).

billing determinant incorporating significantly more than 5 h), when the THI exceeds a defined level or load exceeds a specified threshold. By relying on metered load and the customer's measured demand, load can reduce its demand for capacity and that reduction can be verified without complicated and inaccurate metrics to estimate load reductions. Payments at the customer level would be based on actual metered demand. To the extent that customers enter into contracts with CSPs to manage their payments, measurement and verification can be negotiated as part of a bilateral commercial contract between a customer and its CSP. But the system would be paid for actual, metered usage, regardless of which contractual party takes that obligation.

In the energy market, load could offer a strike price to facilitate PJM dispatch, or load could simply respond to real-time prices and PJM forecasts would learn from experience. Billing would be for actual energy used and there would be no counterfactual estimates of the load reduction. Payments at the customer level would be based on actual metered usage. The same arrangements with CSPs would be possible. If customers do not wish to pay real-time prices, customers could enter contracts with intermediaries for fixed prices and the intermediaries would manage the risk and face the market incentives.

These approaches provide incentives for longer-term investment behavior in addition to response to real-time price signals. Investments in smart meters and appliances as well as smart buildings and factories would enable customers to reduce energy usage as well as automate real-time responses to market conditions. Intermediaries could benefit by packaging groups of customers, managing information and data needs, and sharing benefits with customers.

Although almost all demand-side payments in PJM have been in the capacity market, exactly the same design principles apply in an energy only market with meaningful scarcity pricing. An effective demand-side design would allow customers to avoid paying scarcity prices in market designs where scarcity prices are intended to provide revenues equivalent to capacity market revenues in designs with capacity markets.[30]

---

[30] Despite the characterization of ERCOT as an energy-only market, ERCOT has the equivalent of a capacity market for some demand-side resources, the Emergency Response Service. See Baldick, chapter two.

7. Conclusion

The current treatment of demand-side resources as supply, and the associated set of complex rules governing participation in the energy and capacity markets, should be ended and a transition made to a market-based approach to the demand side of wholesale power markets.

The appropriate end state for demand resources in the PJM markets should be comparable to the demand side of any market. Customers would pay real-time locational prices for energy and capacity. Customers should use energy as they wish, accounting for market prices in any way they like, and that usage will determine the amount of capacity and energy for which each customer pays. There would be no counterfactual measurement and verification. The role of the RTO would be to make as much information as possible available to customers or their intermediaries in real time and to facilitate customers or their intermediaries declaring their load reduction goals in advance for the capacity market and to incorporate those goals in the demand for capacity.

An effective demand side is essential to the competitive functioning of the wholesale power market in PJM. In PJM, and wholesale power markets generally, a new, market-based approach is required in order to maximize the opportunities for the economic participation of demand-side resources. That approach should include treatment of demand as demand, actual real-time locational marginal pricing, improved assignment of capacity costs, better load forecasting by the RTO, provision of real-time information by the RTO, and better coordination with retail tariffs. Improved retail tariffs are also required. Those tariffs are the responsibility of state utility commissions and are not part of the wholesale market design. An improved market design will provide the opportunity for load to benefit from the option to flexibly respond to prices and to receive the benefits of that response.

A transition to this end state would require a commitment to market principles by FERC, by states and state utility commissions, by PJM and by PJM market participants. The infrastructure for a market-based approach to demand-side participation exists and smarter technologies are evolving daily. But there are substantial barriers to progress. As with any significant change to the market design, change is opposed by those who benefit from the current design. The current market design creates an identified and well-organized group of beneficiaries. The beneficiaries of a new market design are less clearly identified and are not well organized.

PART 3

Coupling flexible demand to variable generation

CHAPTER 12

Valuing consumer flexibility in electricity market design[*]

Laurens de Vries[1], Gerard Doorman[2]

[1]Delft University of Technology, Delft, The Netherlands; [2]Statnett, Oslo, Norway

1. Introduction

As the transition to a low-carbon energy system is picking up speed and the cost of renewable energy is decreasing, one of the largest long-term challenges is how to maintain security of supply. Solar and wind energy are becoming cheaper all the time, but how will the lights be kept on when they are not available? Solutions on the consumer side of the market, such as demand flexibility, energy storage behind the meter, and self-generation will play an important role in maintaining the energy balance. However, they will not be enough during periods of a week or longer without much wind and solar generation (a "wind and solar drought" or "Dunkelflaute" in German), when wholesale market solutions such as controllable power generation and large-scale storage will be needed. This raises the question how to design the electricity market in such a way that an optimal mix between these wholesale and decentral solutions is achieved.

Current electricity markets rely on legacy fossil fuel plant for reliability, but their operating hours decline year by year as they are being pushed out of the market by renewable energy. After these plants are phased out, alternative, low-carbon technologies are needed to provide controllable electricity generation. Replacing them will require large volumes of stored energy, e.g., in the form of hydrogen or other synthesized molecules, or perhaps fossil plant with carbon capture and storage or nuclear power, to be available even though it will operate only a limited number of hours per

[*]An initial, short version of this paper was published as "Electricity market design based on consumer demand for capacity" (2017), in ROSSETTO, Nicolò (editor/s), *Design the electricity market(s) of the future.* Proceedings from the Eurelectric — Florence School of Regulation Conference. 7 June 2017, Florence: European University Institute, Robert Schuman Center for Advanced Studies, 2017, Florence School of Regulation, Energy, Electricity, pp. 49—53. Cadmus, European University Institute Research Repository, available at: http://hdl.handle.net/1814/50004.

Variable Generation, Flexible Demand
ISBN 978-0-12-823810-3
https://doi.org/10.1016/B978-0-12-823810-3.00019-4
287

year on average. As the weather affects both electricity demand and renewable energy generation, the year-on-year weather variations cause significant investment risk for these facilities, as their need becomes uncertain. So while controllable generation capacity is essential, its average cost per unit of electricity will be high and the investment risk will also be high.

Demand flexibility and storage behind the meter can help reduce the need for controllable generation capacity and large-scale storage. Electricity consumers may be able to shift when they charge their electric vehicles or heat the reservoirs of their home hot water and heating systems. Industrial consumers may be able to reduce their production occasionally. The market design challenge not only requires the alignment of incentives among the markets for controllable wholesale generation capacity and consumer flexibility but also needs to reflect that solutions like demand flexibility may expire after some time, e.g., when a vehicle really is needed for driving or a building needs to be heated, whereas other, perhaps more expensive options, such as a power plant with CCS, may not have a time restriction on its operating time. In this chapter we propose an electricity market design called *Capacity Subscription*, which provides an economically efficient, transparent, and stable reward for consumer flexibility.

The idea is that consumers *subscribe* to a volume of firm capacity: they pay a fixed amount per month for having access to a certain volume of generation capacity from these controllable electricity generation facilities. Because their monthly payments cover (most of) the fixed costs of these facilities, the electricity can be made available at a cost close to variable cost. There is no need for extreme scarcity prices. In this way, Capacity Subscription removes the uncertainty regarding the demand for controllable generation capacity and spreads its cost to consumers over time. The resulting reduction in risk benefits both the suppliers of reliable power generation and the consumers. Given the relatively high fixed costs of controllable generation capacity, consumers may choose to reduce their consumption at times when renewable energy generation is tight or to install storage devices themselves. Capacity Subscription has a number of advantages:

- Consumers pay directly for the scarce resource: generation capacity;
- Consumers do not face extreme price spikes in the energy market;
- Demand flexibility is internalized in the consumers' decisions;
- Demand is controlled by the consumer in real time and physical shortages are avoided;

- The level of uninterrupted supply is chosen by each consumer individually;
- Producers are remunerated by selling capacity and not dependent on rare scarcity prices;
- The price and quantity of capacity are both market based;
- System adequacy moves in the direction of becoming a private good.

Capacity Subscription provides a market in which all instruments that contribute to system adequacy, from industrial facilities to consumer flexibility, are valued on par. In a power system with much demand flexibility, the current approach of defining reliability in terms of a maximum number or duration of demand interruptions no longer works. If there is enough demand flexibility, demand may be able to adjust to power shortages. Instead, the question becomes how much inconvenience consumers are willing to accept and how to develop an optimal mix of flexibility options. Capacity Subscription allows consumers to choose their preferred level of adequacy and removes the need for a central reliability standard. This makes it superior to the capacity remuneration mechanisms that currently exist, which either ignore consumer flexibility options or rely on artificial instruments such as flexibility markets to give them some value, but without an instrument to optimize the combination of consumer and wholesale solutions. Capacity Subscription can already be introduced in current power markets in which (nearly) all consumers have smart meters. It will facilitate the development of consumer-side flexibility options during the energy transition by providing consumers an inherently strong incentive to become flexible.

This chapter is organized as follows:

- Section 2 describes the techno-economic characteristics of consumer flexibility and their link with system adequacy;
- Section 3 describes the principles of Capacity Subscription;
- Section 4 discusses how such a scheme could function in a low-carbon power system; and
- Section 5 explains how Capacity Subscription facilitates consumer participation in flexibility; followed by the chapter's conclusions.

2. The contribution of flexible demand to system adequacy in a low-carbon system

In a low-carbon energy system, the share of variable renewable energy sources (vRES) is so high that there will be many hours each year in which their production outstrips electricity demand. This results in low electricity prices during these hours, as has already been observed during the Covid-19 crisis in Europe. However, even in the presence of very high shares of vRES there will be a significant periods of time when vRES does not meet electricity demand. Moreover, the relative contribution of vRES to peak demand (i.e., the percentage of their installed capacity that can be relied upon during peak demand moments) declines as their total share increases (Hirth, 2014; Moghanjooghi, 2017). During these times, firm capacity will be needed, such as biomass, hydropower (if available), or gas plant that run on hydrogen.

This firm capacity is inherently expensive as it needs to operate only a limited number of hours per year. Especially the facilities that provide in the very peak hours of demand, the plants that run the fewest number of hours per year, will not only be expensive but also risky from an investor's point of view. The uncertainties about the occurrence of such energy shortages, the electricity demand during these periods, and the levels of the resulting prices make the business case impossible if these facilities need to recover their costs in short-term electricity markets. This market design is also quite unattractive for consumers, as these plants may need electricity prices to peak at levels of up to two orders of magnitude higher than average prices to recover their costs (De Vries, 2007; Doorman, 2005; Doorman and Botterud, 2008).

From the perspective of system cost, it is therefore attractive to lower peak demand for electricity. Expansion of electricity network capacity will reduce the need for flexibility by evening out fluctuations in supply and demand over a larger area, but will not eliminate the need for firm capacity. As discussed in Part 1 of this book, it is increasingly recognized that an important characteristic of a low-carbon electricity system is its "flexibility." Flexible demand—demand response—will reduce the gap between vRES and demand and thereby reduce system cost.

For the market to function efficiently, the market design needs to provide a level playing field for all types of flexibility, from demand response to large-scale commercial storage facilities and back-up generators. There is a need, therefore, for a market that remunerates firm capacity but that also allows consumers to adjust their demand for this capacity based on the degree to which they can be flexible with their peak consumption. Flexible consumers, ones who are able to reduce electricity consumption during shortage hours, should have the opportunity to purchase less firm capacity than inflexible ones. A market design that is based on Capacity

Subscription meets these goals, as will be explained in the following sections. Capacity Subscription provides a stable and economically efficient incentive to consumers to provide flexibility and to producers to provide firm capacity.

Capacity Subscription creates a marketplace, in addition to the wholesale electricity market, for firm capacity. A key feature is that consumers choose their level of demand, so they can make the trade-off between purchasing capacity and being flexible, or even installing their own backup capacity in the form of a battery. The market price in the capacity subscription market reflects the extent of consumers' preference for uninterrupted supply on the one hand, and producers' costs (including risks and hedging) on the other. While traditional market designs exploit consumer flexibility through reactions to price variations, Capacity Subscription incentivizes consumers to internalize the use of flexibility by keeping demand under a predetermined limit.

The creation of an explicit market in which firm capacity is traded and consumers can determine their willingness to pay for this by comparing its cost to their own alternatives has some strong advantages. As it makes the demand for firm capacity explicit and provides a stable revenue stream for back-up units, it reduces investment risk (and therefore the cost of capital) and ensures a sufficient supply of these units. Consumers benefit likewise from more stable prices (compared to an energy-only market), while the linking of consumer and wholesale markets ensures economic efficiency.

3. How capacity subscription works

The main idea of Capacity Subscription is that consumers buy the amount of generation capacity ("subscribe to") that they expect to need during moments of system scarcity. They buy capacity subscriptions from providers of firm capacity (generation and storage). When a consumer buys a capacity subscription of, e.g., 4 kW, he is guaranteed that he can consume electricity up to this capacity level under all conditions. When the energy market is short of generation capacity, e.g., during a period without much solar and wind energy, the system operator activates so-called *Load Limiting Devices* (LLDs)[1] that are installed at each consumer site. These devices force consumers to restrict their consumption to the levels that they contracted, so this consumer may not use more than 4 kW during these hours. In return,

---

[1] More about the LLDs and the role of the system operator in Section 3.4.

he has the certainty that this capacity is available at a price close to the marginal cost of generation. When there is no shortage of generation capacity—most of the time—consumption is unrestricted.

Because physical shortages are avoided, scarcity prices do not occur. A capacity subscription may therefore be considered as a sort of physical option contract: by paying for the capacity subscription, a consumer obtains the right to consume electricity at the system marginal cost at any time. He will not need to pay scarcity prices of one or two orders of magnitude above marginal cost. For generation companies, the benefits are that the demand for reliable capacity is made explicit and that the payments are spread out over time. In fact, this system turns reliable capacity into a product with a steady remuneration, rather than a speculative investment as is the case in an energy-only market. The following sections will detail the rules of the Capacity Subscription concept from the perspectives of consumers, producers, and the system operator.

Limiting demand, physically or economically, is not new and was common in Norway until the 1980s, as described in Box 12.1. Today both Italy and Spain have similar schemes, in which consumers are penalized when they exceed a certain demand level. However, refining these schemes would improve both their effectiveness and user-friendliness.

3.1 The role of consumers

3.1.1 How much capacity?

Crucial questions for consumers that participate in a capacity subscription market are how much capacity they need, when they need it, how their demand peaks coincide with system scarcity, and whether they have means to reduce their need for capacity during system peaks.

The capacity need of a consumer can in principle be based on last year's consumption peak, if measurements per time unit are available. If not, a typical ratio between peak load and annual consumption is a reasonable starting point. How a consumer's individual peak demand coincides with system scarcity is a more challenging question, although a record of historical data is a good basis for an evaluation. It may be expected that few consumers will have the ability and interest to perform such assessments, but apps and websites can be developed to support consumers to make choices that match their preferences with the use of measured data and forecasts. In a system with widespread use of Capacity Subscription, there will be a strong demand for such solutions. This can be expected to lead to commercial interest to develop them, resulting in competition and user-friendly

BOX 12.1 A short look at history

In the early days of Norwegian electricity supply, production often came from small run-of-river plants, where energy had zero marginal cost but capacity was seriously constrained. At the time, lasting up to the 1950s, many consumers had a so-called "seesaw" tariff, where they only paid for capacity. Their demand was limited by a switch that started to turn the power on and off when the subscribed capacity was reached—an obvious signal to reduce demand.

Later, while capacity still was constrained (especially in the local grids), a two-part tariff was common. Consumers paid for a certain amount of capacity. Below this capacity, they paid a (very) low energy price. Above this capacity, however, they typically paid 5—10 times as much. This was common until the mid-1980s. Actual usage and the capacity limit were indicated by black and red arrows, respectively.

Both these tariffs confronted consumers with similar choices as in the case of Capacity Subscription, except that the (crucial) issue of coincidence with system scarcity was not considered.

solutions. Energy service companies (ESCOs), electricity retailers, and demand aggregators all have a business case to make by offering user-friendly ways to consumers to reduce the volume of capacity to which they subscribe and to helping them stay within this limit. Their services might range from mere insight in their own consumption profile and a suitable level of capacity subscription to active demand management services.

At least for household consumers, there should be a minimum mandatory capacity, e.g., based on the previous year's consumption, to avoid that consumers buy zero capacity to save costs. As the individual consumers' preferences will change over time, it is necessary to be able to easily adjust the subscribed limit of the LLD. They must therefore be remotely controlled, by the grid operator or another responsible party.

3.1.2 Demand flexibility

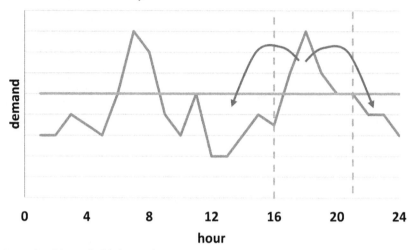

Example of household demand on a typical day. The subscribed limit is given by the brown horizontal line. Between 16:00 and 21:00 there is a scarcity situation, and demand must be shifted, as illustrated by the arrows. A similar peak in the morning however, is unaffected, as there is no LLD activation and demand is unconstrained

A compelling feature of Capacity Subscription is the incentive it provides to keep demand below the subscribed capacity level, which includes the incentive to develop technology for this purpose. When Capacity Subscription is implemented, millions of consumers become interested in controlling their demand, which creates opportunities for companies to develop and sell solutions. While many traditional schemes rely on electricity price variations to incentivize flexibility, Capacity Subscription "internalizes" the incentive in the consumer's capacity payment. When the LLD is activated, the consumer has a strong incentive to keep demand below the subscribed limit, in the least interruptive way possible. If the price of capacity is high, there is an incentive to invest in flexibility in order to reduce the volume of capacity that they buy. Thus, Capacity Subscription triggers demand response in a different way than conventional electricity markets do: if capacity is expensive, a consumer can save money by buying less. However, this reduces his comfort, as he needs to "switch off something" when the LLDs are activated. Flexibility becomes a matter of controlling one's demand in such a way that he stays below a limit, i.e., his subscribed capacity. The more flexible the consumer, the less capacity he can buy without compromising comfort. This mechanism will create a strong "market pull" for flexibility, as consumers will observe the cost of capacity, versus an uncertain risk of occasionally very high prices for energy. The market will also ensure an optimal mix between different types of flexibility, most notably behind small-scale "behind the meter" solutions and wholesale market capacity products.

Part Two of this book discusses how consumers can become more flexible. For customers with electrical heating, heat storage is an option. Others may have less flexibility, but will still have options to shift their most power-intensive appliances in such a way that their peak demand is kept below their procured capacity limit. Batteries may become an increasingly interesting option for managing short consumption peaks. The increasing penetration of electric vehicles provides an interesting potential for "free" battery capacity.

However, for sustained periods of low vRES production, it is hard to see an alternative to generation capacity (possibly from large-scale stored energy such as hydrogen). This is the other side of the Capacity Subscription medal. At northern latitudes, there are occasional periods with adverse weather, e.g., more than a week of cold winter weather at a northern latitude, with low sun and limited wind. During these periods the flexibility provided by electric vehicles, home batteries, and demand shifting will run out to a large degree.

3.1.3 System adequacy becomes a private good
Stoft (2002) identifies two demand-side flaws of electricity markets:
- Lack of metering and real-time billing
- Lack of real-time control of power flow to specific consumers

In the presence of these flaws, reliability is a common good. System adequacy is conventionally ensured by having sufficient generation capacity available, with sufficient being a judgment that is made by the system operator or the regulator. When we cannot distinguish between consumers, we need to design the system in such a way that it satisfies the preferences of those that value reliability the most. This may be a large majority, but we don't really know. If even only a minority of consumers accepts a lower level of reliability, e.g., by being willing to keep demand below a limit, this will create sufficient flexibility to avoid *involuntary* load shedding. The cost of load shedding to consumers can be extremely high, but depends strongly on the time of day, duration, and type of consumer. Ovaere et al. (2018) shows on the hand of multiple studies of the value of lost load—the value that consumers would have been willing to pay to avoid a power shortage—that this ranges from 1300 €/MWh to well over 100,000 €/MWh for some categories of consumers and moments. Shivakumar et al. (2017) estimate the value of lost load for European households, estimating it between 3200 €/MWh and 15,800 €/MWh, depending on the country. As different as these estimates are, they have in common that they are two orders of magnitude higher than the average cost of electricity service.

These high estimates of the value of lost load explain why household consumers traditionally have been considered as price-inelastic: normal electricity prices, even peak prices, did not usually come close to levels at which household consumers would consider changing their behavior. This is no longer true, however, when a combination of new types of flexible load and smart control systems are available. Smart EV charging, home batteries, heating—especially in the presence of a large heat reservoir—and cooling provide this possibility. The cost of shifting these services will be much lower than the above estimates; in fact, one may expect that a large share of their consumption will be shifted toward low-price hours in order to minimize their cost (within the constraints of the users' preferences). Capacity Subscription can, on the one hand, help incentivize them to move away from system peak demand times, and on the other hand spread their loads over time so these flexible loads do not all turn on at the same time.

With Capacity Subscription, consumers weigh the cost of capacity against their preferences for unlimited supply. If the price of capacity is high, industrial consumers will over time redesign their production processes to be able to reduce their need for capacity. Households and services will similarly have incentives to look at ways to reduce demand when necessary. Capacity Subscription thus has the unique feature that it reveals the need for firm capacity in the market as expressed by consumers' preferences for uninterrupted supply. This internalizes system adequacy in the market: it becomes a private good,[2] as a result of which consumers have incentives to be flexible in order to reduce their need to pay for capacity.

---

[2] This is not completely true, as system adequacy also is affected by the transmission and distribution grids. But it is true for the part of adequacy that is determined by the ability to balance generation and demand.

3.2 The role of suppliers

In the future, several types of parties could be capacity suppliers, although the major group probably still will be generators. The major distinction between a supplier and a consumer of capacity is their position during scarcity events. A party that can reasonably expect to inject energy into the system during such events will be a supplier, while one that will withdraw energy from the system will be a consumer. A crucial parameter is the duration of the injection. If Capacity Subscription is to contribute to longer periods of reduced vRES production, it does not look probable today that other parties than generators with storable fuel can deliver such contribution. Still, the market design is generic, and would certainly allow such source, should they be developed in the future.

When a scarcity event occurs, suppliers need to demonstrate their availability by bidding in the relevant markets, day ahead, intraday, and balancing. There needs to be a significant penalty for noncompliance to avoid gaming. Forced unavailability is taken into account by derating. If scarcity events are infrequent, it may be necessary to use random verification checks because the effectiveness of the model crucially depends on the supply capacity being available when needed. Prequalification is probably also necessary to ensure that all bidders are physically able to produce.

The high cost of firm capacity creates new business opportunities for ESCOs, aggregators, and other innovative parties to help consumers manage the flexibility of EV charging, heating, cooling etc., making use of ICT services to minimize transaction costs. Estimates of the electricity price over the next few days, as they are impacted by consumption and by the weather, need to be combined with consumers' preferences in order to optimally charge vehicles and control the indoor temperature. A key challenge is how to minimize the effort on behalf of the consumer while still meeting his preferences (cf. Poplavskaya and De Vries, 2020).

3.3 Activation of the LLDs and the role of the system operator

The main function of Capacity Subscription is to ensure the balance between demand and supply at the system level. In this context, the system operator is the obvious party to activate the LLDs. There may be concerns about the system operator limiting consumer demand. However, the limitation lies in the amount of capacity the consumer has procured, not in the actual activation of the LLDs. The system operator only effectuates the limitation.

While actual activation will only happen close to real time, the system operator may issue advance warnings before the day-ahead market clearing and subsequently throughout the day until (close to) real time. Clear and transparent rules need to be worked out for these advance warnings because they obviously will have strong market impacts. Activation implies that a signal is sent to the LLDs. Consumers need to be "notified" in order to be prepared. Notification can take place via visual warnings on the consumers' premises or through text messages or in other ways, as preferred by the individual consumer. Advance warnings should also be clearly announced through various media platforms. An important issue is the minimum time for prewarning of an actual activation: from the system operator's point of view, it should be as short as possible, as this would limit actual use. On the other hand, consumers may wish to have some time to adapt. On the other hand, this should not be necessary if their demand control is fully automated.

The solution that is presented in Chapter 7 of this book, in which consumers allow an aggregator to access their assets under certain conditions, is optional. With Capacity Subscription, it should not be necessary to have an external party intervene with a consumer's assets, as it is up to the consumer to ensure the limit is observed. Whether he transfers this demand to a third party, e.g., for the sake of convenience, is a choice, as the consumer may also be able to control his load himself if it is "smart" and user-friendly enough. On the other hand, it may be precisely the added value of an aggregator to provide these smart services.

With respect to the capacity that is needed for system balancing (operating reserves), an obvious solution is that the system operator buys this capacity from relevant parties in the same way as consumers buy their capacity. However, consumers who have invested in flexibility may also be able to provide this capacity, for instance in the form of short-term deviations to their vehicle charging rates. TenneT, the Dutch transmission system operator, is already experimenting with this type of service (TenneT, 2020).

3.4 Contract duration and changes to subscription levels

On the one side there are consumers that want to buy capacity based on their preferences and on the other side capacity suppliers that offer their capacity. They meet in a market—probably an organized exchange—in which the price is determined by the intersection of the long-term marginal cost of capacity and the marginal willingness to pay for capacity. The organized exchange could be organized by a power exchange, the system operator, or another relevant party. Large consumers can trade directly with the capacity suppliers, e.g., generators, but for small consumers this does not make sense—instead they can buy from retailers, much like they buy energy in today's markets.

Annual auctions should probably be the primary marketplace. The auctions need to be held well in advance of the season when residual demand (demand minus vRES production) peaks. There is no lead time, i.e., only existing capacity can participate. However, owners of new capacity know that, once it is commissioned, it will receive revenues from selling capacity. All contracts have a validity of 1 year, i.e., until the next major auction. Additional auctions are probably needed to address changes in supply and demand of capacity, but this may also be solved through continuous trade.

Retailers buy the capacity they expect their consumers to demand in the primary auction. They will use price-quantity bids, depending on the price for capacity and their expectations about consumer behavior. They will then offer the capacity they procured to their consumers at a price that ensures a reasonable profit. There will be a secondary market for capacity in which consumers can adjust their contracted volumes, e.g., when they move house or when their family situation changes. Sufficient competition between retailers is necessary to ensure this. Throughout the year they will buy and sell capacity, depending on their total number of consumers and their demand for capacity. Through these mechanisms transparent prices for capacity will occur both at the wholesale and retail levels.

4. Capacity subscription in a low-carbon system

4.1 Conditions for consumers

Perhaps the most fundamental market design question is whether Capacity Subscription should be mandatory for consumers or voluntary, based on either an opt-in or an opt-out scheme? Capacity Subscription has two essential features:

- It makes it possible to limit demand during system scarcity and
- It creates a demand for capacity based on consumers' preferences for uninterrupted supply and capacity providers' bids of capacity.

If the system is not mandatory and the first condition is not met, i.e., it is not possible to limit demand during system scarcity, consumers have no incentive to pay for available capacity. If capacity is to be defined as a product, it cannot be voluntary to pay for it—who would choose this option? Secondly, if not everyone participates, the demand for capacity subscriptions does not signal the need for firm capacity. However, a "light" version that may be desired for household consumers is described in Section 4.2.

An important question for household consumers is how much capacity they should buy. It has already been mentioned that apps and other web-based solutions will be developed to support consumers to make an optimal choice. However, many consumers will not be interested or may not have the technical skills or time. Therefore, it is necessary to have a simple rule-of-thumb default value that is reasonable for most consumers. This needs to be worked out in detail, but a typical value could be 90% of last year's temperature-corrected coincident peak demand (i.e., the consumer's peak demand during periods when the system demand was also peaking).

A third issue is the potential for opportunistic behavior. Instead of buying capacity in the primary auction (cf. Section 3.4), consumers may choose a wait-and-see strategy. They buy a small amount of capacity (the minimum level) and buy more later in the secondary trade if they think it is necessary. Once there is a secondary trade, it is not possible to exclude someone, and this creates a possibility for opportunistic behavior. A possible solution is to make it mandatory to buy a certain percentage (50%—90%) of last year's peak from the retailer after the primary auction, where the retailer ensures the capacity for its customers. This also provides more certainty to the retailer. It is still allowed to trade capacity—however, this requires active engagement, and not too many small consumers will do this. Moreover, if the price of capacity is high, this is a risky strategy, as there probably will be some scarcity leading to LLD activations. If the price of capacity is low, there is not so much to earn from selling it. Small consumers would not be interested in "gaming" the market, but their desire to change would be based on physical change like moving, reconstruction, family changes, etc.

4.2 Consumer risk and social acceptability

Is Capacity Subscription too complex or too expensive for small consumers? A key benefit of Capacity Subscription for consumers is that it allows them to spread out the cost of capacity over time. As a result, they are not surprised by sudden high prices when extreme weather events occur, as they would in an energy-only system that sends price signals to all consumers. See Box 12.2 for a description of how it would work during a prolonged period of adverse conditions. Combined with the gain in economic efficiency from demand response, this means that on average consumers are expected to pay less than in an energy-only market. In case energy poverty is a concern, it is an option to reduce the price of a basic volume of capacity and cross-subsidize it with higher charges to consumers who demand higher volumes of capacity.

BOX 12.2 The perfect storm: a persistent high pressure area in winter

In a decarbonized world, electricity system adequacy will depend much more on the weather. The worst case occurs when weather-driven demand (heating or cooling) coincides with low availability of solar and wind energy for a prolonged period of time. Consider the case of a high-pressure system in northern Europe that stays in place for 2 weeks in January. No wind, limited sun, freezing weather, high heat demand. The weather system is the same in all of northwest Europe, so interconnection only helps to the extent that it provides access to a different weather system—southern Europe—and to storage. However, the stored hydropower in Scandinavia and the Alps is insufficient to meet demand. Demand response runs out after a few days, as homes need heating and cars need to be charged.

In this case, a large volume of infrequently used controllable generation capacity is needed. This may take the form of storage in synthetic fuels such as hydrogen and ammonia, or perhaps fossil fuels with or without carbon capture and sequestration. Other solutions will probably also be developed, but every solution will need to recover its full cost during a few, rare events. In an energy-only market, this means that the electricity price may need to rise to several thousand €/MWh; if the price stays at this level for several weeks, this will create tremendous hardship for both household and commercial consumers. It will also threaten the financial viability of many retailers and thus the market itself.

With Capacity Subscription, the required controllable generation is financed through the consumers' capacity payments. Consumption is limited to the contracted capacity for the duration of the adverse weather event, as a result of which there is no shortage on the wholesale market. The electricity price is set by either the variable cost of the marginal generator or a storage provider. As a result, demand is limited without the hardship of a prolonged price spike. In return for their capacity payments, consumers have the certainty that their minimal needs are securely met at a price based on the marginal cost of generation.

Capacity Subscription benefits consumers by allowing them to select their desired capacity level and thereby control their costs. It corresponds to the recommendation, provided in Chapter 19 of this book, to regulated for consumer control, not by controlling consumers. However, periodically high capacity prices may still be considered a problem for vulnerable consumers. While less economically efficient, an option is to regulate the capacity subscription price for household consumers. If policy makers wish to address energy poverty, a block capacity tariff can be applied that provides a first tier of capacity below cost. Although economists largely agree that social issues in general are better addressed through other means than energy tariffs, political reality may necessitate this.

A final issue is that the system may be too complex for household consumers. A solution is to charge them a default capacity fee that is based on their peak capacity usage in the previous year, without a physical limitation to consumption. There would be no immediate penalty for exceeding the capacity level, but in this case, the next years' capacity subscription would be based on their new consumption peak. This way, consumers do not need to think about a capacity subscription, but they still have a strong incentive for reducing their contribution to the system consumption peak. Consumers who want to reduce their cost have the ability to either pay attention and reduce their consumption during peak hours, or to opt into the system formally by buying a capacity subscription and committing to that level of peak consumption.

Despite these options to make Capacity Subscription simple and cost-effective for consumers, it appears that the underlying complexity and consumers' lack of familiarity with the concept of limiting peak consumption are major barriers to its implementation. The latter argument does not apply to Spain and Italy, where household electricity consumption is limited to a certain peak capacity—not only during peak moments, but permanently. Nevertheless, system operators and policy makers appear hesitant to implement a temporary capacity restriction to a service that has been taken as granted as water from the tap for the past century, even if consumers have the option of tailoring it completely to their needs without paying more than now. Therefore pilot programs should be developed to test consumer acceptance and find out the best way to design the consumer interface. Given the limited response of consumers to time-varying prices that is presented in Chapter 18, it is worth exploring whether Capacity Subscription is more effective at unlocking consumer flexibility. Capacity Subscription may help remove the information asymmetry that is mentioned in Chapter 6 by providing a simple and clear remuneration for lowering peak consumption.

4.3 The role of storage

One of the challenges for the market design of the future is how to remunerate storage facilities. Storage facilities are expected to play an important role as a complement to vRES. While it is unclear which of the many technological options will end up being the most attractive one(s), their general economic characteristics are clear and therefore it is possible to consider how they fit into the market design. Storage facilities need to recover their capital cost from the price difference when they buy and sell energy,[3] while their operational costs consist mostly of energy losses. With respect to security of supply, their main issue is their limited energy capacity.

The average cost of energy storage per unit of electricity that is supplied becomes extremely high for storage facilities that are only used during rare energy shortages. The business case for storage facilities that are needed to maintain system adequacy in an energy-only model during adverse weather events that occur once every 1 to 10 year hinges on such low probabilities that it is unattractive. However, with Capacity Subscription, storage should be allowed to sell capacity. The question is how much capacity a storage facility can supply reliably. In principle, it is equal to the capacity that it can supply continuously for as long as it is needed. For instance, if the reliability standard requires that the worst event for which the electricity system needs to be prepared is a 2-week (336 h) period with scarcely any solar and wind power. In this case, a storage facility with an energy content of x MWh would be allowed to sell $x/336$ MW of capacity subscriptions. This is probably not a market for batteries, which have a high MWh cost, but rather for other storage technologies with a lower MWh cost.

Small-scale storage plays a different role and may indeed be quite attractive with Capacity Subscription. It can help consumers to stay below their subscribed limit, even during longer periods of LLD activation by allowing them to consume more than their capacity limit as long as the storage unit lasts. The figure in Section 3.2 illustrates a typical demand profile: even in a prolonged period of demand limitation, storage can still be charged when demand is below the subscribed limit and discharged when it

---

[3] In addition, storage facilities can participate in several balancing markets, which may add substantial revenues.

is above. If consumers have an electric car, they can use its battery. This is a simpler concept than many of the solutions that have been proposed for central control of (EV) batteries. Moreover, if the consumer also has solar PV, he already has a converter that can also be used for supplying energy from the battery to the household. In addition to batteries, heat storage can also be used to flatten demand below the subscribed limit. As a next step, one could imagine an aggregator combining consumer batteries into a virtual power plant and helping them collectively stay within the limit be matching aggregate battery power with aggregate demand, effectively using spare battery capacity from one consumer to allow another consumer to exceeds his contracted capacity. Such aggregation of batteries already occurs on a significant scale (Vorrath, 2020).

4.4 Capacity subscription and energy prices in a low-carbon system

Capacity Subscription is designed to achieve system adequacy in an economically efficient manner by providing incentives to consumers to choose their own optimal mix of controllable generation capacity and/or storage that they purchase from the market, self-generation, self-storage, and flexible behavior. Capacity Subscription will provide system adequacy regardless of the design of the short-term market, but the economic efficiency of the system does of course depend strongly on this. Some form of real-time pricing for consumers appears to be a necessary feature of a future electricity market design. Extreme scarcity prices, such as the ones that theoretically are needed in an energy-only market, should not need to occur in a market with Capacity Subscription, however, because the cost of controllable generation is to a large extent covered by the capacity subscription payments (similar as in a capacity market). Such high prices are also not needed to incentivize consumers, as the capacity limits of their subscriptions signal to them when they need to shift load or invest in another form of flexibility. In a conventional power system, the electricity price is therefore expected to be close to the marginal cost of generation at all times, as actual generation shortages are prevented.

However, how will electricity be priced when the marginal cost of generation is close to zero much of the time, and storage and demand response are marginal the rest of the time? Storage units should be expected to sell at opportunity cost, which can be quite high during an extreme shortage period. During more limited demand spikes, the willingness to pay of demand may set the price, which may still be higher than what consumers who paid for a capacity subscription would consider as fair. A capacity subscription provides a sort of option contract to consumers: in exchange for a fixed payment, they receive a right to consume up to the level of their subscription. If the system works well, there should be no energy shortages and the consumers should receive the power for a price that is close to the cost of the provision of capacity, e.g., by generation or storage. However, it is not a real option contract because the price at which the provider of capacity sells electricity is not specified.

A solution could be to provide the consumers with explicit option contracts so they have a guaranteed maximum price linked to their capacity subscription. In this variation, the capacity subscription would provide consumers with the right to consume up to the volume of subscribed capacity and pay no more than the option strike price. This would protect them against unforeseen shortages, e.g., due to an investment cycle or unexpected unavailability of large generation and storage facilities. In an interconnected market, the option variant would also protect consumers against importing high shortage prices. As the sellers of the capacity subscription compete for the subscription price, the requirement to sell the energy that they produce at a maximum price would be priced into the capacity price and therefore this should not distort market efficiency. This variation of capacity subscription could be considered as a decentral version of the well-known reliability options model. (Pérez-Arriaga, 2001; Vázquez et al., 2002). Box 12.3 describes how a related concept is also emerging in the field of distribution network pricing.

BOX 12.3 Can capacity subscription also be used as a basis for grid tariffs and the handling of grid constraints?

The cost structure of the electricity networks is almost exclusively a function of peak usage, as the variable costs are low in comparison with the capital costs. Therefore the principle that underlies Capacity Subscription can be applied to network tariffication as well in the form of a capacity-based network tariff that is based on the contribution of the network user to the annual network peak load. A first step in this direction has already been made in the Netherlands, where the usual volumetric (kWh-related) network tariffs for small consumers have been replaced with a fixed network tariff. This is the same for all small consumers, however, while a link with peak consumption would provide a further improvement. Several large Dutch distribution companies are currently considering basing their household network tariffs on capacity "bandwidths," in

Continued

BOX 12.3 Can capacity subscription also be used as a basis for grid tariffs and the handling of grid constraints?—cont'd

which the tariff increases stepwise when a household moves to a higher bandwidth. The number and size of the bands—or whether not to have a tariff that increases continuously with peak consumption—and the penalty for exceeding the contracted bandwidth are key design parameters of this proposal.

An obvious solution is that the subscribed capacity level is also the basis for the grid tariff. An option is then to allow the DSO to activate the LLDs in case of local grid constraints. The downside for the consumer is increased activation of the LLDs (at least at some locations) and more complexity. Still there is a potential here that should be further explored. Currently, the Dutch DSOs are not considering the implementation of LLDs but a fine for exceeding the contracted network capacity (perhaps only after two warnings). As this could save the cost of implementing LLDs in case the existing smart meters do not have the functionality to curtail consumption, the use of a penalty instead of physical curtailment could also be considered for the capacity subscription market.

5. Conclusions

Capacity Subscription is designed to ensure an adequate volume of controllable electricity generation and storage capacity in a power system with a high share of variable renewable energy. An important feature of Capacity Subscription is that demand for capacity is based on the individual consumers' preferences for uninterrupted supply. Consumers can adjust the amount of capacity to which they subscribe to their other options such as self-generation, storage, and demand flexibility. Therefore it helps consumers to make an optimal trade-off between their own flexibility and the cost of flexible resources that they buy in the market. This unique feature of this market design is expected to contribute significantly to the economic efficiency—and therefore lower consumer cost—of a low-carbon electricity system. It is the only market design that actually lets the consumer himself decide how and how much he will spend on the reliability of power supply.

The consumer demand for capacity subscription provides wholesale suppliers of flexible generation capacity a market-based, but stable revenue for their capacity. Consumers, on the other hand, are provided the certainty that they have access to the electricity that they need at affordable prices. Thus capacity subscription reduces risk for both sides of the market. For generators, this means that they can provide the required capacity at a lower cost of capital. For consumers, this means that they do not need to worry about the possibility of infrequent but high price spikes. While we focused in this chapter on small consumers, Capacity Subscription is intended to be implemented for all consumers.

Capacity Subscription is designed to give the consumer full control: the consumer decides how much to pay for firm capacity, versus becoming more flexible or investing in his own (battery) capacity. Consumers may delegate these decisions to an aggregator who may be better able to choose between these options and to provide the required operational interventions when the capacity limit is enforced, but fundamentally, this is not necessary.

Grid tariffs that are based on the same principle are being investigated by some DSOs. For similar reasons, they may prove a better alternative to volume-based grid charges in a power system with a high share of flexible, and therefore coincident, demand.

References

De Vries, Laurens, 2007. Generation adequacy: Helping the market do its job. Utilities Policy 15 (1), 20−35.
Doorman, G.L., 2005. Capacity subscription: solving the peak demand challenge in electricity markets. IEEE Transactions on Power Systems 20 (1), 239−245.
Doorman, G.L., Botterud, A., 2008. Analysis of generation investment under different market designs. IEEE Transactions on Power Systems 23 (3), 859−867.
Hirth, L., 2014. The economics of wind and solar variability: How the variability of wind and solar power affects their marginal value, optimal deployment, and integration costs. Technical University Berlin: Doctoral dissertation.
Moghanjooghi, H.A., 2017. Model-based Analysis of Generation Resource Adequacy in Energy-Only Markets, pp. 1−94865.
Ovaere, M., et al., 2019. How detailed value of lost load data impact power system reliability decisions. Energy Pol. 132, 1064−1075.
Pérez-Arriaga, I.J., 2001. Long-term Reliability of Generation in Competitive Wholesale Markets: A Critical Review of Issues and Alternative Options. IIT Working Paper IIT-00-098IT.
Poplavskaya, K., Lago, J., De Vries, L., 2020. Aggregators today and tomorrow: from intermediaries to local orchestrators? In: Sioshansi, F.P. (Ed.), Behind and beyond the Meter, Digitalization, Aggregation, Optimization, Monetization. Elsevier.
Shivakumar, A., Welsch, M., Taliotis, C., Jakšić, D., Baričević, T., Howells, M., Gupta, S., Rogner, H., 2017. Valuing blackouts and lost leisure: estimating electricity interruption costs for households across the European Union. Energy Res. Soc. Sci. 34 (May), 39−48. https://theconversation.com/why-did-energy-regulators-deliberately-turn-out-the-lights-in-south-australia-72729.

Stoft, S., 2002. Power System Economics — Designing Markets for Electricity. IEEE Press, Wiley-Interscience.

TenneT, April 23, 2020. Equigy-platform biedt Europese consumenten toegang tot de duurzame energiemarkt van morgen. Press release.

Vázquez, C., Rivier, M., Pérez-Arriaga, I.J., 2002. A market approach to long-term security of supply. IEEE Transactions on Power Systems 17 (2), 349—357. https://doi.org/10.1109/TPWRS.2002.1007903.

Vorrath, S., May 29, 2020. Ausgrid Seeks Thousands of NSW Homes to Join Expanded Virtual Power Plant. Renew Economy. Available at. https://reneweconomy.com.au/ausgrid-seeks-thousands-of-nsw-homes-to-join-expanded-virtual-power-plant-16010.

CHAPTER 13

Variable renewables and demand flexibility: Day-ahead versus intraday valuation

Reinhard Madlener[1,a], Oliver Ruhnau[2]
[1]Institute for Future Energy Consumer Needs and Behavior (FCN), School of Business and Economics / E.ON Energy Research Center, RWTH Aachen University, Aachen, Germany; [2]Hertie School, Berlin, Germany

1. Introduction

The volatility and uncertainty related to electricity generation from renewables, and in particular from wind and solar energy sources, has profound implications for day-ahead (DA) and intraday (ID) electricity markets. First, the volatility of these variable renewable energy sources (VRES) leads to a lower market value relative to the base price, already in the DA market. This is because the power supply from VRES features near-zero marginal cost, depressing prices whenever it is available. Second, the uncertainty of renewable power production leads to balancing cost: VRES need to adjust their day-ahead positions at adverse prices in the ID and imbalance (IB) markets, which further reduces their market value.

Three main mitigation strategies can be applied to counteract low market values and high balancing cost (Fig. 13.1):

- improved forecasting,
- enhanced marketing and trading, and
- the use of flexibility resources.

Among these flexibility resources are flexible conventional power plants, storage units, and demand-side flexibility, also referred to as demand response (DR). The latter seems particularly promising in the context of the expected increase in potentially flexible electricity demand for heating, transportation, and the production of synthetic gases.

This chapter describes the market effects of volatile and uncertain renewable power supply and the corresponding mitigation options. A framework is provided to understand and analyze the mechanisms behind

---

[a] The authors gratefully acknowledge helpful comments received from Anselm Eicke.

Variable Generation, Flexible Demand
ISBN 978-0-12-823810-3
https://doi.org/10.1016/B978-0-12-823810-3.00005-4

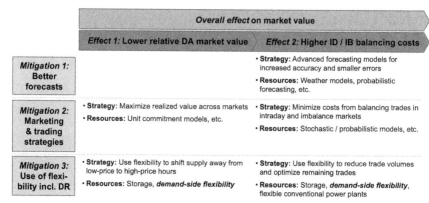

Figure 13.1 Effects of volatile renewable power supply, mitigation strategies, and resources available. *(Source: Garnier, E., Madlener, R., 2020. Valuation of Demand Response for Day-Ahead and Intra-day Trading of Renewable Electricity, FSR Symposium "Variable Generation, Flexible Demand", Fiesole (Florence), Italy, February 19, 2020.)*

DA and ID markets and their interrelatedness. The effect of price and volume uncertainty in DA and ID markets is evaluated for renewable power generators in terms of their market value and balancing cost, including the economic implications of forecasting and trading optimization. Moreover, the role and value of demand-side flexibility is evaluated in the context of renewable volatility and uncertainty. Some key aspects in this chapter are illustrated with empirical analyses that are based on, but not limited to, the example of Germany.

The remainder of the chapter is organized as follows:

- Section 2 describes a framework for the understanding and analysis of DA and ID markets;
- Sections 3 and 4 discuss trading on the DA and ID markets from the perspectives of renewable power generators and demand response aggregators, respectively;
- Followed by the chapter's conclusions (Section 5).

2. Trading in day-ahead and intraday markets

The value of supply and demand in electricity markets is time-dependent, in a double sense: it depends on the time of delivery and on the (lead) time of trading (Hirth et al., 2016). This chapter focuses on electricity spot markets, namely the DA and ID markets. These two consecutive energy-only markets concern the same product, viz. electricity delivered at a certain time, but that differs by the lead time it is traded.

The DA market is the primary market for matching electricity supply and demand. In Europe, this market is organized in coupled pricing zones, for which electricity can be traded for individual hours or blocks of multiple consecutive hours. Bids can be submitted up to 12 p.m. on the day before delivery, which are then cleared at a single price emerging from the intersection of the aggregated supply and demand curves (pay-as-clear). Further trade can happen outside the power exchange (over-the-counter, OTC), but the DA price often serves as a reference for such bilateral deals.

In contrast, the ID market fulfills the secondary function of adjusting hourly DA positions to updated, quarter-hourly ID schedules. In Germany, the ID market consists of the ID auction, which—despite the name— happens on the day before delivery at 3 p.m., and of a continuous trading period until 30 min before delivery. Whereas the ID auction mainly serves to balance deterministic deviations between hourly and quarter-hourly schedules, continuous trading helps to balance stochastic deviations due to supply and demand forecasts as well as unplanned power plant outages. In the continuous ID market, the price varies from trade to trade over time, i.e., there is not a single price. In addition, prices are different for buying and selling in the continuous ID market, and the difference is referred to as the "bid-ask spread." Most European countries feature a similar market design, but some organize the ID market in several consecutive auctions instead of continuous trading.[1]

The distinct roles of the DA market and the ID market are reflected in the trading volumes at the European power exchange EPEX SPOT, which covers eight central-western European countries. While the primary DA trading accounts for 400−500 TWh of trades, the sum of the secondary ID markets, including both the ID auction and the ID continuous trading, reached a maximum of only 82 TWh in 2018 (Fig. 13.2).

Around half of the DA trading volume (200−250 TWh) relates to Germany, Austria, and Luxembourg, which up to September 2018 shared one bidding zone. For comparison, the overall annual electricity consumption in these countries is around 620 TWh, which implies that 30%−40% of all consumption is traded at the DA market. Concerning the ID trading volume, the majority relates to Germany (50 TWh in 2018), and 90% of this volume is traded continuously whereas only the remaining 10% is traded at the ID auction.

---

[1] For an example of how this is treated in Spain, refer to Chaves-Ávila and Fernandes (2015).

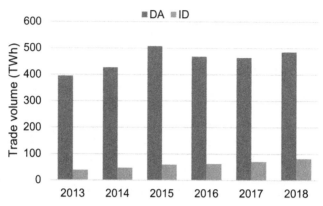

Figure 13.2 Trading volumes on the EPEX SPOT day-ahead (DA) and intraday (ID) electricity markets for 2013—18, covering Austria, Belgium, France, Germany, Great Britain, Luxembourg, the Netherlands, and Switzerland. *(Source: Own illustration, based on data from EPEX SPOT.)*

Within the continuous trading period, most trading occurs shortly before gate closure (Koch and Hirth, 2019). Hence, we conclude that continuous ID trading is used for balancing stochastic deviations shortly before delivery, and—within the broader context of ID markets—this is more important than balancing deterministic deviations in the ID auction. Even though being substantially smaller than the DA trading volume, the total ID trading volume has doubled at EPEX SPOT and even tripled in Germany from 2013 to 2018. This trend has been examined in more detail by Koch and Hirth (2019), who find that trading volumes particularly increased for quarter-hourly products and during nighttime, indicating the implementation of more and more 24/7 trading.

In economic terms, DA and ID markets can be described by their supply and demand curves (Henriot, 2014; Pape et al., 2016). As illustrated in Fig. 13.3, adjusted forecasts of load and renewable generation, as well as power plant outages, can be represented as a shift of the residual demand curve: market participants realizing that they will produce less or consume more than their position in the DA market have to buy (back) additional electricity, and vice versa. The dispatchable supply can be divided into plants that committed production in the DA market (with marginal cost below the DA price) and plants that were not (with higher marginal cost). In the ID market, committed plants are willing to buy back electricity when the ID price falls below their marginal cost, whereas uncommitted plants are willing to sell electricity when the ID price rises above their marginal

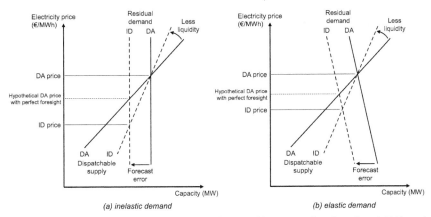

Figure 13.3 Impact of a forecast error on the equilibrium in the day-ahead (DA) and intraday (ID) electricity markets with inelastic (a) and elastic (b) demand. *(Source: Own illustration.)*

cost. However, technical restrictions limit such short-term supply adjustments and hence reduce the liquidity of the ID market as compared to the DA market, which leads to a steeper supply curve in the ID market (Fig. 13.3).

Every short-term shift of the residual demand curve, whether caused by forecast errors, plant outages, or other reasons, will lead to a price spread between the ID and DA markets. For the example provided in Fig. 13.3, think of unexpected additional renewable generation causing a decline in the ID price. This price decline is aggravated by the steeper supply curve in the ID market. For comparison, Fig. 13.3 illustrates the hypothetical case of perfect foresight: if the additional renewable power generation was already known in the DA market, this would have also caused a decline in the DA price ("merit order effect," cf. Sensfuß et al., 2008). It can be seen that, because of the steeper supply curve in the ID market, the ID price drops below this hypothetical DA price with perfect foresight. Likewise, an unexpected shortage of electricity will lead to a stronger rise in the ID price compared to the merit order effect in the DA market. Due to the efficient market hypothesis (Fama, 1970), the expected price spread between the ID and DA markets should be zero, which is empirically supported by Narajewski and Ziel (2019) for the example of Germany. Nevertheless, DA and ID markets offer scope for strategic behavior, which may further increase price volatility in the ID market (Rintamäki et al., 2020).

The resulting increased price volatility in the ID market, however, can be moderated by demand response. This is illustrated in Fig. 13.3b, where demand response is indicated as a price elasticity of demand, leading to a nonvertical, downward-sloping demand curve. As a result, part of the unexpected additional renewable power supply is compensated by an increase of demand as a response to falling prices. Note that this effect would also have been occurred in the hypothetical case of a DA market with perfect foresight, but is more significant in the ID market because of the stronger decline in the ID price (due to the lower ID liquidity). Hence, increased demand-side flexibility can partly attenuate the price-driving effect of inflexibility on the supply side.

Empirically, the reduced flexibility in the ID market becomes apparent in the increased volatility of ID prices. This is illustrated in Fig. 13.4 for the example of Germany. As there is no single price in the continuous ID market, the volume-weighted average ID price over the entire trading period (ID average) and over the last 3 h of trading (ID3) are considered here.[2] Indeed, it can be observed that price volatility increases with

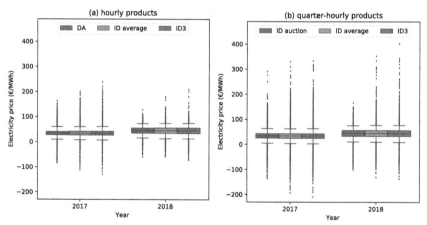

Figure 13.4 Price distribution for hourly (a) and quarter-hourly (b) products in the German electricity spot market 2017—18. The day-ahead (DA) and intraday (ID) auction prices are compared with the volume-weighted average ID price over the entire trading period (ID average) and over the last 3 h of trading (ID3). Boxes depict the range between the first and third quartile (interquartile range), whiskers indicate the 5%—95% quantiles, and points indicate observations exceeding these. *(Source: Own illustration, based on data from EPEX SPOT.)*

[2] Note that the price of single trades will be even more volatile than the volume-weighted average.

decreasing lead time: the ID average price and in particular the ID3 price are more volatile than the prices in the DA and ID auctions, both held on the day before delivery. In a more detailed econometric analysis, (Kiesel and Paraschiv, 2017) identify updated forecasts of renewable power generation as a key driver of intraday market prices. Goodarzi et al. (2019) investigate the impact of renewable energy forecast errors on both imbalance volumes and electricity intraday spot market prices.

Four more insights on the ID price volatility are provided in Fig. 13.4:

1. While the interquartile range (depicted as boxes) does not change significantly, outliers (depicted as points) become more frequent and more extreme.

2. It can be observed that volatility is lower for hourly products (Fig. 13.4a) than for quarter-hourly products (Fig. 13.4b). This seems reasonable as, because of the no-arbitrage condition, the price of the hourly products should equal the mean price of the related quarter-hourly products.

3. There are less extreme negative prices, which may be explained by renewable electricity being curtailed when negative prices fall below a certain threshold.

4. The mean price increased and the frequency of extreme prices decreased in 2018. This can be traced back to an increase in CO_2 prices, pushing inflexible coal power plants out of the market in that year (Agora Energiewende, 2019).

While the ID market helps to balance stochastic deviations from the DA market positions, actual generation and consumption may still deviate from the adjusted ID schedules, e.g., due to forecast errors and plant outages becoming apparent only after the market gate closure. In Europe, such deviations are balanced in real time by the transmission system operators and financially settled at the imbalance settlement price, hereafter referred to as the IB price. In this context, Koch and Hirth (2019) argue that trading at the ID market substitutes for balancing by the transmission system operators: the better market parties adjust their schedules in the ID market, the less deviations are settled at the IB price.[3]

---

[3] Indeed, they find for the case of Germany that more active ID trading (Fig. 13.2) was one of the reasons why the utilization of balancing reserves declined despite of an increase in uncertain renewable electricity generation.

3. The perspective of renewable power generators

From the perspective of renewable power generators, trading implications can best be evaluated in terms of their market value, i.e., the revenues they can earn per MWh of electricity generation (Pape, 2018). Based on the prices at the DA and ID markets at time t, p_t^{DA} and p_t^{DA}, as well as imbalance price, p_t^{IB}, the market value, MV, can be expressed as:

$$MV = \frac{\sum_t g_t^{DA} \cdot p_t^{DA} + \left(g_t^{ID} - g_t^{DA}\right) \cdot p_t^{ID} + \left(g_t - g_t^{ID}\right) \cdot p_t^{IB}}{\sum_t g_t} \tag{13.1}$$

Where g_t^{DA} and g_t^{ID} are the committed generation volumes after trading in the DA and ID market, and g_t is the actual generation.

This definition of the market value explicitly considers the uncertainty related to renewable power generation. The effect of this uncertainty can be isolated by calculating the market value under the assumption of perfect foresight, MV^*, when trading in the DA market:

$$MV^* = \frac{\sum_t g_t \cdot p_t^{DA}}{\sum_t g_t} \tag{13.2}$$

On this basis, balancing cost can be defined as the difference between the market value with and without uncertainty (Ruhnau et al., 2020):

$$BC = MV^* - MV = \frac{\sum_t \left(g_t - g_t^{DA}\right) \cdot p_t^{DA} + \left(g_t^{DA} - g_t^{ID}\right) \cdot p_t^{ID} + \left(g_t^{ID} - g_t\right) \cdot p_t^{IB}}{\sum_t g_t}$$

$$= \frac{\sum_t \left(g_t^{DA} - g_t^{ID}\right) \cdot \left(p_t^{ID} - p_t^{DA}\right) + \left(g_t^{ID} - g_t\right) \cdot \left(p_t^{IB} - p_t^{DA}\right)}{\sum_t g_t}$$

$$= \frac{\sum_t \left(g_t^{DA} - g_t^{ID}\right) \cdot \left(p_t^{ID} - p_t^{DA}\right) + \left(g_t^{ID} - g_t\right) \cdot \left(p_t^{IB} - p_t^{DA}\right)}{\sum_t g_t} \tag{13.3}$$

Note that this definition is not limited to costs arising in the balancing mechanism but also includes costs from balancing activities in the ID market.

Trading the perfect forecast in the DA market serves as a benchmark for the calculation of balancing cost for two reasons:

1. ID and IB prices and related revenues per MWh of renewable electricity are more volatile than in the DA market, implying that renewable power generators can reduce their risk by trading DA, and
2. ID and IB prices tend to worsen from the perspective of renewable power generators as compared to DA prices.

Both points are connected to the lower liquidity and steeper bid curve in the ID market. For the simplified example of Fig. 13.3, additional generation needs to be sold at an ID price that is lower than the DA price, even when accounting for the price decline that would have been occurred in the DA market in the hypothetical case of perfect foresight. Similarly, a lack of generation needs to be bought back at an ID price which is over-proportionally higher than the DA price. Both lead to a decline in the market value and an increase in balancing cost. Against this background, balancing cost can be interpreted as "opportunity costs for trading the forecast error at potentially worse prices in the ID market or IB mechanism" (Pape, 2018, p.186). Note, however, that while prices tend to worsen, they can also improve occasionally. For example, if some unexpected additional renewable power supply coincides with a larger power plant outage, the additional supply may be sold at an ID price that exceeds the DA price.

To illustrate the importance of balancing cost, they are estimated for the example of the national wind and solar generation in Germany. Balancing costs are distinguished into the cost of balancing deterministic and the cost of balancing stochastic deviations. The cost of balancing deterministic deviations (ramps) is evaluated based on the hourly DA price and the quarter-hourly ID auction price, assuming that the hourly average of the DA forecast is sold at the DA auction, whereas quarter-hourly deviations from the hourly average are traded at the ID auction. The cost of balancing stochastic deviations (forecast errors) is evaluated using the DA price and the ID3 price. Hence, the assumption is made that (1) all stochastic deviations are balanced in the ID market and not by the imbalance mechanism[4] and (2) all forecast errors are balanced during the last 3 h before ID gate closure.

---

[4] Hence, it is assumed that $g_t^{ID} = g_t$, and Eq. (13.3) simplifies to

$$BC^{DA-ID} = \sum_t \left(g_t^{DA} - g_t^{ID}\right) \cdot \left(p_t^{ID} - p_t^{DA}\right) \Big/ \sum_t g_t.$$

The role of the imbalance mechanism and the optimal timing of trade is discussed afterwards.

The results of the balancing cost estimation are presented in Fig. 13.5a. While the cost of balancing stochastic deviations are in the same order of magnitude for wind and solar power (0.34−0.67 €/MWh), the cost of balancing deterministic deviations are substantially higher for solar power (0.31−0.43 €/MWh) compared to wind power (0.03 €/MWh). This can be explained by the diurnal ramps in the generation profile of solar power. It should be noted that stochastic balancing cost are not evenly distributed throughout the year. In fact, 1% of all yearly observations account for around 25% and 40% of the stochastic wind and solar balancing cost, respectively. This is when large forecast errors lead to extreme spreads between the ID auction and the ID3 prices.[5]

For comparison, the market value of wind and solar power under the assumption of perfect foresight was around 30 €/MWh in 2017 and 40 €/MWh in 2018 (Fig. 13.5b). This exceeds balancing cost by two orders of magnitude, making the latter seem almost negligible. However, support schemes such as market premia and contracts-for-differences are often designed such that renewable power generators do not bear the risk of variances in the DA market value. In contrast, they are fully accountable for balancing cost and hence there is a strong incentive to optimize these.

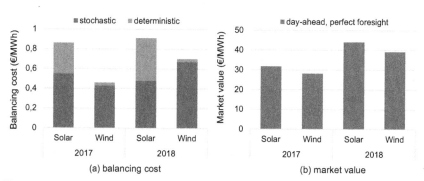

(a) balancing cost   (b) market value

Figure 13.5 Balancing cost in 2017 and 2018 in Germany due to deterministic (ramps) and stochastic (forecast errors) deviations between the DA auction and ID gate closure (a), compared to the hypothetical DA market value with perfect foresight (b). *(Source: Own analysis, based on data from ENTSO-E and EPEX SPOT.)*

---

[5] As illustrated in Fig. 13.3, forecast errors drive the spread between DA in ID prices, and balancing costs are the product of forecast errors and the price spread (Eq. (13.3)).

The above estimation of the balancing cost neglected the balancing mechanism. In fact, there is a remaining forecast error at ID gate closure, leading to inevitable deviations that are financially settled at the IB price. As for the ID market, forecast errors of renewable power generators drive the IB price in such a way that it tends to worsen from the generator's perspective. Depending on the design of balancing mechanisms, IB prices may be even more volatile than ID prices. The IB price hence constitutes an economic incentive for balancing in the ID market (Koch and Hirth, 2019).

Balancing cost can be reduced through forecasting and optimized bidding strategies. Technically, DA and ID bids are based on short-term electricity generation forecasts. Multiple studies present forecasting models for solar and wind power generation using different approaches, e.g., fundamental equations, statistical methods, artificial intelligence techniques, or a combination of these (Foley et al., 2012; Inman et al., 2013). Concerning the economic implications of forecasting, it is an intuitive assumption that trading based on the most accurate, unbiased point forecast would lead to optimal results. However, as discussed by Ruhnau et al. (2020), there are two exemptions to this rule:

1. The "asymmetry effect" concerns the case when upward and downward deviations lead to different costs, e.g., by design of the balancing mechanism. As a result, it is beneficial to bias trading so as to minimize the probability of the more costly deviations. The optimal quantity can be derived from probabilistic forecasts (e.g., Pinson et al., 2007).
2. The correlation effect refers to correlation between forecast errors and the market price spread, e.g., between the DA and the ID market. In fact, if there were no correlation between these two metrics in the sense that prices worsen from the perspective of renewable power generators, the average balancing cost would be zero. As shown in Ruhnau et al. (2020), forecasting models differ by their correlation with the market price spread and hence the most accurate forecast is not necessarily economically optimal; a forecasting model with larger forecast errors but lower correlation with market prices may lead to a better economic outcome.

Balancing cost can further be optimized through the timing of trading. In this regard, the above-introduced framework of the market value and the balancing cost simplifies ID trading: it implicitly assumes that there is only one trade at one price. In fact, as pointed out in Section 2, ID trading is continuous and agents must decide when to trade, and they may trade multiple times. This optimization of continuous trading implies a trade-off between the (expected) cost of uncertainty (which decreases over the trading period) and the (expected) cost of adverse prices (which tends to increase over the trading period). In addition, trading early based on an updated forecast means risking to trade double in the sense that subsequent forecast updates may require trades in the opposite direction, which would have netted out with the early trade (Henriot, 2014; Garnier and Madlener, 2015).

Some renewable power generators may be part of a larger portfolio including controllable generators (power sources) and consumers (power sinks), and storage units. In this case, the portfolio operator of such a "virtual power plant" (VPP) may dispatch these assets as an alternative to trading at the ID market and settling deviations at the IB mechanism. The logic and economics of such self-balancing is discussed in the subsequent section for the case of demand response.

4. The perspective of demand response

Here, DR is understood as a short-term adjustment of electricity demand to changes in timely resolved electricity prices: demand is reduced when prices are high and increased when prices are low. Often, however, DR comes in the form of a temporal shift, rather than an overall reduction or increase, i.e., a demand reduction at time t is compensated by an equivalent increase at another time t' (demand shifting).

Traditionally, consumers have hardly responded to short-term price changes, but the electricity demand was taken as "given" (Chapter 5). However, as discussed throughout this volume, this is expected to change, for multiple reasons:

1. There is an increased requirement for flexibility by renewable power generators, making DR more valuable (Part One of this volume).
2. Information and communication technologies make DR more accessible, also for small consumers (inter alia Chapter 14).
3. Electricity is expected to be increasingly used for heating and transportation, which entail flexibility in the form of thermal storage units and vehicle batteries (inter alia Chapter 8). In addition, excess electricity may be flexibly used to produce synthetic fuels (Ruhnau et al., 2019).

Regulatory and policy aspects as well as business model availability have an influence on the realizable DR potentials, too (IRGC, 2015).

In this context, conventional research has often narrowly focused on the response of individual energy users (e.g., private households) to smart meters, in-home displays, dynamic pricing, or peer comparison effects

(Goulden et al., 2014). However, behavioral uncertainty (change in attitudes, structural change) and the recognition of energy use as a complex socioeconomic technical landscape raises caution regarding the predictability of consumer behavior at the individual level. Aggregators pooling individual users and appliances—to VPPs (Garnier and Madlener, 2016) or within citizen/clean energy communities (CECs, Gui and MacGill, 2018)—can mitigate the challenge of predicting DR by focusing on aggregates rather than individual behavior, and enable the trading of distributed small-scale flexibilities as aggregate entities on the exchange-based electricity markets.[6]

Focusing on DA and ID markets, the value of DR can be calculated based on the prices in these markets, and based on how much electricity consumption is increased (d_t^+) and reduced (d_t^-):

$$MV^{DR} = \frac{\sum_t \left(d_t^{DA+} - d_t^{DA-} \right) \cdot p_t^{DA} + \left(d_t^{ID+} - d_t^{ID-} \right) \cdot p_t^{ID}}{d_{max} \cdot T} \tag{13.4}$$

Here, the value is normalized by the maximum demand that can be shifted in energy terms, d_{max}, and the examination period, T (e.g., the value is calculated per kWh of demand-side flexibility per year of operation).

Based on this formula, two DR strategies can be distinguished: DA activation, which is valued at the DA price, and ID activation, which is valued at the ID price. Of course, a mix of these strategies, i.e., trading DR in both markets, is conceivable, and DR could be traded at multiple occasions over the continuous ID trading period. In the case of demand shifting, the net of the overall increase and the overall decrease is zero. Hence, DR profits from volatility in market prices, which is driven by renewables and their forecast error, as discussed above.[7]

Next, the market value of DR is quantified for a stylized example in the German market. For this, it is assumed that 1 kWh demand is shifted every

---

[6] While aggregating demand eases market integration, it can be challenging to establish a trustable and working relationship with customers (Chapter 7), and Burns and Mountain describe the limitations of what average customers are able to understand or do (Chapter 18).

[7] For a (decomposition) analysis of the evolution of electricity price levels and volatility in light of rising shares of VRES on the German day-ahead market see, e.g., Khoshrou et al. (2019). Note that DR can also be used to avoid balancing costs in the balancing mechanism and to provide balancing services, as discussed by Koliou et al. (2014), a topic which is beyond the scope of this chapter.

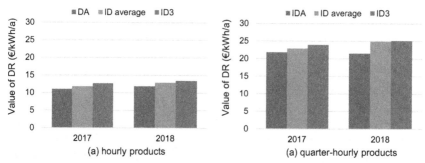

Figure 13.6 The value of DR in the German electricity spot market 2017–18. Activation strategies are compared for hourly (a) and quarter-hourly (b) products in the day-ahead (DA) and intraday (ID) auction as well as in continuous ID trading, using the volume-weighted average ID price over the entire trading period (ID average) and over the last 3 h of trading (ID3). *(Source: Own illustration, based on data from EPEX SPOT.)*

day, away from the time with the highest price and to the time with the lowest price. This assumption implies perfect foresight on market prices and that DR is available all the time, but it can only be activated once a day. The results are presented in Fig. 13.6 for DA and ID activation, and for hourly and quarter-hourly products.

As expected, the market value of DR is driven by price volatility as analyzed in Fig. 13.4: it is twice as high for quarter-hourly products (21−25 €/kWh/a, Fig. 13.6a) than for hourly products (11−13 €/kWh/a, Fig. 13.6b), and it slightly increases with decreasing lead time. Note that, in order to capture the higher price volatility of quarter-hourly products, the DR application must be able to adjust demand by 1 kWh in a quarter-hour, which implies a power adjustment of 4 kW rather than 1 kW for the hourly products.

For the above definition and estimation of the market value, it was implicitly assumed that the flexibility of DR is sold at the ID and DA markets. In contrast, Garnier and Madlener (2020) examine the case of DR being integrated into a portfolio of renewable power generation units, which is jointly optimized at these marketplaces. In this case, the DA activation of DR can increase the sales revenues of renewables (i.e., the DA market value), and the ID activation of DR can reduce the balancing cost of renewables. As compared to the isolated DR optimization, the integrated optimization can yield additional benefits by avoiding the bid-ask spread and other transaction costs.[8]

Fig. 13.7 illustrates the different activation strategies. In the DA strategy, the portfolio operator has different amounts of excess power available for delivery (or sale) across a certain time window. The flexibility of shifting some of the demand into times of low market prices creates economic value, as some supply volumes are freed for sale in the market at times of high prices. Relevant are the price differences and the bid-ask spread between current time t and some other time slot M (cf. Fig. 13.7).

In the ID strategy, in contrast, valuation depends on both prices and forecast error dynamics, and in particular on the sign of the forecast error. An operator is either short, having to purchase ID for the current time slot t, or long, being able to sell volumes ID for the current time slot t. These two settings enable the operator to exercise two distinct strategies (cf. Fig. 13.7):

(a) Shifting consumption from the current long (short) position to another slot M with a long (short) position with a lower price.

(b) Enlarging (or netting) an open position by moving loads from the current long (short) position to a slot E (N) with a short (long) position. Evidently, such shifts are only possible if there are time slots available with both long and short positions.

Note that option (a) is similar to the DA valuation, whereas option (b) needs some further differentiation. In particular, using DR for position netting will avoid the bid-ask spread and other transaction costs (see Garnier

---

[8] Wozabal and Rameseder (2020) determine optimized bidding strategies of integrated trading of renewable electricity in the Spanish day-ahead and all (six staggered, auction-based) intraday markets, treating both electricity prices and wind power production as stochastic. The approach enables the decision-maker to account for the interdependency of the decisions in these markets and the information available from the opening of the day-ahead market until the physical delivery of the electricity. The combined trading strategy is proven to outperform more limited trading strategies, and stochastic optimization is shown to dominate trading based on deterministic planning. In related work, Rintamäki et al. (forthcoming) investigate the potential of strategic, coordinated bidding in day-ahead and intraday markets by a flexible producer aiming at raising profits through exerting market power (achieved through induced transmission network congestion or lack of competitive generation capacity). Simultaneous market clearing aimed at reducing balancing costs reduces, but not entirely eliminates, the potential for strategic bidding (for the case of Nordic market data); the potential role of DR in such a setting is not investigated, but seems to be an important topic for future research.

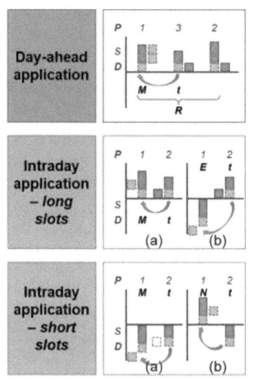

Figure 13.7 Day-ahead and intraday strategies of the system operator using DR. *(Source: Garnier, E., Madlener, R., 2020. Valuation of Demand Response for Day-Ahead and Intra-day Trading of Renewable Electricity, FSR Symposium "Variable Generation, Flexible Demand", Fiesole (Florence), Italy, February 19, 2020)*

and Madlener (2020) for a formal and more detailed analysis, both analytical and empirical).

In the context of portfolio optimization, it can be shown analytically that quite likely ID activation of DR will yield more value. Three value determinants can be identified that drive the relative advantage of ID activation: ID prices are more volatile, there is a (larger) bid–ask spread in the ID market[9], and other transaction costs are also higher (Garnier and Madlener, 2020).

---

[9] Hagemann and Weber (2013) find an intraday bid–ask spread of 3 €/MWh compared to only 0.25 €/MWh day-ahead. In addition, because of the higher average bid–ask spread and the higher price volatility intraday, it seems plausible that also BAS volatility might be higher intraday.

Apart from market-side value creation effects of DR, costs should not be ignored. The variable costs of shifting loads arise from the opportunity costs of customers and from activation costs. Obviously, both opportunity costs and activation costs of DR will reduce its net benefit, but it is often hard to quantify these exactly. Moreover, they can be quite heterogeneous, both across segments (industry, service sector, private households) and actors (behavioral differences, differences in the appliance stock). Activation costs are also dependent on whether the appliances are "smart," so that they can be activated remotely and maybe even automatically controlled in a sophisticated manner. However, additional investment costs may occur for such a smart infrastructure.

Finally, it should be noted that whether DR is activated DA or ID will also influence costs. DA activation implies better predictability for the DR providers, which can lower activation and opportunity costs (better anticipation). The extent of the advantage compared to ID activation depends on the timing of the latter. On the other hand, intraday activation can be done for quarter-hourly time slots—enabling a fine-tuned temporal shifting of demand (e.g., within quarter-hourly intervals), which possibly reduces opportunity costs. Whether activation costs are higher for the DA or ID activation strategy hence needs case-by-case assessments.

5. Conclusions

In this chapter, trading in the DA and ID markets was analyzed, first from a general perspective and then from the perspectives of renewable power generators and DR, respectively. As a key feature of continuous-trade ID markets, we have identified the increased price volatility at shorter lead times and higher temporal granularity. This increased volatility can be traced back to the limited flexibility of market participants and hence the reduced liquidity in the ID market.

The increased price volatility in the ID market is amplified by renewables: their actual production inevitably deviates from hourly DA schedules, due to (deterministic) ramps and (stochastic) forecast errors. Because renewable deviations are correlated with the market price spread, renewable power generators face balancing cost, which they can optimize through enhanced (less correlated) forecasting and optimized trading strategies. At the same time, price volatility drives the value of DR and hence triggers its adoption. Indeed, we have discussed that activating DR based on

ID prices is likely to be more economical. An increased activation of DR in the ID market will in turn moderate the ID price volatility.

Put differently, the renewables create a demand for short-term flexibility, which can be supplied by DR. This supply and demand of flexibility is matched in the ID market based on price volatility. Of course, DR is not the only source of flexibility, but it will compete with dispatchable electricity generation and storage. In this context, we have argued that a competitive advantage arises from integrating renewables and DR into one portfolio or VPP, i.e., matching flexibility supply and demand outside of the market, thus avoiding transaction costs.

As the volatile and uncertain electricity generation from renewable energy sources grows, the demand for short-term flexibility increases. In the future, DR may become a key resource to supply this flexibility demand. In any case, the ID market provides a critical price signal to both renewable power generators and flexibility suppliers, and its importance is likely to continue to grow.

Bibliography

Agora Energiewende, 2019. Die Energiewende im Stromsektor: Stand der Dinge 2018. Rückblick auf die wesentlichen Entwicklungen sowie Ausblick auf 2019. Agora Energiewende.   https://www.agora-energiewende.de/veroeffentlichungen/die-energiewende-im-stromsektor-stand-der-dinge-2018/.

Chaves-Ávila, J.P., Fernandes, C., 2015. The Spanish intraday market design: a successful solution to balance renewable generation? Renew. Energy 74, 422—432.

Fama, E.F., 1970. Efficient capital markets: a review of theory and empirical work. J. Finance 25 (2), 383—417.

Foley, A.M., Leahy, P.G., Marvuglia, A., McKeogh, E.J., 2012. Current methods and advances in forecasting of wind power generation. Renew. Energy 37 (1), 1—8.

Garnier, E., Madlener, R., 2015. Balancing forecast errors in continuous-trade intraday markets. Energy Sys. 6 (3), 361—388.

Garnier, E., Madlener, R., 2016. The value of ICT platform investments within distributed energy systems. Energy J. 37 (SI2), 145—160.

Garnier, E., Madlener, R., 2020. Valuation of Demand Response for Day-Ahead and Intra-Day Trading of Renewable Electricity, FSR Symposium "Variable Generation, Flexible Demand". Fiesole (Florence), Italy, February 19, 2020.

Goodarzi, S., Perera, H.N., Bunn, D., 2019. The impact of renewable energy forecast errors on imbalance volumes and electricity spot prices. Energy Pol. 134, 110827.

Goulden, M., Bedwell, B., Rennick-Egglestone, S., Rodden, R., Spence, A., 2014. Smart grids, smart users? The role of the user in demand side management. Energy Res. Soc. Sci. 2, 21—29.

Gui, E.M., MacGill, I., 2018. Typology of future clean energy communities: an exploratory structure, opportunities, and challenges. Energy Res. Soc. Sci. 35, 94—107.

Hagemann, S., Weber, C., 2013. An Empirical Analysis of Liquidity and its Determinants in the German Intraday Market for Electricity, EWL Working Papers 1317. Chair for Management Sciences and Energy Economics, Essen, Germany.

Henriot, A., 2014. Market design with centralized wind power management: handling low-predictability in intraday markets. The Energy J. 35 (1), 99—117.

Hirth, L., Ueckerdt, F., Edenhofer, O., 2016. Why wind is not coal. On the economics of electricity. The Energy J. 37 (3), 1—27.

Inman, R.H., Pedro, H.T., Coimbra, C.F., 2013. Solar forecasting methods for renewable energy integration. Prog. Energy Combust. Sci. 39 (6), 535—576.

IRGC, 2015. Demand-Side Flexibility for Energy Transitions. Ensuring the Competitive Development of Demand Response Options. International Risk Governance Council, Lausanne, Switzerland.

Kiesel, R., Paraschiv, F., 2017. Econometric analysis of 15-minute intraday electricity prices. Energy Econ. 64, 77—99.

Koch, C., Hirth, L., 2019. Short-term electricity trading for system balancing: an empirical analysis of the role of intraday trading in balancing Germany's electricity system. Renew. Sustain. Energy Rev. 113, 109275.

Koliou, E., Eid, C., Chaves-Ávila, J.P., Hakvoort, R.A., 2014. Demand response in liberalized electricity markets: analysis of aggregated load participation in the German balancing mechanism. Energy 71, 245—254.

Koshrou, A., Dorsman, A.B., Pauwels, E.J., 2019. The evolution of electricity price on the German day-ahead market before and after the energy switch. Renew. Energy 134, 1—13.

Narajewski, M., Ziel, F., 2019. Econometric modelling and forecasting of intraday electricity prices. J. Commod. Mark. 100107.

Pape, C., 2018. The impact of intraday markets on the market value of flexibility — decomposing effects on profile and the imbalance costs. Energy Econ. 76, 186—201.

Pape, C., Hagemann, S., Weber, C., 2016. Are fundamentals enough? Explaining price variations in the German day-ahead and intraday power market. Energy Econ. 54, 376—387.

Pinson, P., Chevallier, C., Kariniotakis, G.N., 2007. Trading wind generation from short-term probabilistic forecasts of wind power. IEEE Trans. Power Syst. 22 (3), 1148—1156.

Rintamäki, T., Siddiqui, A.S., Salo, A., 2020. Strategic offering of a flexible producer in day-ahead and intraday power markets. Eur. J. Oper. Res. 284 (3), 1136—1153.

Ruhnau, O., Bannik, S., Otten, S., Praktiknjo, A., Robinius, M., 2019. Direct or indirect electrification? A review of heat generation and road transport decarbonisation scenarios for Germany 2050. Energy 166, 989—999.

Ruhnau, O., Hennig, P., Madlener, R., 2020. Economic implications of forecasting electricity generation from variable renewable energy sources. Renew. Energy 161, 1318—1327.

Sensfuß, F., Ragwitz, M., Genoese, M., 2008. The merit-order effect: a detailed analysis of the price effect of renewable electricity generation on spot market prices in Germany. Energy Pol. 36, 3086—3094.

Wozabal, D., Rameseder, G., 2020. Optimal bidding of a virtual power plant on the Spanish day-ahead and intraday market for electricity. Eur. J. Oper. Res. 280, 639—655.

CHAPTER 14

The value of flexibility in Australia's national electricity market

Alan Rai[1], Prabpreet Calais[2], Kate Wild[2], Greg Williams[2], Tim Nelson[3]

[1]University of Technology Sydney (UTS), Sydney, NSW, Australia; [2]AEMC, Sydney, NSW, Australia; [3]Griffith University, Nathan, QLD, Australia

1. Introduction

Schweppe et al. (1988) envisaged the wholesale electricity market would eventually share the characteristics of other commodity markets, namely active demand- and supply-side participants. Schweppe et al. (1988) imagined technological advances would enable a two-sided market for electricity by the dawn of the 21st century.

When Australia's National Electricity Market (NEM) was established in December 1998, electricity generation was provided by a relatively small number of centralized generators. Electricity effectively flowed one way, from generators to consumers. Technologies of the type considered by Schweppe et al. (1988) as enabling active demand-side participation were prohibitively expensive for all but the largest of customers. For mass-market customers, consumption was measured using analogue, consumption-based (accumulation) meters and tariffs were simple two-part structures with a fixed daily charge and a volumetric tariff. Therefore, electricity markets were necessarily designed with greater emphasis on the supply side of the market. Power system operation, including the operation of the real-time market (i.e., dispatch), was about forecasting demand and scheduling and coordinating generation to efficiently serve forecast demand, holding some capacity in reserve for unexpected events or "contingencies." Sioshansi (Chapter 5) notes the historical focus on the supply side partly reflected a view that continued economies of scale in both the generation and transport of electricity would eventually mean electricity would soon be too cheap to even meter. For those convinced of such a scenario, why bother with a meter at all and why bother to spend time and effort to understand the demand side?

Variable Generation, Flexible Demand
ISBN 978-0-12-823810-3
https://doi.org/10.1016/B978-0-12-823810-3.00021-2

329

As discussed in other chapters in this book, technological limitations to the development of an active "two-sided" wholesale electricity market for mass-market customers were also prevalent internationally; for example, in the North American markets of ERCOT (Baldick, Chapter 2) and PJM (Bowring, Chapter 11), as well as in Germany (Bauknecht et al., Chapter 15).

However, advances in technology and declining costs—enabled by automation, digitalization, and the Internet of Things—are changing electricity markets and beginning to enable Schweppe et al. (1988)'s vision to be realized for all customers. Today, the generation mix is increasingly renewable and dispersed, networks are facing more dynamic two-way power flows, and the nature of a "consumer" is fundamentally changing. Power flows are increasingly weather-dependent and less predictable than historically. These challenges are likely to intensify given the expected greater uptake of variable renewable energy (VRE) in the NEM; under Australian Energy Market Operator's (AEMO's) "Central" scenario, the NEM's VRE penetration rate[1] is projected to almost quadruple from 19% today to 72% by 2040 (AEMO, 2019a). This projection assumes no additional electricity sector emissions reduction targets beyond the status quo.[2]

This chapter has three purposes. One, it provides estimates of the value of being dispatchable and flexible, based on prices in South Australia, which has one of the highest VRE penetration rates worldwide. The key finding here is that the "flexibility premium" has grown as VRE penetration has risen. Moreover, this premium can increasingly be earned by both utility- and smaller-scale generators (e.g., residential "prosumers").

Second, this chapter discusses some of the benefits of active demand-side resources and a two-sided market. In short, a more active demand side can, when coupled with an increasing penetration of zero-emissions generation, help resolve the energy trilemma: improved reliability, improved affordability, and lower emissions. Moreover, while the growth of VRE generation creates challenges for maintaining system reliability and security, participation by the demand side is needed to help resolve the energy trilemma.

---

[1] In this article, the "penetration rate" of a generator (or type of generators, such as VRE plants) is defined as the proportion of total electricity generation produced by that (type of) generator.

[2] Under the status quo, the NEM's emissions reduction target is its pro rata share of the economy-wide target to reduce emissions by 26–28% on 2005 levels by 2030.

And third, this chapter outlines some of the key enablers of a more active demand side, including policy-driven enablers (design considerations for a two-sided market), discussing some of the trials of two-sided markets currently on foot.

The balance of the chapter is organized as follows:

- Section 2 provides estimates of the value of dispatchability and flexibility in the NEM, based on the experience in South Australia;
- Section 3 outlines the enablers of a two-sided NEM;
- Section 4 discusses the benefits of more active demand-side resources and of a two-sided market; and
- Section 5 provides the chapter's conclusions.

2. The value of dispatchability and flexibility in the NEM

2.1 Comparing three key prices

"Dispatchability" is commonly considered as the extent to which a resource (either a demand-side or supply-side resource) can be relied on to "follow a target" in relation to its load or generation (Joskow, 2011). "Dispatchability" therefore incorporates notions of controllability and flexibility, and is best thought of as a "spectrum of controllability," with different technologies at various points along this spectrum. VRE plants are less dispatchable than conventional plants: Certain generation technologies are more dispatchable than others: wind and solar PV generators' output can be controlled down much more easily and cheaply than they can be controlled up (primarily because their "fuel" is "use it or lose it"), while gas-fired thermal plant and hydro plant can be controlled up more easily vis-à-vis VRE plant. There are also differing degrees of dispatchability within particular plant types, e.g., thermal plant.[3]

In the NEM, the value (or premium) of being dispatchable can be inferred by comparing the dispatch-weighted price (DWP) received by dispatchable generators, such as coal, gas, and hydro, against the DWPs of VRE plant. A DWP is the volume-weighted price received by a generator. The gross, uniform-price auction feature of the NEM means that each generator that generates during a particular 5-min interval is paid the same

---

[3] For example, black coal-fired plants are more flexible than brown coal-fired plants. Black coal plants can operate between 40% and 100% of their maximum output and can ramp in a few hours. By contrast, brown coal-fired plants operate between 70% and 100%, with ramps over more hours than black coal plants (Rai and Nunn, 2020b).

price (ignoring electrical losses).[4] Therefore, differences in DWPs across generators reflect the *time* (and associated levels of electricity demand) at which each power station generates its output.

There are three DWPs worth comparing:

- Time-weighted prices (TWPs), also known as settlement prices, an equally weighted average of the price in each of the six 5-min dispatch intervals that comprise a 30-min trading interval in the NEM. TWPs are DWPs for generators with a constant level of output; for example, a high-capacity factor (baseload) plant such as a brown coal-fired plant or a nuclear plant, which is technically or economically unable to adjust its output over time.
- While the constant-output assumption implied by TWPs is unrealistic, as all generators have some ability to vary output, TWPs nonetheless serve as a useful benchmark for a hypothetical *inflexible* plant.
- The DWP received by a generator that is able to positively correlate its output with demand. "Demand" could refer to either operational (i.e., grid-sourced) demand, or *residual demand*, defined as operational demand less supply from utility-scale VRE plant. The latter demand is met by generators that are both *dispatchable and flexible*. The DWPs earned by these plant are dubbed *firming prices*, as the plant that service residual demand are, by the definition of residual demand, effectively firming the output of VRE plant.
- DWPs received by VRE (i.e., nondispatchable) plant.

2.2 Application to South Australia

South Australia's (SA's) utility-scale VRE penetration rate, comprised almost entirely of wind farms, was over 50% in 2019, compared to 9% in 2008 (Fig. 14.1).[5]

SA also has one of the highest utility-scale[6] VRE penetrations globally (Fig. 14.2). Including small-scale VRE (i.e., systems sizes of less than

---

[4] In the NEM, price differentials between coincident-output generators do occur, due to different amounts of electrical losses associated with locational differences of coincident-output generators. That is, generators located in areas with higher export losses receive lower prices than coincident-output generators located in areas with lower losses.

[5] Unless noted otherwise, all years in this chapter refer to the year to June 30.

[6] Utility-scale is defined as plant sizes of 30 MW or more. Small-scale relates to plant sizes of 100 kW or less.

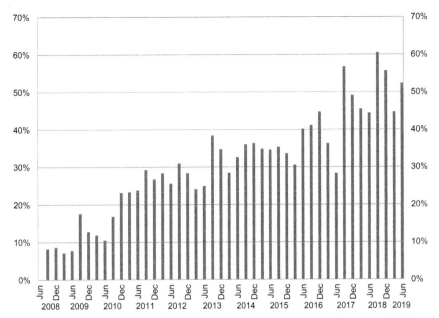

Figure 14.1 Penetration of utility-scale VRE generation in SA. *(Source: AEMO.)*

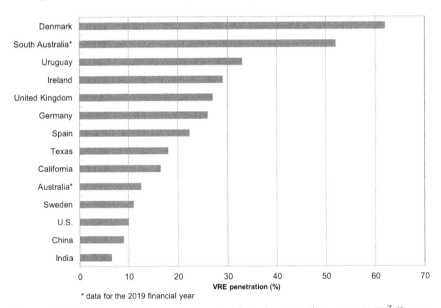

* data for the 2019 financial year

Figure 14.2 Utility-scale VRE penetration in selected geographic areas, 2018.[7] *(Source: AER, IEA.)*

[7]The geographic areas in Fig. 14.2 are inconsistent as both countries and intracountry regions are included. As intracountry interconnections are typically greater than between countries, VRE penetration rates can be, and often are, higher within countries than between countries. Even within a given region (e.g., Texas), certain areas have higher VRE penetration. For example, West Texas has wind penetration regularly above 100% i.e., it exports to the rest of the ERCOT market.

100 kW) in Fig. 14.2, would make SA's overall VRE penetration rate during 2019 around 60%.

The following two trends have been observed in SA (Rai and Nunn, 2020b):

• wind output has typically been negatively correlated with operational demand. For example, between January 2013 and December 2018, monthly correlations ranged from −0.36 to −0.08, and

• high pairwise correlations between different wind farms' output. For example, during the 2018 calendar year, these correlations ranged from 0.35 to 0.95, with the highest correlations seen for those wind farms geographically the closest.

Together, these two findings mean there is more aggregate supply available at times when wind farms generate, resulting in low spot prices at these times. This in turn means wind farms can, and do, earn DWPs that are lower than either TWPs or firming prices.

The firming premium (i.e., the difference between firming prices and wind farm DWPs) is higher than the TWP premium, due to the increased volatility in residual demand. For example, over 2019, the firming price averaged $144/MWh versus an average TWP of $110/MWh and wind farm DWP of $83/MWh (Fig. 14.3).

In addition, the firming premium has increased over time, both absolutely and relative to the TWP premium. For example, the firming premium was around $61/MWh (i.e., $144/MWh less $83/MWh) during 2019, compared to a premium of $15/MWh during 2014 (Fig. 14.4). This $46/MWh increase exceeded the $20/MWh increase in the TWP premium over the same period (from $7/MWh to $27/MWh).

Expressed as a mark-up on wind farm DWPs, the firming premium was 73% over 2019, while the TWP premium was 32%. Both of these premia have more than doubled since 2014, with the firming premium increasing by more (from 28% to 73%) than the TWP premium (from 13% to 32%).

So, the value of being dispatchable and flexible (i.e., the firming premium) has grown over time, coinciding, as expected, with a rising VRE penetration rate. This is likely to become even more pronounced with increasing investment in solar PV, which unlike wind, is almost completely coincident in its production profile. It is worth noting the firming premium can be earned by supply side and, where there are more cost-effective, demand-side resources. Hence the need to consider ways to unlock the potential flexibility in demand.

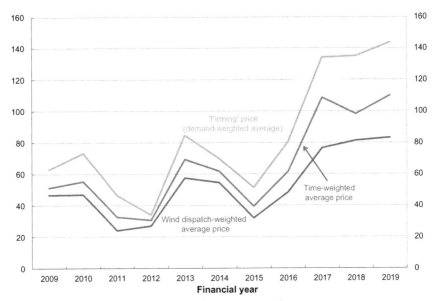

Figure 14.3 Three DWPs in SA ($/MWh).[8] *(Source: AEMO.)*

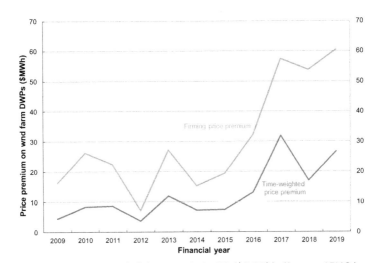

Figure 14.4 Dispatchability premia in SA ($/MWh). *(Source: AEMO.)*

[8]Unless noted otherwise, all dollar values in this chapter refer to Australian dollars.

2.3 A role for demand flexibility

As noted in Section 1, technologies that enable active demand-side participation have existed since the start of the NEM, for the largest customers (e.g., C&I and other nonresidential customers). These customers have faced signals to be more flexible from both:

- the wholesale market, by being exposed to the spot price for some or all of their demand and
- networks via demand (i.e., $/kVA-based prices) and/or time-of-use (time-varying $/MWh prices) charges.

The first signals for residential demand flexibility came via distribution network controlled-load tariffs on air conditioners, electric hot water systems, and pool pumps. Such tariffs, coupled with installation of ripple controls, have existed for electric hot water systems in Australia since the 1960s (APQRC, 2014; Swanston, Chapter 22). Outside of these controlled-load tariffs, minimal signals for residential demand flexibility exist, as most customers remain on time-invariant, two–part fixed-volumetric rate tariffs. However, this is changing, with the emergence of innovative new distribution network tariffs, such as a "solar sponge" tariff (discussed in Section 2.3.1), as well as increasing numbers of spot price pass-through retail contracts, providing varying degrees of exposure to spot electricity markets, providing increasing opportunities for valuing flexible residential demand.

2.3.1 New network tariff structures and new retail business models

A "solar-sponge" tariff for residential and small business customers is currently under trial by South Australian Power Networks (SAPN), to inform SAPN's 2020–25 tariff structure statement. Participation in the trial is limited by SAPN to 7000 customers (SAPN, 2019).

This time-of-use (ToU) tariff incentivizes electricity consumption during the middle of the day, when electricity is typically cheap in SA due to the high output from rooftop solar PV (and increasingly also from utility-scale PV). Fig. 14.5 shows the ToU residential tariff versus average wholesale spot prices in S.A, by time of the day.

The solar sponge tariff provides the following opportunities for demand flexibility:

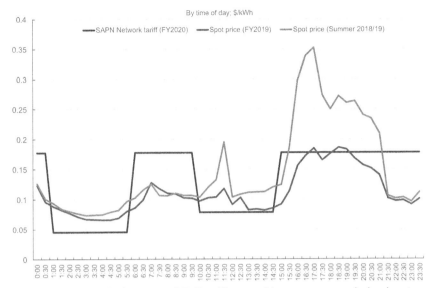

Figure 14.5 SAPN's "solar sponge" ToU tariff versus SA average spot wholesale prices. Note: Summer 2018/19 refers to the 4 months to 31 March 2019. *(Source: AEMO, SAPN.)*

- more expensive to consume in the morning (6 am—9.30 a.m.) and afternoon/evening (3 p.m.—12.30 a.m.), and
- cheaper to consume in the early hours (1 am—5.30 a.m.) and middle of the day (10.30 a.m.—2.30 p.m.).

However, during the summer, the wholesale spot market provides *greater* benefits for demand flexibility than the network tariff. For example, over summer 2018/19, the peak to off-peak spot price ratio was around 4.8:1, whereas the corresponding price ratio for the solar sponge was around 4:1.

That is, the wholesale spot price over summer provided a 20% sharper signal for demand flexibility than the network tariff. A sharper price signal can create more volatility for customer bills if demand is not managed properly, an issue discussed in more detail in Section 3.3.

A final point to note is that Fig. 14.5 is based on spot price averages and so smoothes out the volatility inherent in 5-min prices, and in turn reduces the full value of demand flexibility, especially in the summer. For example, the minimum 5-min spot price in SA in summer 2018/19 was -$850/MWh, and the maximum was $14,500/MWh. Demand that could flex and take advantage of these spot price differences would have realized significant cost savings (or in turn profits if paid to be flexible).

The "solar sponge" tariff is similar to the "Sunshine tariff" offered by Western Power Distribution to residential customers in the South West of England during 2016, and similar residential tariffs in parts of North America such as California (Faruqui, 2018).

Retailers are increasingly offering dynamic tariffs that incentivize demand flexibility. For example, new-entrant retailers like Flow Power and Amber Electric offer contracts which pass the wholesale electricity spot price directly through to consumers.[9] Often these prices and products are only accessible for customers on flexibility-enabling technologies, such as a residential battery and/or a smart meter (AEMC, 2019). Amber Electric, in coordination with Energy Locals, currently offers residential customers in New South Wales (NSW) and SA the following offer:

- 30-min wholesale spot price charges for energy usage,
- 30-min wholesale spot price payments for energy generation,
- network tariff structure pass-through. This tariff could be two-part, ToU, or capacity-based, as determined by the relevant distribution network, and
- $10 per month fixed charge to recover Amber Electric's operating costs.

Amber Electric noted the tariff would likely become even more dynamic over time as more dynamic, cost-reflective, network tariffs are introduced, and also when 5-min settlement is introduced.[10]

The next section discusses what a two-sided market could look like in a future where all customers, and many loads, are digitally enabled and measured.

3. Designing a two-sided electricity market

3.1 Context for a two-sided market

Before discussing the design of a two-sided market (Faruqui and Lessem, 2020), it is worth providing some context for the increased focus on two-sided markets, namely:

---

[9] See https://flowpower.com.au/faqs/(Flow Power), and https://www.amberelectric.com.au/how-it-works (Amber Electric) for more details.

[10] In November 2017, the AEMC made a rule that uses 5-min dispatch prices as the basis for settlement, starting from 1 July 2021. In April 2020, the AEMC received a request to delay implementation of 5-min settlement by 12 months. At the time of writing, the AEMC was yet to make its decision on delaying implementation.

- The long-running view that the NEM does not (fully) value demand response. Historically, the focus was on peak-time demand reduction (so-called "negawatts"), but "demand flexibility" has become the preferred term vis-à-vis "demand response," as the former phrase recognizes that *increased* demand can at times be just as valuable to the system as decreased demand (Rai, 2020b).
- Growing instances in those NEM regions with high PV penetration rates—SA and Queensland—of low daytime demand; for example, in SA, minimum demand now occurs in the middle of the day rather than overnight, with new record lows continuing to be set (Rai et al., 2019). Such low demand pushes down spot prices, leading to thermal plant decommitment, increasing the risk of system-wide blackout were a credible contingency to occur.[11]
- Concerns that distributed energy resources (DER) are currently not adequately integrated into the NEM, resulting in inefficient and inequitable outcomes; namely, the creation of significant cross-subsidies from non-DER to cum-DER customers due to existing network tariffs being designed for a pre-DER world (Rai, 2020a).
- Decreased costs of providing demand flexibility due to technological advances, digitalization, and system improvements, especially for small-scale, mass-market, customers (ESB, 2020).

3.2 Characteristics of a two-sided electricity market

Parties obtain access to the electricity network via a connection point. A connection point is where:

- connection to, and disconnection from, the network occurs
- flows of electrical energy and quantities of other energy services are measured and accounted for
- the party with operational control and obligations relating to energy flows into and out of the network is established and changes from time to time.

For example, at a connection point a party may provide energy to, or receive energy from, the system. Metering arrangements at the connection point will measure the quantity of electricity provided or received, and these data will be used to determine the amount of money the relevant

---

[11] The risk is of inadequate system strength when the credible contingency is a thermal unit outage. In SA, there is also the risk of insufficient inertia when the credible contingency is SA being islanded.

party receives from, or owes to, the market as a result (ESB, 2020). Once connected to a power system, a customer can freely use electricity taken from the grid through their connection meter whenever they like (as long as they pay for it!). A connection to the grid provides a level of flexibility not seen in other commodity markets.

The NEM takes better advantage of valuing flexibility through organized exchange than some other electricity markets, by virtue of having a high market price cap and a low price floor ($14,700/MWh and -$1000/MWh, respectively, for the year to 30 June 2020) and real-time markets that allow for relatively late rebidding, which enables, at least in theory, greater flexibility on both the load and generation side. However, the value of this flexibility does not, by and large, reach down to small customers, and therefore it is likely that the NEM currently does not maximize the value of flexibility, though it is difficult to estimate the precise amount of underutilization.

In the NEM, the most common form of residential and small-business tariff type offered by retailers remains the two-part tariff, offered either "flat" or inclining block.[12] For example, 53% of small businesses were on two-part electricity tariffs in 2018, while more than two-thirds of residential customer offers had two-part tariff structures (AEMC, 2018).

The prevalence of flat and block tariffs means most households and small businesses in Australia have no way of knowing when it is cheap or expensive to consume electricity or export it back to the grid. In addition, existing measures fall short of fully valuing demand flexibility as follows:

- Controlled-load tariffs provide an opportunity for valuing flexibility, but these tariffs are set-and-forget in relation to when the loads turn on and off,[13] and can also be disempowering for consumers as the loads are controlled by distribution network businesses, not consumers.

---

[12] A two-part tariff has a time- and volume-invariant daily supply charge (cents-per-day), and a time-invariant volumetric charge (cents-per-kWh). Flat and block tariffs differ with respect to the structure of the volumetric charge: in the former, the per-kWh rate is fixed, in the latter, it varies by consumption "block" (e.g., a different per-kWh rate is applied to the first 500 kWh consumed, with a different rate applied to the next 500 kWh block, and so on).

[13] For example, most electric hot water loads in NSW, Victoria, and Queensland distribution networks are controlled to come on at night, due to concerns night-time demand would otherwise fall below the minimum stable generation levels for coal-fired plants in these states.

- The level and structure of network tariffs are set for a period of time (typically, a financial year). Even the innovative "solar sponge" tariff (see Section 2.3) fixes both prices and peak-to-off-peak price ratios. These relativities might provide an accurate signal of system need on average, but can at times provide the wrong signal as compared to spot prices.
- Finally, most residential consumers pay for their electricity in arrears on a quarterly basis, which can create a further disconnect between consumption and real-time market conditions.

Making a two-sided market for electricity would provide the opportunity to fully unlock and value demand flexibility, empowering consumers that choose to participate and improving efficiency.

In order to maximize the value of demand flexibility, a two-sided market should ideally have the following three characteristics:

1. All market participants, big and small, should be able to gain exposure to wholesale spot prices, as these prices best reflect real-time system needs. Currently, utility-scale generators and large customers can maintain or increase their spot exposure by reducing the extent of their hedged positions, but smaller customers typically have their loads fully hedged by default with limited choice regarding reducing their hedge position and increasing their spot exposure.
2. All participants should be subject to a common set of obligations with respect to how they participate in the wholesale market. Moreover, obligations should be placed on functions and activities, rather than participant categories or technologies. This said, the obligations should be proportionate to the size of the customer; for example, individual households may be exempted from some of the obligations placed on utility-scale generators.
3. The price of electricity at each connection point should incorporate the congestion and losses associated with importing or exporting electricity to or from that connection point, where the benefits of doing so exceed the costs.

It is worth noting these three characteristics, especially the last, represent the ideal end state, and we are currently quite far away from that end state. Currently, congestion is not priced at the transmission level let alone at the distribution level, and so a lot would need to change in order to facilitate locational marginal prices (LMPs) at both the transmission and distribution level.

Indeed, this ideal end state may *never* happen, given our caveat that this only be achieved if the "benefits exceed the costs." The ESB's work on a post-2025 fit-for-purpose market framework is considering whether LMPs should be implemented at the transmission level.[14] It remains to be seen if LMPs can be successfully implemented in the NEM at the transmission level, a necessary (albeit not sufficient) precedent for having LMPs at the distribution level.

3.2.1 Exposure to wholesale prices

Enabling electricity consumers to face wholesale prices does *not* mean all consumers have to be fully exposed to 5-min spot prices. What it does mean is consumers should be offered a spectrum of options regarding the extent of that exposure; for example, time-of-use pricing at one end of the spectrum, and spot prices (or real-time prices, RTP) at the other end (Fig. 14.6).

Different points on this spectrum have different tradeoffs between risk (volatility of electricity bills over time) and reward (the potential size of bill savings vis-à-vis a guaranteed-price bill). For example, RTP-based retail

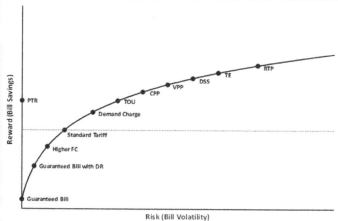

Figure 14.6 Spectrum of retail pricing choices. *(Source: Faruqui, A., Lessem, N, 2020. Enhancing Rate Design Choices for Ontarians, 11 June.)*

---

[14] http://www.coagenergycouncil.gov.au/energy-security-board/post-2025.

tariffs would enable flexible demand to capture the wholesale market benefits noted in Section 2.2 in the context of South Australia (see Fig. 14.5).

Enabling parties to gain wholesale price exposure also means default retail tariff structures need to be set above the current default structure: flat-volumetric rate tariffs. Instead, the default structure should be a TOU tariff, with the structure based on the conditions in that distribution network (ideally, even further localized, at the subdistribution network). For example, a "solar sponge" tariff should be the default in SA (and in parts of Queensland) given SA's high PV penetration rate.

It is therefore important that network businesses continue to design and offer a broad and innovative range of controlled-load tariffs and associated DREDs, such as the "solar sponge" tariff noted in Section 2. For this reason, electricity network businesses should *not* be precluded from providing incentives for demand flexibility and from providing tariff structures more volatile than two-part tariffs, provided the potential for anticompetitive behavior (i.e., the potential for regulated businesses to predatory-price its unregulated rivals) is minimized.

Being exposed to wholesale prices also doesn't mean households and small businesses will have to become expert energy traders. Automation, digitalization, and the Internet of Things means demand is able to respond to price signals without requiring end users to either become energy traders or to physically or consciously control their demand, and without sacrificing consumer comfort. Rather, small-scale consumers and prosumers would be able to choose if and how they participate in the wholesale market—either directly or through a third party like a retailer as they do now or an aggregator of demand-side flexibility. In most cases, a third party would trade and settle all electricity bought and sold in the wholesale market on behalf of its customers.[15]

There are some signs of this occurring in the NEM today. For example, electricity retailer Pooled Energy remotely controls the operation of household pool pumps, optimizing their use based on water quality and

[15] Any move to a two-sided market where customers are buying and selling energy into the wholesale market (either themselves or through a third party) will require careful consideration of the consumer protection framework. This may range from general consumer protection laws to more energy-specific legislation to provide additional protections given that electricity is an essential service. These consumer protections would deal with any information asymmetry between the customer and the trader buying and selling energy on their behalf.

electricity costs from wholesale spot prices, enabled by control systems.[16] Pooled Energy's control systems use machine learning and artificial intelligence techniques and technologies to optimize the value of demand flexibility associated with pool pumps and this logic could be applied to other appliances (e.g., home air-conditioning/heating).

Unlocking demand flexibility also offers rewards to retailers and other parties acting on behalf of the customer. Demand flexibility may at times be cheaper than traditional financial derivative contracts for hedging wholesale price risk, such as cap contracts (effectively an option to limit prices to an agreed strike price, currently $300/MWh). Enabling more customers to participate in a two-sided market is likely to result in innovation on pricing and service offerings, further to the offerings currently made by Pooled Energy, Amber Electric and Flow Power; potential innovations include coupling dynamic pricing with demand response-enabling devices (DREDs) and advisory services on how to minimize bills or maximize the value of flexibility. For example, SmartestEnergy[17] offers price-DRED bundling for small businesses and households in the United Kingdom, and has aspirations to do the same in Australia, as do other European-based retailers and aggregators.

3.3 Existing trials of two-sided electricity markets in the NEM

Moving closer to a two-sided market would be a shift for the NEM and any design for a two-sided market should include a transitional approach to maximize benefit while also minimizing adverse impacts on consumers and market participants. The design and sequencing of this transitional approach, and the design of a two-sided market more generally, should learn from the multitude of existing trials and processes in relation to two-sided markets.

Enabling the participation of end users within a two-sided market will also require new network operational models for determining and applying network constraints as well as optimizing the flow of electricity across both distribution and transmission networks. The framework must allow the aggregation of DER to nodes in the transmission network to take account of network constraints (ESB, 2020).

---

[16] For more information see: https://pooledenergy.com.au/.

[17] For more details, see https://smartestenergy.com/about-us/about-us.

These trials explore the ability for distributed generation, storage, and controllable loads to provide flexibility. The proceeding discussion focuses on a select subset of existing trials; for a comprehensive list, see Appendix C of ESB (2020).

3.3.1 The wholesale demand response mechanism rule change

A wholesale market-based demand response mechanism (DRM) is being introduced for the NEM (AEMC, 2020). Under the mechanism, large customers would be able to actively participate in central dispatch and be rewarded for the value they provide to the system. A new category of registered participant, a demand response service provider (DRSP), would be created, and would be able to engage directly with a customer, without the involvement of that customer's retailer, to bid that customer's demand response into the wholesale market.

The wholesale DRM is to apply for large customers, as the costs of applying the same requirements on small customers would be prohibitive (e.g., determining baselines for millions of individual loads; the requirement for demand response to be scheduled in central dispatch). In addition, as the service provided by DRSPs to customers is not considered a sale or supply of energy, there is no legal requirement for a DRSP to be an authorized retailer, and therefore existing energy-specific consumer protections for small customers would not apply to demand response provided through a third-party DRSP (AEMC, 2020).

This said, the learnings from the wholesale demand response changes will be used to inform the design of a two-sided market. For example, scheduling requirements placed on large customers participating in the wholesale DRM can inform the proportionality of such requirements on small customers participating in the wholesale DRM (as noted above, costs of applying the same requirements on large customers as for small customers may be prohibitive).

3.3.2 Virtual power plant (VPP) trials

A VPP broadly refers to an aggregation of resources (chiefly, rooftop PV, battery storage, and controllable loads) coordinated to deliver services for power system operations and electricity markets. There are four key VPP trials currently in place that can help inform various aspects of the design of two-sided markets, as follows:

1. AEMO's trials—these trials test a selection of VPPs' ability to provide frequency control ancillary services (FCAS) (AEMO, 2019b)

2. **VPP trial by Tesla in SA**—similar to AEMO's trial, the Tesla trial seeks to gain insights into the ability of end users to participate in wholesale markets through aggregators; in particular, the capability of the aggregators to provide FCAS through use of the end users' resources.
3. **VPP trial by Simply Energy**—this focuses on the role and interaction of VPPs, networks, and other market participants through distribution service operator (DSO) and distribution market operator (DMO) platforms. This trial tests the ability to send signals to VPP aggregators to modify behavior based on network conditions and price.
4. **Advanced VPP**—this trial tests dynamic export limits to maximize VPP (integrated solar-and-battery systems) output to the SA grid. It utilizes a Tesla- and CSIRO-developed platform for DER participation in the wholesale market.

3.3.3 Trials of new distribution network pricing structures

In the pre-DER world, efficient network price signals focused on managing peak demand (e.g., peak shaving) as a means of maintaining power system reliability and security whilst maintaining affordability. In contrast, in today's age of decarbonization and the "prosumer," efficient price signals are needed on both withdrawals (i.e., consumption and demand) and injections (i.e., supply and production), to manage import and export congestion.

The importance of such price signals is growing: rooftop PV capacity is projected to double by 2030, and uptake of other DERs, chiefly electric vehicles (EVs) and home batteries, are likely to also accelerate (Rai et al., 2019). In addition, the emergence of digital load-control technologies such as Google Home and Nest may result in demand that was once thought to be price-inelastic in the short term, becoming more price-elastic.

SAPN's "solar sponge" tariff is the first ToU tariff focused on managing both import congestion/peak demand and export congestion/high renewables output. Similar tariffs are likely to emerge in Queensland, where rooftop PV penetration is around 50%, well above the 40% threshold where reverse power flows occur with associated power quality issues. Such tariffs may provide a cheaper way of controlling PV output than existing measures like imposing inverter limits and curtailing PV generation when output is too high relative to demand (Rai, 2020b).

3.3.4 ARENA's distributed energy integration program (DEIP)

This work is considering changes to the existing distribution network access regime for rooftop PV and other distributed generation, to minimize the costs imposed on all electricity consumers from unconstrained exports. Reforms to existing network pricing—in particular, the inability of distribution networks to charge a negative price for exports—are also being considered (ARENA, 2019).

3.3.5 Open Energy Networks (OpEN) project

A joint initiative between AEMO and Energy Networks Australia (ENA) to explore a new network operational model that optimizes the flow of electricity across electricity distribution and transmission networks, OpEN is considering how to optimize DER, while managing distribution and transmission network constraints (AEMO-ENA, 2019).

For example, under a proposed hybrid framework, the distributor would be responsible for managing and communicating network constraints and the need for network services, while AEMO would manage a market platform that provides a standardized interface for DER to participate in the wholesale energy market. These inputs could be incorporated into a security constrained economic dispatch, which would be calculated at a nodal level and aggregated to the whole system. The market platform could also be utilized to facilitate participation of DER in network-level markets and services.

3.3.6 Standalone power systems (or microgrids)

Standalone power systems (SAPS) provide an opportunity for two-sided markets to be established, and for these learnings to be applied to a potential two-sided market for NEM-connected customers. This is especially the case for SAPS with a high VRE penetration, which can and does create a mismatch between generation and load. Flexible demand is often used to resolve mismatches, along with battery storage and back-up diesel plant. Price signals for demand flexibility in SAPS are likely to be greatest for so-called third-party SAPS—SAPS where the provider of the system is not the local distribution network service provider (LNSP)—as customers in these SAPS are not required to be on the same retail tariff structures as NEM-connected customers. As LNSP-led SAPS are economically regulated (i.e., the assets in these SAPS form part of an LNSP's regulated asset base), customers in LNSP-led SAPS are required to be on the same retail tariff structures as NEM-connected customers (Rai et al., 2020).

3.3.7 Decentralized energy exchange (deX) trials

GreenSync, a technology provider, has developed a platform which allows electricity end users to interact directly with networks and market operators on a local area basis. The deX platform creates the ability for all users signed up to the platform to respond to the needs of the system, akin to a two-sided market. There have been several trials run using the deX platform, including an 18-month trial with distribution network Evoenergy in the Australian Capital Territory.[18]

3.4 Incorporating insights on consumer decision-making

In addition to learning from these trials, it is also imperative that a two-sided market design be informed by learnings from Australia's experience with full retail contestability and retail price deregulation. In particular, the reregulation of retail prices in the NEM reflects failures with price deregulation in practice, such as offers which imposed, *de facto*, significant financial penalties on consumers for not paying their bills on time (Ben-David, 2020). More fundamentally, the reregulation of retail prices reflects a failure by market designers to test their underlying assumptions around consumer decision-making and behavior—chiefly, people's ability to make rational, long-term decisions with respect to choices around their energy contracts and offers—before introducing key market reforms.

Difficulties encountered by consumers in comparing offers, and the risk of making wrong choices, could magnify under a two-sided market given the multidimensionality of contracts—the dimensions of reliability, flexibility, and price. Therefore, two-sided market trials can and should also provide learnings about how consumers evaluate and decide between competing offers when faced with a broader range of options vis-à-vis offers under the existing one-sided market. These learnings can reduce the risk that an economically elegant solution fails to work in practice or is unsustainable due to a misunderstanding (or a lack of understanding) about how people make decisions in a similar way to how deregulated retail prices ultimately came to be reregulated (Ben-David, 2020).

---

[18] See: https://dex.energy/ and https://arena.gov.au/projects/decentralised-energy-exchange/.

4. Benefits of a two-sided electricity market

As noted in Section 3, signals for demand flexibility by small customers (i.e., residential and small business customers) have historically been limited. A two-sided market provides the means by which demand flexibility by *all* customers can be adequately and appropriately rewarded. The benefits of two-sided electricity markets include:
- Lower retail and wholesale prices
- Improved reliability and security
- Increased customer empowerment
- Reduced emissions
  Each of these is briefly discussed below.

4.1 Lower retail and wholesale prices

Extending wholesale price signals to all connection points will reveal the value of demand flexibility according to the needs in each location. As customers respond to these price signals for demand flexibility, a two-sided market should improve the allocative and productive efficiency of wholesale prices (AEMC, 2019). As noted in Section 3, automation, digitalization, ubiquitous batteries, and the Internet of Things mean demand is more able to respond to price signals, without requiring end users to become energy traders and without sacrificing consumer comfort.

For example, the market will be more productively efficient and reduce wholesale prices where demand response is cheaper than peaking (gas-fired or hydro) generation. Employing greater use of demand flexibility can also result in more efficient network investment spend by efficiently managing the level of both export- and import network congestion. As noted in Section 3.2, an efficient network charge in today's age of decarbonization and the "prosumer" needs to be agnostic to both imports into, and exports from, connection points.

In the absence of well-designed retail price signals, which are a critical aspect of any two-sided market, consumers will bear the costs of DER not being integrated efficiently into the electricity system. At best, poor integration of DER could mean that the electricity system would not realize all the benefits that the significant investment in DER could provide. At worst, consumers would bear additional costs through poor planning decisions, an inability to realize benefits from their DER investment, and potentially inefficient investment such as over building network capacity relative to long-term requirements.

4.2 Improved reliability and security

A two-sided market can enhance both power system reliability and security. A reliable power system has sufficient generation, demand response, and network capacity to supply customers with the energy that they demand in a manner that satisfies applicable reliability standards for network and generation.[19] A secure power system is able to continue operating within defined technical limits (e.g., limits on both the level of, and changes in, voltage and frequency) even if a major system element, like a large generator or customer, disconnects from the system.

In Australia, most (96%) customer supply interruptions are due to network outages caused by storms, high winds, floods, or other network faults. Of the remaining 4% of interruptions, the vast majority are due to system insecurity; just 0.3% of customer supply interruptions occur due to insufficient generation; that is, due to a lack of generation reliability.

Despite reliability-related events comprising a small share of supply interruptions, reliability remains front of mind for policymakers and governments, due to concerns about the future risks to reliability from higher VRE penetration rates (Rai and Nunn, 2020a; Simshauser and Gilmore, 2020) given the historical poor correlation between VRE output and electricity demand, as well as concerns about the reliability of coal-fired plant, especially Victorian brown coal plant.

A two-sided market can enhance generation and network reliability by:
- enabling more price-responsive resources to respond to wholesale price spikes induced by periods of generation inadequacy and
- enabling more resources to respond to resolve real-time network limitations, such as export- and/or import-induced congestion. Residential storage (batteries or EVs) can play a useful role in resolving both import and export congestion.

Furthermore, as noted in Section 4.1, a two-sided market can reduce the cost of achieving a given level of reliability.

Similarly, a two-sided market can enhance system security by:
- enabling demand, along with supply, to respond to deviations in frequency and voltage. AEMO's VPP trials (discussed in Section 3.2)

---

[19] The generation reliability standard is set by the independent Reliability Panel, and requires at least 99.998% of forecast customer demand to be met each year, for each region in the NEM. Reliability standards for transmission and distribution electricity networks are set by state and territory governments, with reference to the extent and duration of load interruptions.

reveal demand-side resources can participate in FCAS markets, and can provide frequency raise services faster and more reliability than thermal plant and

- utilizing the greater information provided about real-time DER performance, a key enabler of a two-sided market. This information can improve the design of load shedding schemes and minimize the risk of load shedding exacerbating, rather than mitigating, the effect of system contingencies. The OpEN project, discussed in Section 3.2, provides the opportunity for this information transfer between network businesses and AEMO to occur.

A two-sided market can also reduce the cost of maintaining system security. As noted above, demand-side resources can often provide a faster frequency and voltage response than traditional supply-side resources (i.e., thermal plant).

4.3 Customer empowerment with respect to choosing desired levels of reliability

Finding the right balance between the cost of reliable supply and the actual levels of reliability for society collectively is inherently difficult due to the heterogeneity across electricity consumers. In general, a higher value is placed on reliability or uninterrupted supply where the cost of adapting to supply interruptions is higher. For example, aluminum smelters typically place a much greater value on reliability than, say, retail clothing stores, reflecting the higher costs to smelters of load interruptions (e.g., pot lines freezing). Similarly, amongst residential customers, households with life-saving equipment place a much greater value on reliability than other households. The value of reliability is also a function of time and space.

Demand flexibility and two-sided markets provide individual consumers with the opportunity to nominate their preferred level of reliability at points in space and time. As discussed in Section 3, to an extent this happens already: controlled loads are available to customers who tolerate some interruption to continuous on-demand electricity supply for specific, nominated, loads (Swanston, Chapter 22). However, as discussed in Section 3, a two-sided market would provide customers with the opportunity to flex other loads and to do them in a more dynamic manner than existing controlled loads.

Customers could flex their entire supply at the connection point, or for individual loads (e.g., pool pumps, air conditioners or hot water systems, or in adjusting the extent of rooftop PV export vis-à-vis self-consumption). In return for a willingness to flex demand, customers should be offered bill savings, as currently occurs with controlled-load tariffs.

Customers themselves, or a party acting on their behalf (e.g., distribution network businesses offering controlled-load tariffs) can translate these preferences into demand bids and automatically adjust the level of supply a customer receives such that generation always meets demand while ensuring all customers receive their desired level of reliability.

Automation and digitalization would provide the means to achieve a more dynamic and individualized provision of reliability. If customers revealed higher tolerances for a reduction in service than implied by the existing reliability standard, then this would reduce the cost of electricity for all consumers. And if consumers desired a higher level of reliability, then the resulting cost increase would be an efficient response reflecting the higher desired level of reliability.

4.4 Reduced emissions

A two-sided market provides the opportunity to reduce emissions by:
1. enabling demand-side resources to substitute for thermal generation in both energy and ancillary service markets and
2. reducing the risk of renewables curtailment by demand increasing at times where VRE output would otherwise be curtailed in order for CO_2-e emitting plant to run.

VRE curtailment across the NEM has risen over time, from close to zero prior to 2018 to 7% for the 3 months to end-March 2020 (Fig. 14.7). To date, most of this curtailment has been in SA and Queensland, the two regions with the highest VRE penetration rates. Curtailment of rooftop PV output has occurred in Victoria and NSW (AEMO, 2020). In all these cases, curtailment has occurred in order for emissions-intensive plant to run,

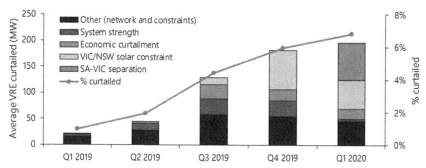

Figure 14.7 VRE curtailment by quarter, for those regions where curtailment occurred. *(Source: AEMO (2020).*

due to concerns about system insecurity from a combination of high VRE output and low demand. This is also an issue in other regions with high VRE penetration rates, such as California (Sioshansi, Chapter 1) and Texas (Baldick, Chapter 2).

Unlocking and enabling more flexible demand can reduce these instances of VRE curtailment, by demand increasing in high-VRE output times. The increase in demand can lessen the risk of system insecurity, resulting in higher VRE output displacing some of the output from emissions-intensive plant, leading to lower emissions.

Additionally, by providing a better price signal to all connection points than is the case currently, future investment in VRE can be optimized to provide for the needs of the system, possibly minimizing curtailment.

5. Conclusions

When the NEM was established, technologies enabling active demand-side participation were prohibitively expensive for all but the largest of customers. Electricity markets were thus necessarily designed with greater emphasis on the supply side of the market. However, technological advances and declining costs are fundamentally changing electricity markets and increasingly enabling a more active demand side.

Today, the generation mix is increasingly renewable and dispersed, networks are facing more dynamic two-way power flows, and the nature of a "consumer" is fundamentally changing. Today's power flows are increasingly weather-dependent and less predictable than historically. These challenges are likely to intensify given the projected quadrupling in VRE penetration rates over the next 2 decades.

This chapter has shown the value of demand flexibility has grown over time as VRE penetration has risen. The incentive to flex one's demand, when coupled with increasing VRE penetration, can help resolve the energy trilemma: improved system reliability and security, improved affordability, and lower emissions.

A two-sided market provides the means by which to incentivize demand flexibility at both the transmission and distribution level. And incorporating insights on consumer decision-making, drawing on observed experiences with full retail contestability and retail price deregulation in the NEM, will help make the design for a two-sided market both fit-for-purpose and sustainable. Hooray for flexibility!

Bibliography

AEMC, 2020. Wholesale Demand Response Mechanism — Rule Determination, 11 June.

AEMC, 2019. How Digitalization Is Changing the NEM — The Potential to Move to a Two-Sided Market, 14 November.

AEMO, 2020. Quarterly Energy Dynamics — Q1 2020, 23 April.

AEMO, 2019a. Draft 2020 Integrated System Plan for the NEM, 12 December.

AEMO, 2019b. NEM Virtual Power Plant Demonstrations Program — Final Design. July.

AEMO-ENA, 2019. Open Energy Networks: Required Capabilities and Recommended Actions. Interim Report. 22 July.

APQRC, 2014. Ripple Injection Load Control Systems. Australian Power Quality & Reliability Centre. Technical Note 14.

ARENA, 2019. DEIP Access and Pricing Package: Report on Workshop 2, 9 December.

Ben-David, R., 2020. Response to ESB Consultation on Two Sided Markets, 17 May.

ESB, 2020. Moving to a Two-Sided Market, Consultation Paper. April.

Faruqui, A., 2018. Modernizing Distribution Tariffs for Households: A Presentation to Energy Consumers Australia, 9 November.

Faruqui, A., Lessem, N., 2020. Enhancing Rate Design Choices for Ontarians, 11 June.

Joskow, P., 2011. Comparing the costs of intermittent and dispatchable electricity generating technologies. Am. Econ. Rev. 101 (3), 238—241.

Rai, A., 2020a. Network Tariffs in an Increasingly Distributed, Decentralized, and Decarbonized Power System. IAEE Energy Forum, 3rd quarter, pp. 11—14.

Rai, A., 2020b. Demand Response in a Distributed, Decentralized, Decarbonized and Digitalized NEM: Much More than "Negawatts", 16 March.

Rai, A., Nunn, O., 2020a. On the impact of increasing penetration of variable renewables on electricity spot price extremes in Australia. Econ. Anal. Pol. 67, 67—86.

Rai, A., Nunn, O., 2020b. Is there a value for dispatchability in the NEM? Yes. Electr. J. 33 (3), 106712.

Rai, A., Esplin, R., Nunn, O., Nelson, T., 2019. The times are a changin': current and future trends in electricity demand and supply. Electr. J. 32 (6), 24—32.

Rai, A., Rozyn, C., Truswell, A., Nelson, T., 2020. Regulating off-the-grid: stand-alone power systems in Australia. In: Sioshansi, F. (Ed.), Behind and Beyond the Meter: Digitalization, Aggregation, Optimization, Monetization, pp. 317—339.

SAPN, 2019. Tariff Information, South Australian Power Networks.

Schweppe, F., Caramanis, M., Tabors, R., Bohn, R., 1988. Spot Pricing of Electricity. Springer US.

Simshauser, P., Gilmore, J., 2020. Is the NEM Broken? Policy Discontinuity and the 2017—2020 Investment Megacycle. EPRG Working Paper 2014. University of Cambridge.

CHAPTER 15

Demand flexibility and what it can contribute in Germany

Dierk Bauknecht, Christoph Heinemann, Matthias Koch, Moritz Vogel
Oeko-Institut, Energy & Climate, Freiburg, Germany

1. Introduction

In Germany, the share of renewables in power generation has increased significantly over the past 2 decades. As a consequence, the system implications of renewable generation have come to the fore. This includes the increasing demand for flexibility.

The chapter presents the role demand flexibility can play in this context and what its challenges and limitations are. It presents key insights from a range of research projects on flexibility in the German context.[1] The chapter does not look at any specific type of demand flexibility or one specific regulatory issue in detail, but provides an overview on key issues that need to be considered in a political strategy for introducing demand flexibility. Even though the chapter focuses on Germany, the general insights are relevant for other systems as well.

The chapter is structured as follows:

- Section 2 presents a quantitative analysis of how the demand for flexibility increases with a rising share of renewables in Germany and how different flexibility options can contribute to cover that demand, including demand flexibility. This includes a generic analysis of different flexibility functions, as well as a quantitative scenario analysis for the German power system.
- Section 3 complements these quantitative results with a more qualitative discussion on what needs to be taken into account when developing demand flexibility. What are the pros and cons of demand flexibility compared to other flexibility options? What are the trade-offs between

---

[1] For an overview: https://www.oeko.de/forschung-beratung/themen/energie-und-klimaschutz/flexibilitaet-im-stromsystem-herausforderungen-und-ansaetze/ (in German).

Variable Generation, Flexible Demand
ISBN 978-0-12-823810-3
https://doi.org/10.1016/B978-0-12-823810-3.00011-X

demand flexibility and efficiency? And which role can new electricity consumers such as electric vehicles play?

• Section 4 highlights regulatory issues to develop demand flexibility and looks at why demand flexibility still plays a minor role, despite the high renewables share that has been reached in Germany.

This is followed by the chapter's conclusions.

2. Demand for flexibility and flexible demand

The increasing share of RES-E[2] in the German electricity sector results in a change of flexibility demand. As described in Chapter 5, traditionally, flexibility has been mainly provided by conventional power plants as well as pumped water storage plants. Along with the increasing share of variable RES-E from wind and solar additional flexibility is needed to integrate the electricity produced by RES-E—topics covered in Part One of this volume.

The main questions within this discussion are:

• When is additional flexibility needed in the system?
• What role can flexible demand play to meet that flexibility need?

It is obvious that an increasing share of variable and uncertain RES-E increases the demand for flexibility within the system to match supply and demand for electricity. However, in order to assess in more detail what kind of flexibility is needed and when it is needed, it is suggested to take a differentiated view on flexibility that distinguishes between:

• Different system contexts;
• Different flexibility functions;
• Different values of flexibility; and
• Different flexibility options.

In terms of system contexts, the demand for additional flexibility depends on:

• The share of RES-E and
• The kind of flexibility that is already available in the system and that can be exploited, before new flexibility options and new technologies are set up.

In terms of flexibility functions, it is important to note that different flexibility options can serve different purposes.

• Some flexibility options can be used to cover a capacity shortfall;
• Some flexibility options can be used to accommodate surplus RES-E generation;

---

[2] RES-E: Renewables for electricity generation.

- Some flexibility options can be used to accommodate surplus RES-E generation and use it to replace conventional generation or capacity; and
- Moreover, new flexibility options can be used to replace existing flexibility in the system.

The different flexibility functions are illustrated in Fig. 15.1, including an overview of which flexibility options can be used to fulfill which functions. Fig. 15.1 shows a stylized load duration curve of the residual load (i.e., the load that cannot be covered by renewable generation), taking into account existing flexibility options that are used to the extent possible to balance demand and supply (typically conventional generation and existing pumped hydro plants). The left-hand side of the graph shows the hours that cannot be covered by existing options (neither renewables, nor conventional generation, nor pumped storage). The right-hand side of the curve

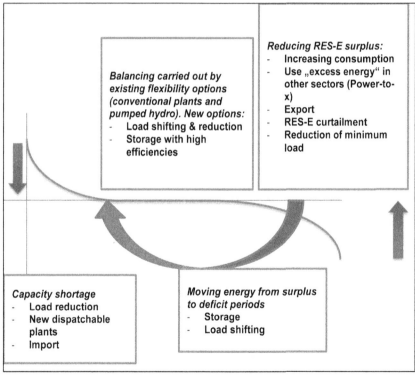

Figure 15.1 Different flexibility functions and options related to different sections of the residual load duration curve. *(Based on Bauknecht, D., Heinemann, C., Koch, M., Ritter, D., Harthan, R., Sachs, A., Vogel, M., Tröster, E., Langanke, S., 2016. Systematischer Vergleich von Flexibilitäts- und Speicheroptionen im deutschen Stromsystem zur Integration von erneuerbaren Energien und Analyse entsprechender Rahmenbedingungen. Öko-Institut; energynautics, Freiburg, Darmstadt, 106 pp. https://www.oeko.de/fileadmin/oekodoc/ Systematischer_Vergleich_Flexibilitaetsoptionen.pdf. (Accessed 22 September 2020).)*

shows surplus RES-E generation that cannot be accommodated by existing pumped storage options or by turning down conventional generation. In the flat section of the curve, RES-E generation matches demand, or any difference between RES-E generation and demand is balanced by the existing flexibility of conventional generation or pumped storage.

As can be seen in Fig. 15.1, demand flexibility can play a role in various ways.

- Load reduction as a specific form of demand flexibility, where demand is not shifted, but reduced at specific times, can be used to cover a capacity shortage;
- Both load reduction and load shifting can be used to replace flexibility that has so far been provided by the conventional power plants in place;
- Load shifting can be used to use power that would otherwise be curtailed, thereby avoiding a capacity shortage in other time periods; and
- Increasing consumption or adding new consumers can use surplus power, thereby avoiding the curtailment of renewables. This will be discussed in more detail in Section 3 of this chapter.

Looking at the four functions in more detail:

- In the flat section of the above graph, additional flexibility can replace flexibility that is already in place and provide flexibility at higher efficiencies (if ramping of conventional plants is reduced or if new flexibility options have higher efficiencies than the ones replaced) or enable conventional power plants to operate at lower costs (if flexibility increases the share of base load). Those will be the main effects of additional flexibility for some time to come, which means that additional flexibility is not yet essential. If new flexibility mainly has the effect to provide flexibility at higher efficiencies, then the system value of this additional flexibility is relatively low. While the shift toward base load plants that results from additional flexibility can reduce system costs, it can also entail higher CO_2 emissions, especially if the shift takes place from gas to coal. In this first phase, low cost flexibility potentials with high operating efficiencies like some demand flexibility or increasing the flexibility of CHP plants can be an attractive flexibility option, compared to for example storage. New options can also compensate for the decreasing share of conventional generation, which currently provides the bulk of the flexibility.
- Looking at the right-hand side of the above graph, if RES-E shares increase, flexibility can also be used to accommodate additional RES-E that would have been curtailed without that flexibility available. In

this case flexibility can be used to replace conventional generation that has relatively high marginal costs and CO_2 emissions with RES-E generation at very low marginal costs and without CO_2 emissions. The system value of additional flexibility increases accordingly.

- Looking at the left-hand side of the above graph, new flexibility options can also provide missing capacity that is required to cover demand and that would normally be provided by additional conventional capacity. If the system is faced with a capacity shortfall, then new flexibility options do not just replace conventional generation but can also be set up instead of conventional capacity. In this case, the system value of flexibility increases further. The capacity shortfall can be driven by RES-E that replaces conventional generation, yet without reducing the residual peak capacity accordingly. However, it can also be driven by a demand increase or the nuclear phaseout as in the case of Germany. In this case, options like demand flexibility can help to replace new conventional generation capacity.

Besides contributing to the demand–supply balance, flexibility options including demand flexibility can also be used for the management of the electricity grid and for providing system services. In this case, the system value depends on the extent to which there are bottlenecks in the grid. If flexibility options are used only for grid management, then they will typically only be used a few hours per year. Hence, flexibility options should be mainly market driven, and on top of that should be used for grid management purposes if required.

Fig. 15.2 shows the residual load duration curve that has been discussed above for Germany for three scenarios: 40% RES-E, 61% RES-E, and 75% RES-E.[3]

For all scenarios the assumption is that there are no new conventional plants added. It shows that for a 40% RES-E share basically all of the RES-E generation can be balanced by options that are already in operation. There is only a very limited surplus in a few hours. For 61% RES-E, both the deficit and the surplus increase, but are still limited to a relatively small number of hours per year. Both show significant peaks, with only relatively small amounts of energy. At 75% the problem becomes obviously more

---

[3] The scenarios were originally defined to represent the years 2020, 2030, and 2050. In 2019, renewables actually already reached a share of 42.1% of power generation in Germany, see https://www.umweltbundesamt.de/en/topics/climate-energy/renewable-energies/renewable-energies-in-figures. This means that flexibility demand increases faster than expected.

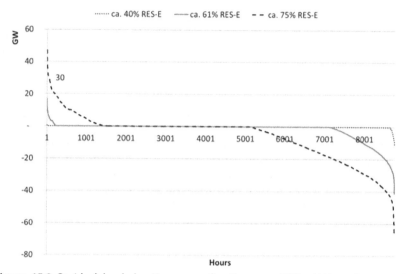

Figure 15.2 Residual load duration curve for Germany (40%, 60%, and 75% RES-E). *(Based on Bauknecht, D., Heinemann, C., Koch, M., Ritter, D., Harthan, R., Sachs, A., Vogel, M., Tröster, E., Langanke, S., 2016. Systematischer Vergleich von Flexibilitäts- und Speicheroptionen im deutschen Stromsystem zur Integration von erneuerbaren Energien und Analyse entsprechender Rahmenbedingungen. Öko-Institut; energynautics, Freiburg, Darmstadt, 106 pp. https://www.oeko.de/fileadmin/oekodoc/Systematischer_Vergleich_ Flexibilitaetsoptionen.pdf. (Accessed 22 September 2020).)*

relevant, both in terms of surpluses and deficits. Please note that import and export flows are not yet taken into account in this analysis and would further reduce the demand for flexibility as illustrated in Fig. 15.3.

Fig. 15.3 shows what different flexibility options and their technical potential within Germany can contribute to cover a capacity deficit in the scenario 2030 (approx. 60% RES-E). The capacity deficit is used here as an important indicator for the required flexible capacity. The deficit could also be covered by new gas turbines, but this analysis looks at what other options can cover.

The technical potential for demand flexibility is based on the assumption that the flexibility that is currently available is not extended, for example, by additional product storage options, which would be relevant especially for industrial flexibility, so that companies are able to a larger extent to produce more in some periods and less in others.

The capacity deficit is shown on the left. It corresponds to the deficit shown in Fig. 15.3 for the 60% scenario. It can be seen that interconnectors are a very important option to reduce the capacity deficit. The analysis also

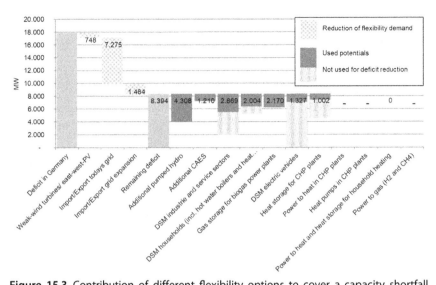

Figure 15.3 Contribution of different flexibility options to cover a capacity shortfall (approx. 60% RES-E). *(Based on Bauknecht, D., Heinemann, C., Koch, M., Ritter, D., Harthan, R., Sachs, A., Vogel, M., Tröster, E., Langanke, S., 2016. Systematischer Vergleich von Flexibilitäts- und Speicheroptionen im deutschen Stromsystem zur Integration von erneuerbaren Energien und Analyse entsprechender Rahmenbedingungen. Öko-Institut; energynautics, Freiburg, Darmstadt, 106 pp. https://www.oeko.de/fileadmin/oekodoc/ Systematischer_Vergleich_Flexibilitaetsoptionen.pdf. (Accessed 22 September 2020).)*

shows the potential of different types of demand flexibility: DSM (demand-side management) in the industry and service sector (see Chapter 16 for more details on industrial demand flexibility in Germany) and DSM in households.

A distinction must be made between two types of flexibility options: on the one hand options which can use their total installed capacity at all times (e.g., pumped hydro); and on the other hand options which cannot provide their total flexibility at all times because of other purposes they have to fulfill (see "Not used for deficit reduction" in Fig. 15.3). While storage options can provide their full potential to reduce the capacity shortfall, demand flexibility options are less flexible. Since their flexibility depends on their power consumption profile (or heat demand profile in the case of CHP with heat storage), these options may not necessarily provide their full potential at the time when the maximum power is missing in the overall system.

Yet it may still be more cost-efficient at least in the medium to short term to use demand flexibility than to build up additional storage capacity. And, for example, using EV batteries in CA is expected to be far less costly than conventional storage—see Chapter 8. Demand flexibility on its own would not be able to provide the flexible capacity that is needed, but it can make a relevant contribution.

3. The pros and cons of demand flexibility

Having discussed the potential contribution of flexible demand based on a quantitative analysis, this is complemented in this section by a qualitative discussion of the pros and cons of demand flexibility.

Table 15.1 provides an overview of the advantages and disadvantages compared to other flexibility options.

The rest of this section highlights two issues:

- Flexibility that can be provided by new demand and the potential pitfalls and
- The trade-off between flexibility and efficiency which is particularly relevant for the demand side.

Table 15.1 Pros and cons of demand flexibility.

Pros	Cons
Consumers can become active participants in power markets.	How many consumers actually consider it as an advantage to become active participants in power markets? How many consumers would rather prefer to continue inflexible consumption? Some consumers can provide more flexibility than others, and especially for households inflexibility should not be penalized. There should be a focus on automated processes such as heat pumps so that any interference with consumption processes is reduced to a minimum.
Demand flexibility provides flexibility of existing hardware, instead of requiring new hardware. This means new investment is mainly needed for "control" and no major innovations are needed. This also means that there are less natural resources needed, e.g., compared to storage in lithium-ion batteries.	As opposed to other options like storage plants, the flexible capacity DSM can provide is time-dependent, as a certain demand profile has to be met, and this may not correlate with the profile of flexibility demand. Demand flexibility competes with the process that needs power, which leads to opportunity costs (see also Fig. 15.3).

Table 15.1 Pros and cons of demand flexibility.—cont'd

Pros	Cons
Demand flexibility can provide flexible capacity without power generation of spinning reserve. As a result, there is a relatively high efficiency of flexibility, as there is no energy storage or conversion.	Demand flexibility has a relatively low "storage volume" and can only provide short-term flexibility. For RES-E shares at and even above 60%, the analysis in Section 2 shows that there is mainly short-term demand for flexibility, so that demand flexibility can contribute. At and above 75%–80%, long-term flexibility like power-to-gas-to-power will become more relevant, when there are long periods of RES-E surplus and deficit that cannot be covered by demand flexibility.

Source: Oeko-Institut.

3.1 New flexibility from new demand?

New demand from sectors that have so far been based on energy sources other than electricity is often seen as a way to provide flexibility to the electricity sector. This refers to the right-hand side "Reducing RES-E surplus" in Fig. 15.1.

Sector coupling and new demand (often labeled as Power-to-X, i.e., E-Mobility, Power-to-gas, Power-to-heat, etc.) can be seen from the perspective of the electricity sector and are then seen as an additional flexibility option for the use of surpluses. However, the primary objective of sector coupling should not and cannot be to provide flexibility to the power sector. Rather, the main objective of sector coupling is to enable the use of renewable energies in other sectors. This should be done in a flexible way to the extent possible. The intelligent loading of electric vehicles is a case in point as covered in Chapter 8.

Yet an additional consumer always means that additional inflexibilities are introduced. For the sustainable use of renewable energy in other sectors, it will not be sufficient to rely on curtailed power from renewables. Looking at Fig. 15.2, it is evident that the surplus on the right-hand side will not be sufficient to operate a relevant part of the new consumers that are being discussed. The surplus does increase with a rising share of renewables, yet so does the energy shortfall on the left-hand side, which means that a major part of the curtailed power will need to be used to cover that shortfall, for example via storage.

Therefore, the idea that so-called surplus electricity from renewables can be used for new consumers is misleading. Rather, a key challenge of sector coupling is to enable the development of additional renewable capacities. If additional consumers are introduced too early, there is a danger of increasing conventional power generation. This important requirement tends to be forgotten when these new consumers are mainly regarded as a flexibility option for the power sector.

3.2 Demand flexibility and efficiency

When talking about the demand side, it should be kept in mind that a key issue is not just demand flexibility but also demand reduction and hence energy efficiency. What is important here is that there can be trade-offs between demand-side flexibility and demand reduction. Such trade-offs can be found on different levels:

- First, on the level of individual appliances we observe the following trade-off: Increasing the efficiency of appliances tends to reduce the available flexibility. And the use of demand flexibility is often associated with an increase in consumption because flexibility leads to efficiency losses. However, efficiency losses from other options such as storage are typically more significant.
- Second, on the level of individual consumers the following question arises: Should people invest in flexibility or rather in demand reduction, i.e., should they for example buy a battery storage to increase self-consumption from PV, or should they rather invest in more efficient appliances to reduce their power consumption. From the perspective of the energy transition, the second strategy arguably provides better value for money.
- Third, on the system level, lower demand means that less renewable capacity is needed to reach a certain RES share. This in turn reduces the need for flexibility. Therefore, rather than making demand more flexible, it could be reduced in the first place.
- Finally, from an efficiency point of view, demand flexibility should always mean load shifting rather than merely increasing the load at specific times of high wind generation (see right-hand side in Fig. 15.1).

These arguments do not imply that demand flexibility is not relevant. However, they indicate that efforts to develop demand flexibility should not be made at the expense of demand efficiency; and in a renewable system demand efficiency can be one important way to reduce the need both for generation *and* flexibility.

Night storage heating (NSH) systems are presented here as an example for how various objectives can compete. This technology is less efficient than heat pumps that generate the same amount of heat with less use of electricity. Yet for this very reason storage heaters can in principle offer a higher flexibility potential. A comparison of these two technologies reveals that the advantages associated with reducing consumption outweigh the disadvantage of the lower flexibility potential. For this comparison three cases have been analyzed and compared in terms of system costs and CO_2 emissions: flexible use of NSH, replacement of NSH by heat pumps, and flexible use of heat pumps (Heinemann et al., 2014).

Fig. 15.4 shows in a scenario for the year 2020 that the greatest positive effect can be achieved by replacing NSH by electric heat pumps (the two cases on the right), both in terms of system costs and CO_2 emissions. Therefore, reducing the load by introducing a more efficient technology has much greater positive effects on the electricity system than making the NSH as an inefficient technology more flexible. While making NSH and especially heat pumps more flexible can significantly reduce system costs (variable costs of electricity generation), it also leads to a slight increase in CO_2 emissions compared to the inflexible version. This is because the electricity system in this scenario still contains a large proportion of conventional power plants. In the current German system introducing additional flexibility leads to coal plants replacing gas plants and hence increasing emissions. Although the relative effects are only small in this example, it illustrates the general trade-off between efficiency and flexibility.

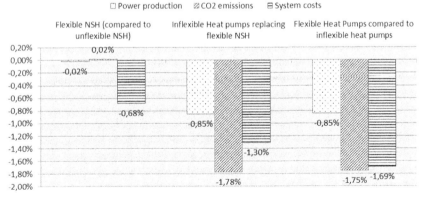

Figure 15.4 Efficiency versus flexibility: The case of night storage heaters and heat pumps in 2020. *(Source: Oeko-Institut.)*

4. Regulatory obstacles: why hasn't it happened yet?

When designing the regulatory framework for flexibility in general and demand flexibility in particular, the following needs to be considered:

On the one hand, new flexibility options such as demand flexibility have a relatively low system benefit in systems like the German system at 40% RES-E, since new flexibility has to compete primarily with the flexibility already available in the system. New options can mainly provide flexibility that is more efficient than the one already in operation (flat section in the curve in Fig. 15.1), but they cannot significantly increase the integration of RES-E, which would have a higher system benefit. Therefore, the development of flexibility options is restricted not only by the regulatory framework but also by this still relatively low system benefit. Nevertheless, efficient options can in principle already compete with existing options, as long as the regulatory framework does not prevent this. However, specific support for these options is not needed.

On the other hand, however, flexibility options have to be developed in time to be available when the need for flexibility increases. This is not so much of matter of technology development, but of getting the infrastructure and the processes in place. The fact that Germany had an RES-E share of 42.1% already in 2019 and thus higher than the 40% share assumed for 2020 in the scenario in Section 1 confirms this argument. Are market price signals and especially price spreads sufficient for the development of flexibility that is needed in the future? Or are further measures needed?

In principle, the development of flexibility options should be left to the competitive market, i.e., the development and use of flexibility should be price-driven and thus depend on the need for flexibility on the one hand, and on the costs of different options and their ability to cover demand on the other hand. RD&D programs for options that are not as advanced yet are also needed, but this is less relevant for demand flexibility.

However, the regulatory strategy to develop flexibility cannot merely rely on market prices. Given the existing framework conditions, the market price could lead to an ineffective or inefficient development of flexibility. This is because the various options have so far operated under different conditions in different regimes and have only partly been used as flexibility at all. This is particularly true for demand flexibility. As the regulatory framework for providing flexibility is not in place for some options or is geared toward other objectives, such as flat load profiles in the case of industrial consumers, just relying on market prices would be counterproductive. Even though it can be argued that different flexibility options should operate on a level playing field, it cannot be sufficient to just expose the different options to the same market price signals. Rather, it is necessary to remove the many technology-specific barriers for flexibility so that the various flexibility options are able to respond to the price signals according to their capabilities.

Table 15.2 provides an overview of some important aspects that need to be addressed in this context for demand flexibility.

Table 15.2 Regulatory framework for demand flexibility: key issues.

Structure of grid fees	The structure of network tariffs is geared towards incentivizing flat demand curves. Providing flexibility can therefore increase grid fees and thus becomes unattractive. This is a problem especially for larger consumers.
Standardized load profiles	As long as electricity suppliers balance households with a standardized load profile, it is difficult to introduce time-varying electricity prices to exploit demand flexibility.
Role of aggregators	The relationship between flexibility aggregators and power suppliers needs to be organized.
Processes and incentives to use demand flexibility for grid management (instead of grid expansion)	Providing flexibility becomes more attractive if it can be used both for balancing generation and demand and for grid management. Especially for DSOs, small-scale demand flexibility can play an important role. But also for TSOs, if TSO–DSO cooperation improves and if acceptance of grid expansion remains low. The EU calls for market-based procurement of flexibility. Especially for demand flexibility, regulated redispatch is difficult, as costs are mainly opportunity costs. This means that market-based redispatch needs to be established.
Capacity mechanism	If a capacity mechanism for power generation is set up, demand flexibility should also be included.

Source: Oeko-Institut.

5. Conclusions

In a renewable energy system a mix of different flexibility options is needed and demand flexibility will certainly become more relevant and play an important role within this mix. Demand flexibility is always restricted by the demand that needs to be covered, yet it is also a flexibility that does not require any major innovations and can be based on the infrastructure that is already in place, rather than developing a new infrastructure, i.e., based on storage. This is relevant not the least to reduce the amount of natural resources needed.

Yet the debate in Germany has also highlighted a number of issues and trade-offs that should be kept in mind. As with other developments in the context of the energy transition, "the more the better" and "the earlier the better" do not always apply, even for options that are needed to support an energy transition based on renewables.

One trade-off that has been highlighted in this chapter concerns flexibility and efficiency. The increasing demand for flexibility should not imply that demand reduction is no longer relevant. Wasting energy to use electricity when there is a lot of wind is not a sensible way of providing flexibility. Renewable electricity does have marginal costs close to zero, but it does not come for free. And it will certainly be very challenging to build up the renewable capacity needed for a carbon-free energy system. In that sense, there is no surplus energy. In the short term, developing flexibility at the expense of efficiency can even lead to higher CO_2 emissions.

Nevertheless, it is important to develop demand flexibility and set up the right regulatory framework and the incentives in due time. As demand did not play a major role in providing flexibility in the old power system, demand flexibility faces a number of specific barriers that need to be addressed. It is only then that it can compete on a level playing field with other flexibility options.

Another confusion that should be avoided is the following: New consumers such as electric vehicles (sector-coupling, Power-to-X) should not be considered as a flexibility option for the power sector. Rather, they introduce a new flexibility problem: More renewable generation is required for them to be sustainable and new consumers come with a specific demand profile that is more or less inflexible. The objective should be to make these new consumers as flexible as possible so that this additional inflexibility is reduced to the extent possible. This is certainly more attractive than introducing these new consumers and combining them with additional storage such as batteries. So flexibility does play a role for new consumers, but it is important to understand what it is about.

References

Bauknecht, D., Heinemann, C., Koch, M., Ritter, D., Harthan, R., Sachs, A., Vogel, M., Tröster, E., Langanke, S., 2016. Systematischer Vergleich von Flexibilitäts- und Speicheroptionen im deutschen Stromsystem zur Integration von erneuerbaren Energien und Analyse entsprechender Rahmenbedingungen. Öko-Institut; energynautics, Freiburg, Darmstadt, 106 pp. https://www.oeko.de/fileadmin/oekodoc/Systematischer_Vergleich_Flexibilitaetsoptionen.pdf. (Accessed 22 September 2020).

Heinemann, C., Bürger, V., Bauknecht, D., Ritter, D., Koch, M., 2014. Widerstandsheizungen: ein Beitrag zum Klimaschutz und zur Integration fluktuierender Erneuerbarer? Energiewirtschaftliche Tagesfr. 64, 45−48.

CHAPTER 16

Industrial demand flexibility: A German case study

Sabine Löbbe[1], André Hackbarth[1], Heinz Hagenlocher[2], Uwe Ziegler[2]

[1]School of Engineering/Distributed Energy Systems and Energy Efficiency, Reutlingen University, Reutlingen, Germany; [2]AVAT Automation GmbH, Tübingen, Germany

1. Introduction

A "virtual power plant (VPP) is a group of distributed generations, energy storage systems, and controllable loads, which are aggregated, optimized, coordinated and controlled so that it can function as one dispatchable unit in power system operations as well as one tradable unit in electricity wholesale markets. The primary goal of VPP is to provide capacity and ancillary services to the grid operations, sell power to electricity wholesale markets, enhance the overall system economics and reliability, promote efficient optimization in resources, and facilitate renewable integration to the grid. Its main framework includes information communication technology (ICT) infrastructure, smart metering and controlling at the customer sites, modeling, forecasting, scheduling, optimization at DERs [distributed energy resources], and real-time coordination with grid and market operations." (Wang et al., 2019).

The main focus of this chapter is to demonstrate the viability, profitability, and strategic approaches for different actors, who are willing to exploit flexible industrial demand in such VPPs. The main focus is on value creation for small and medium-sized enterprises (SMEs) on the one hand, and on cooperative business models for small and medium-sized utilities on the other hand.

The results are based on a research project which aimed at connecting SMEs to a VPP via a specific communication platform. Unlike other VPPs, the focus is on cooperative load management, i.e., participation, data, and control sovereignty for the SMEs, applied for real industrial processes and plants. Based on these results, business model opportunities for utilities concerning such a VPP solution are developed.

Variable Generation, Flexible Demand
ISBN 978-0-12-823810-3
https://doi.org/10.1016/B978-0-12-823810-3.00009-1

The chapter's main contributions are to exemplify viable business models and processes in value chain ecosystems for small and medium-sized utilities in the VPP market. Progressing digitalization and cooperative approaches are identified as game changers to integrate flexible industrial demand in a VPP.

The remainder of the chapter is organized as follows:

- Section 2 provides an overview of the regulatory environment for VPPs in Germany concerning flexible demand in the industry;
- Section 3 introduces the research project;
- Section 4 elaborates on the project findings regarding potential for flexibility and arising value in the industry in a specific example; and
- Section 5 derives typical business models for utilities followed by the chapter's conclusions.

2. Regulatory framework for flexible demand in Germany

In 2010, the German government adopted the "Energy Concept for an Environmentally Friendly, Reliable and Affordable Energy Supply" with the goal of completing the transition to a sustainable and greenhouse gas—neutral energy supply by 2050 (BMWi and BMU, 2010). Against this background, the Federal Ministry for Economic Affairs and Energy issued the "Electricity Grid Action Plan" in 2018 to accelerate the expansion of the grid, to improve the feed-in of renewable electricity into the grid, and to modernize the existing grid for better utilization of its capacity. At the same time, the grid infrastructure could be developed into a smart grid (BMWi, 2018). A smart grid would enable the implementation of an energy system which records the behavior of all market players by measuring their feed-in and consumption quantities in real-time and automatically communicates with the energy market (BDEW, 2013). This process is accompanied by the "Law on the Digitalization of Energy System Transformation," which stipulates the mandatory installation of smart meters (BMWi, 2020). Real-time monitoring with smart meters makes it possible to create behavioral incentives for consumers—such as (SMEs)—and to exploit their flexibility potential by actively participating in the electricity market (Albersmann et al., 2016). Accordingly, the "Electricity Market 2.0" adopted by the Federal Government in 2016 (Electricity Market Act) focuses on the market-based development of flexibility potentials, such as flexible consumers, flexible producers, and storage facilities (BNetzA, 2017). Thus, only the best technologies and solutions for meeting the demand for flexibility are intended to prevail in the competition on energy-only and flexibility markets.

All these developments represent a challenge, especially for smaller energy supply companies. One possible approach to exploit the opportunities inherent in the system is the operation of a VPP—especially regarding small amounts of demand flexibility which solitarily do not meet the minimum quantity that can be traded on energy markets (Loßner et al., 2015).

3. The research project

Against this background, the "Virtual Power Plant as a Cooperation Model" (VK KOOP) research consortium investigated the cooperative integration of SMEs in a VPP. The specific objective of the project was to develop a cooperation scheme based on the model of "industrial symbiosis" and the necessary tools for operating such a VPP. The approach was intended to enable SMEs to participate in the electricity markets while guaranteeing data and plant sovereignty via cooperative load management and to evaluate business models for them as well as for small and medium-sized utilities. Thus, the scope of the project was:

- to develop and design a communication interface between the VPP and the respective companies,
- to develop a methodology for the identification and utilization of flexibilities in companies while keeping their freedom of decision, and
- to investigate business opportunities for municipal utilities as operators of such VPPs with cooperative load management.

Funded by Deutsche Bundesstiftung Umwelt (DBU), the following institutions carried out the study from October 2016 to November 2019: ebök, Reutlingen University, AVAT, and Patavo, four SMEs from different industry sectors (mechanical engineering, food production, building material production, tool production), one sewage treatment plant, one municipal utility, and three supporting stakeholders. The model plants brought 68 different electricity consuming processes and generation units into the VPP—diesel generators, combined heat and power plants, physical storages (materials), electroplating plants, pumps, ovens, heating and cooling devices, and several further machines—which were assessed individually regarding their flexibility potentials.

In this regard, major contributions from Sebastian Wieners, Alexa Münz, Ansgar Sutterer, and René Hötzel are particularly acknowledged.

4. Case study: potential for flexibility in the industry

4.1 Requirements of industrial companies imposed on the electricity industry

Industrial companies offer attractive load shifting potentials (see Sioshansi, Chapter 7 in this book) which can be used in a VPP to generate revenue in electricity markets, in particular if the size of the company's demand flexibility does not suffice to solitarily participate in the energy markets. The potential revenues from flexibility marketing increase with traded volume and capacity as well as congruency of the forecasted and actual values when managing the balancing groups. However, bottlenecks in the electricity grid can define stability criteria in certain cases.

These framework conditions usually cannot be aligned perfectly with the flexibility potential of industrial companies. That is, production processes in the company generally have to be changed in order to exploit the entire flexibility potential. This is a great challenge for industrial companies, since their main focus is on achieving the production targets. Postponements or even down-times of production must be avoided and production times must be scheduled and determined in advance. In addition, industrial companies usually demand full control of their processes and are very reluctant to allow a third party manage their critical operations. Furthermore, they demand that the (demand response) measures do not compromise security of supply. Thus, electricity market revenues, which are made possible by a flexible production regime, must be highly financially advantageous for companies to be comfortable and confident in the arrangements and convince them to participate. However, energy costs often play only a minor role in companies' overall cost optimization and additional costs of flexibilization may occur (e.g., personnel costs due to shift changes, material costs, etc.), making the endeavor even more complicated as described in the Enel X example in Chapter 7.

Thus, the following requirements are demanded by all parties involved, e.g., flexibility marketers and industrial companies: a fully automated operation with minimum personnel deployment, low system costs and, of course, high data protection and secure communication.

The aim of a VPP architecture, hence, is to take the concerns of industrial companies into account by offering freedom of decision and acceptable compromises in the flexibility marketing options, in order to achieve profits for the parties involved and to pursue the goals of energy system transformation at the same time. These monetary and environmental benefits can be used to increase motivation and acceptance among SMEs in order to exploit their latent flexibility potential.

In the following, the developed VPP architecture is described, its implementation is illustrated, and first experiences are reported.

4.2 Architecture of the VPP

The architecture displayed in Fig. 16.1, which is based on the models of "industrial symbiosis" and "decentralized intelligence," was chosen in order to resolve the partially contradictory requirements of the actors involved (see Section 4.1) and to achieve a win-win situation (or win-win-win, if potential benefits for the grid are included).

The VPP with cooperative load management, which was developed in this research project, consists of two levels:

- On the lower level an optimization box is decentrally installed at each of the participating industrial companies. This intelligent box uses weather and energy exchange price (electricity and gas) forecasts as well as current storage levels and calculates overall cost-optimized schedules. It also takes into account technological constraints of the production process, production planning specifications, and energy procurement contracts. The person in charge of production can enter restrictions regarding plant availability (e.g., maintenance times, etc.) via a graphical interface, which are then taken into account in the optimization algorithm. The box communicates with the production units and is able to control them directly.

- The upper level consists of an aggregation layer across all optimization boxes, i.e., the schedules generated by the optimization boxes are centrally aggregated at this level. If the aggregated schedules deviate from a global optimum (e.g., as a result of grid restrictions), a superordinate

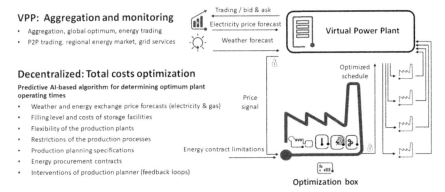

Figure 16.1 Architecture of the VPP. *(Source: AVAT.)*

optimization takes place at this level. Furthermore, the interfaces to the marketer as well as to the service providers for weather and energy price forecasts are implemented in this layer.

The optimization process is performed chronologically as follows:

- Forecast data for the upcoming days for weather as well as electricity and gas prices are obtained **centrally** and forwarded to the optimization boxes.
- Based on these data, future production is **decentrally** optimized at the company level to minimize total costs while taking company-specific conditions (e.g., requirements of the operating personnel) into account. The calculated flexibility in operation, which is based on the combination of generators, consumers, and production plants in conjunction with heat and process storage, is exploited. In general, plant sovereignty remains with the industrial company at all times. This also applies to sensitive company data.
- In a further step, a scheduling proposal is transmitted to the aggregation level, where a superordinate optimization is carried out if needed. This second optimization step may become necessary if the proposed schedules do not correspond to the global optimum of energy marketing.
- If necessary, the updated price signal is sent back to the optimization boxes, where processes are reoptimized and a modified schedule proposal is created. The procedure is repeated until a trading optimum is reached.
- The target schedules created after the completion of such iteration are finally used to control the individual production units.

At the aggregation level, this schedule is marketed on the spot market, or more precisely on the day-ahead market, as it currently is the option with the highest potential revenues for marketing flexibility in the German energy markets with this VPP approach (in comparison to, e.g., the provision of balancing power or trading in the intraday market)—although different results might occur under different circumstances (see Madlener and Ruhnau, Chapter 13).

Similar to the Enel X approach described in Chapter 7, the flexibility of consumers can be used in our VPP approach to reduce their energy costs. In addition, however, the flexibility of almost any combination of storage with electricity consumers and generators (within and across companies) can be used to increase revenues. Not only will incentives be generated for the customers, but specific schedules will be automatically generated in the optimization box, which are sent to the power consumers and generators as commands and set points at execution time.

4.3 Process flow modeling using the example of a cement mill

In the course of the project, solutions for several SMEs were investigated. However, in the following, the approach and the results of the optimization procedure are demonstrated by focusing on the example of an energy-intensive cement plant with considerable demand flexibility (which does not necessarily have to participate in a VPP scheme to participate in the wholesale market).

The plant consists of several process chains with combinations of electricity consumers and storage units. The electricity consumers are for the most part rock mills. The storages are silos in which the raw materials or the ground rock are stored. The flexibility during operation results from the storage function of the silos: The mills do not have to run permanently, especially when the plant is only working at partial capacity, and can thus be operated variably during the course of the day, taking into account other boundary conditions.

The sequence of processes consists of four steps:

4.3.1 Step 1: Plant configuration

In a first step, the plant configuration, i.e., the modeling of the production process and process flow—represented by a combination of generators, consumers, and storages—is carried out. Fig. 16.2 exemplifies a production line. In reality, the plants correspond to the mills and the storages to the silos. On this basis and with the respective parameters and constraints of the components, the specific model (digital twin) is created, utilizing the

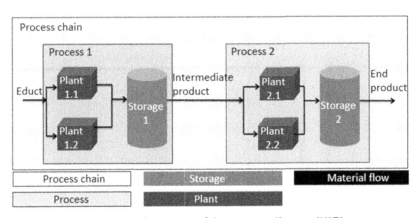

Figure 16.2 Abstraction of the process. *(Source: AVAT.)*

graphical interface developed for the optimization box. This detailed process modeling is usually carried out only once, but can be adapted if plant conditions and constraints change.

4.3.2 Step 2: Integrated planning

In a second step, the production manager can define specifications regarding the runtimes of the units in the optimization period, e.g., by setting revision times, planned operating times, and pause times. The planned production quantities are also communicated to the optimizer in this step (Fig. 16.3).

4.3.3 Step 3: Optimization

In the third step, the optimization is performed by the algorithm implemented in the optimization box. Optimization criteria are electricity costs, personnel costs, and storage costs. For instance, by taking into account the constraint "staff availability between 6:00 a.m. and 6:00 p.m." and the restriction "duration of production process is 4 h" the algorithm calculates an optimal process operation between 6:00 a.m. and 10:00 a.m., as illustrated in Fig. 16.4.

4.3.4 Step 4: Acceptance of the optimized schedule

In a final step, the production manager can accept or reject the result of the optimization, which is displayed in the graphical interface as illustrated in

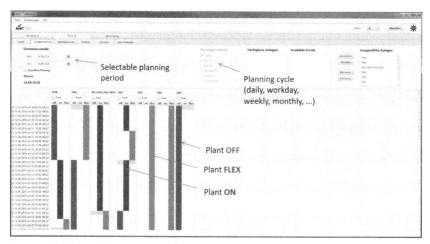

Figure 16.3 Specification of the operating times (screenshot of graphical interface). *(Source: AVAT.)*

Costs

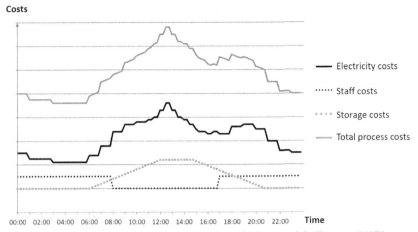

00:00 02:00 04:00 06:00 08:00 10:00 12:00 14:00 16:00 18:00 20:00 22:00   **Time**

—— Electricity costs

······ Staff costs

* * * * Storage costs

—— Total process costs

Figure 16.4 Optimization principle (simplified example). *(Source: AVAT.)*

Fig. 16.5. If the optimization proposal is rejected, it is possible to switch to the production planning or to change parameters in order to start a new optimization and to obtain a modified schedule.

As already mentioned, a superordinate optimization at the VPP aggregation level may follow in a further step, if necessary. The final schedule is then transmitted to the marketer and placed accordingly. At the same time, the optimization box implements the schedules and transmits specific requirements and target values to the corresponding production units.

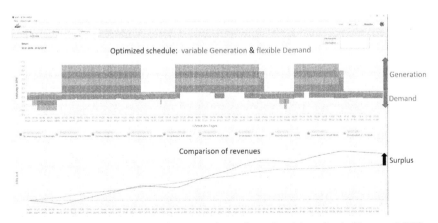

Figure 16.5 Optimization result (screenshot of graphical interface). *(Source: AVAT.)*

4.4 Experiences with the developed VPP architecture

The five model plants (four SMEs and a sewage treatment plant) were satisfied with the developed system. Above all, the data protection (guaranteed by the two-step optimization) and the possibility to intervene in production planning or to reject the schedules was seen as a major improvement over established systems. The expectation that previously unused flexibility potential can be exploited by taking company-specific constraints into account in the overall cost optimization was also fully met. The simple integration of the optimization box into the existing infrastructure of the company, made possible by a wide range of interfaces, has also been proven successfully.

Owing to the high degree of process abstraction and generalizability of the underlying optimization processes, the developed scheme can be applied to all participating SMEs, although they belong to different industries (metal processing, wastewater treatment, building materials industry, electroplating). Hence, it could easily be used by any industrial customer with sufficient load flexibility.

As shown in Section 4.3, the optimization box was integrated and successfully tested at a cement mill during the research project. The optimization box is currently converted into a commercial product and will be offered by AVAT (SE^{2}OPIMIZER and SE^{2}DIRECTOR).

5. Typology of business models for utilities

The solutions of cooperative load management for industrial companies developed in the previous section can be used in (existing) VPPs. The planning, construction, and operation of such VPPs is of interest to specialized service providers and to utilities who wish to offer comprehensive solutions to their industrial customers. According to a survey, 50 VPPs were in operation in Germany in 2016 (Albersmann et al., 2016).

The main focus of the study was to investigate which type of business models are feasible for small and medium-sized utilities which are intending to use the unique VPP approach developed in this project. Accordingly, several business model types were developed based on a theoretically and empirically founded analysis of the value chain of VPPs as well as on the basis of qualitative interviews and workshops with several utilities. The challenges for energy supply companies resulting from an integration of a VPP into the existing business model and its effects on company development were identified in order to derive promising business models for the operation of such a VPP.

5.1 The value-added ecosystem of a VPP

For an appropriate assessment of potential business models, it is necessary to schematically work through the individual processes of a VPP, to integrate the roles and tasks of individual actors and, thus, to describe the value creation ecosystem with its value-added steps. Fig. 16.6 illustrates the summarized results. The central roles are defined as follows:

- **Industrial company**: VPP customer who makes its flexibility potential available to the VPP for marketing and is remunerated in return for this provision.
- **Pool manager**: The pool manager is responsible for the administrative and organizational activities within the VPP. These include customer acquisition and integration, ongoing support of the customer interface, and the monitoring and settlement of customer contracts.
- **Aggregator**: The aggregator is responsible for carrying out the practical activities required to operate a VPP. Main tasks are to ensure the technical functionality of the VPP, the processing and maintenance of incoming data, and the optimization of the schedules of the plants which are bundled by the pool manager. Thus, the aggregator plays a central role within the value-added ecosystem and forms the heart of the VPP.
- **Trader**: The trader is responsible for marketing the optimized schedules of the aggregator on the respective energy markets.

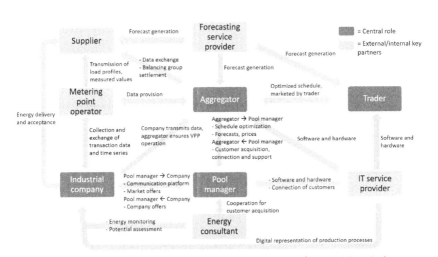

Figure 16.6 Value-added ecosystem. *(Source: Reutlingen University.)*

- **IT service provider**: The IT service provider is assigned by the pool manager to build a digital model of the companies' production processes (digital twin) and to connect the industrial company with the VPP via information technology.
- **Energy consultant**: Energy consultants are likely included in order to identify flexibility potential in industrial companies through energy monitoring.
- **Forecasting service provider**: The forecasting service provider generates performance, consumption, and price forecasts on the basis of which optimization, marketing, and balancing activities are offered and carried out.
- **Metering point operator**: The metering point operator installs and operates the metering equipment for determining the data relevant to balancing and acts as a central data hub.
- **Supplier/balancing group manager**: The supplier is responsible for the acceptance or supply of energy quantities from and to the VPP customer, including balancing energy.

Other theoretically involved roles and actors, such as the distribution or transmission system operator and the balancing group coordinator, were not included in the analysis for reasons of simplicity and clarity.

Each of the displayed roles and tasks can be performed by one or more stakeholders—i.e., energy providers or specialized service providers. Depending on the number of collaborating actors and the allocation of tasks among them, different types of business models with different responsibilities concerning the management of processes or the handling of communications and settlements arise for small and medium-sized utilities (see Section 5.3).

Thus, the roles of stakeholders in the value-added ecosystem have to be linked to the activities in the cooperative VPP concept. For this purpose, the main activities have been sequenced and categorized into different stages of the value chain (Fig. 16.7). The special feature of the cooperative concept is the feedback loop between the activities "global schedule optimization" and "exchange of offers via communication platform":

- Initially, companies receive a price signal, e.g., a forecast for the day-ahead market, on the basis of which they optimize their schedules—consisting of demanded loads and generation capacity—making use of their flexibility.

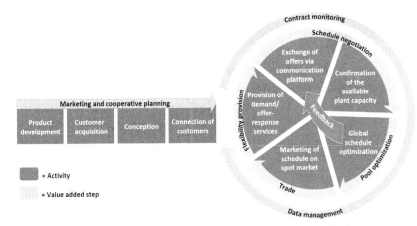

Figure 16.7 Value chain. *(Source: Reutlingen University.)*

- These electricity volumes are aggregated with those of other connected VPP participants in the "global schedule optimization." The specific market price is determined for the aggregated schedule.
- This price is then returned as an improved forecast into the "exchange of offers via communication platform" activity. As a result, the schedules in the companies are recalculated if necessary. If the improved forecast differs significantly from the original one, this process can be repeated several times.
- After the convergence of the iteration, each company can decide whether to accept the offer or not.
- If the offer is accepted, the aggregated electricity quantities are released for procurement or sale.

5.2 Potentials and challenges of a VPP integration

To analyze the opportunities and challenges that arise for small and medium-sized energy supply companies from such a VPP with cooperative load management, the strengths and weaknesses of the energy supply companies must be assessed with regard to the opportunities and risks of such business activity. To this end, it is first necessary to clarify the potentials or challenges that energy providers face regarding the marketing of this specific VPP approach to SMEs and its implementation:

- The clarification of customer interest regarding cooperative load management: The approach of decentralized optimization is in principle considered to have promising potential, since it allows to retain the data and control sovereignty within the industrial company—a unique selling proposition. The increasing attention of the industry concerning climate neutrality, energy monitoring, and energy efficiency will likely have a positive effect on acceptance. However, a cultural change is still needed within the industrial companies, since production planning in cooperative load management is influenced by external actors (price signals and associated interactions), so that SMEs might be reluctant.

- The clarification of a promising marketing strategy for cooperative load management: For instance, in our case, only the day-ahead market was identified as being economically attractive. Furthermore, not all flexibility potentials are economically viable, i.e., adjusting companies' production processes to forecasted electricity prices might not be worthwhile.
- The clarification of the own and the needed resources in order to operate a VPP: For instance, energy portfolio management and access to trading must be available. In a scenario-based profitability analysis, it was concluded that a utility without well-established portfolio management and trading access cannot achieve a positive business case without partners, as the costs for the establishment and operation of the necessary processes are prohibitively high (personnel, hardware, and software). This is particularly true for small and medium-sized energy supply companies, which generally do not reach the volume limit of approximately 1 TWh/a, above which in-house portfolio management is worthwhile.

The following SWOT (strengths, weaknesses, opportunities, and threats) analysis gives further insight with regard to the evaluation of a (municipal) electricity supply company's entry into the VPP market.

The typical strengths/opportunities include:

- Congruency with the goals and strategies of the utility of becoming an innovator in the transformation of the energy system, i.e., to strategically shift away from selling energy commodities toward innovative, digital energy solutions, to master an increasingly decentralized generation structure, to develop renewable generation without subsidies (in Germany the so-called "post-EEG age"), to exploit the potential of sector coupling and digitization.
- Availability of flexibilities from own generation units as well as the potential for coupling with the heat sector (operation of a proprietary heat network with integrated storage units).
- Sufficient customer potential within the area of the municipal utility.

- Using the VPP organizational unit as internal coordinator and driver of innovation (based on a professionalization of communication, objectives, competencies, and responsibilities within the company) and process optimization from a holistic perspective.
- Knowledge of own strengths and weaknesses, ability to cooperate (cooperation culture), and ability to manage and control cooperations or critical processes that are outsourced to service providers.

Typical weaknesses/risks include:

- Insufficient connection to goals and strategies of as well as insufficient acceptance within the own municipal utility.
- Lack of ability for customer-oriented solution sales.
- Insufficient profitability due to a lack of customer access and too limited customer portfolio (insufficient technical potential for optimizing the VPP pool).
- Limited human and infrastructural resources, causing, e.g., hurdles in the energy procurement structure, such as insufficiently equipped energy portfolio management or full supply contracts.
- Insufficient endowment of the organizational unit responsible for the VPP regarding power and influence over other organizational units, such as energy procurement, sales, marketing, or the grid dispatch center.
- Requirement of cooperation with external parties in the value chain.
- Strongly risk-averse strategy, especially in the environment of municipal shareholder structures.

5.3 Three types of business models for operating a VPP as energy supplier

Based on (1) the value and process analysis, (2) the analysis of existing business strategies of selected small and medium-sized utilities currently offering VPP services, and (3) the business model typologies developed by Gassmann et al. (2013), three potential business models were identified under the presumption of a flexibility marketing via the spot market. The respective business models for small and medium-sized utilities differ in terms of the responsibilities for the provision of services, i.e., whether they are performed internally ("make" column in Fig. 16.8) or by partners/service providers ("buy" column). The VPP customer is treated as a service provider as well.

Orchestrator: A municipal utility as "orchestrator" administers specialized service providers from the energy and IT sector. The energy

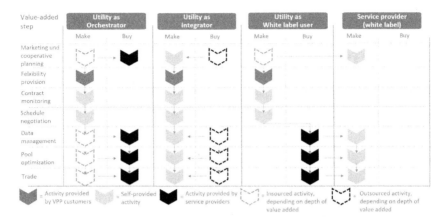

Figure 16.8 Business models for municipal utilities. *(Source: Reutlingen University.)*

supply company has the activities that lie outside its own core competencies taken over by specialized service providers and coordinates their respective services. Thus, the costs for operational, adaptation, and improvement measures are outsourced to third parties. This enables the energy supply company to focus on its core competencies and benefit from the special know-how and the potential for economies of scale on the part of cooperation partners. Above all, the orchestrator is able to benefit from competitive advantages through a flexible selection of highly specialized service providers whose specific innovative strength can be utilized (Gassmann et al., 2013). However, the coordination of service providers is time-consuming and high transaction costs are incurred. Professional management of the cooperation and service provision processes is therefore essential. In general, there is a risk that the municipal utility is not able to achieve a sufficient pool size to economically operate the VPP (see, e.g., Poplavskaya and de Vries (2020)).

Integrator: While the orchestrator essentially combines the services of cooperation partners, the "integrator" combines the activities of third parties with its own service provision. Through this integration of external services, the energy supply company uses and expands its own core competencies in the energy industry, hereby achieves greater independence from third parties, and retains control over specific key activities and resources. It optimizes the value added to meet its own needs and processes. The self-execution of the majority of activities requires a high implementation effort to create the necessary structures and to acquire the needed expertise. Likewise, high fixed costs are incurred due to the high expenditure on personnel and operations, which is why a sufficient and scalable customer pool size is absolutely essential. A high depth of value creation also facilitates the expansion of the portfolio of by-products that build on already established structures, such as white label solutions. This enables the municipal utility to develop growing expertise in the VPP sector and to strive for a firm position in the VPP niche market.

White label user: In contrast, the so-called "white label user" purchases a ready-to-use VPP product, whose technical and operational aspects are handled by the white label provider. To the outside world, the energy supply company portrays itself as an innovative VPP operator without the need to possess the necessary capacities and skills. This enables the energy supply company to concentrate on customer acquisition and support activities. The use of white label products thus enables a relatively simple expansion of the product portfolio, with only limited and low-investment structural or capacity-related adjustments having to be made. The integrated customer plants are managed by the white label provider in a larger pool, with economies of scale leading to economic efficiency and reducing the risk for the energy supply company. On the one hand, this approach puts the energy supply company in a close relationship of dependency, so that limited opportunities for further individual development can be expected. On the other hand, a withdrawal from the VPP business can be arranged quickly and in general also with a certain degree of flexibility.

6. Conclusions

A cooperative VPP is a promising and marketable application for SMEs in the industrial sector, as it enables the optimization of own energy supply and demand. Additional revenues are achievable, although profitability is limited in the current German market, as the revenue potential is mainly restricted to the spot market (where price spreads currently are relatively low), so that participation might be uneconomical for SMEs with relatively small load flexibilities. Furthermore, the revenue potential correlates with the scalability of the pool size. Hence, the number, size and aggregated load curve of VPP participants matters for all partners of the value chain.

On the side of utilities, VPPs open up the possibility for positioning themselves as service providers and an energy transition pioneers and, thus, to grow in a new market. However, establishing a scalable VPP without the participation of cooperation partners is difficult to implement for small and medium-sized energy supply companies. Depending on the strengths, weaknesses, opportunities, and risks as well as the strategy and objectives of the energy supply company, different paths can be taken to enter the VPP market.

In all three business models presented, the energy supply company (municipal utility) can build upon their customer proximity as key resource. The cooperation partners in the VPP—such as aggregators, IT service providers, and energy consultants—can rely on the utility's regional problem-solving ability and customer access.

An energy supply company with well-established energy procurement departments or portfolio management with a high degree of value creation (real net output ratio), financial, technical, and personnel capacities as well as sufficient customer potential will be more interested in operating a VPP as an orchestrator or integrator, while an energy supply company with outsourced procurement is likely to be more interested in a white label solution.

All three business models open up further opportunities for cross-selling—further products, customer segments, and other cross-connected businesses, including sector-coupling—and might be a basis for regional expansion. A VPP offers a structure to tap market potential of unsubsidized PV generation and to digitize the own business. Hence, it offers a range of strategic approaches to ensure the future viability of municipal utilities.

Furthermore, the increasing sensitization of the industry regarding climate issues offers future business potential for VPP solutions. The attractiveness of the overall VPP market, however, depends largely on the development of the marketing potential of electricity demand/supply flexibility and, thus, on the revenues for the provision of capacity in the balancing energy market and for electricity supply in the energy-only market. Since the further expansion of renewable energies will increase the volatility on the spot markets, the outlook is not too bad. Further marketing potentials are, for example, the introduction of grid flexibility markets for the efficient elimination of grid bottlenecks or the establishment of an internal peer-to-peer trading platform (Herrmann et al., 2018). Thus, the quality and speed of developments regarding cooperative VPPs largely depend on future regulatory framework conditions—a common finding in energy markets.

Bibliography

Albersmann, J., Dütsch, G., Theile, H., Erken, E., 2016. Markt und Technik virtueller Kraftwerke. PricewaterhouseCoopers. November 2016.

BDEW — Federal Association of the Energy and Water Industry, 2013. BDEW-Roadmap: Realistische Schritte zur Umsetzung von Smart Grids in Deutschland. Berlin, 11.2.2013. Available online: https://www.bdew.de/media/documents/Pub_20130211_Roadmap-Smart-Grids.pdf.

BMWi — Federal Ministry for Economic Affairs and Energy, 2018. Aktionsplan Stromnetz. Berlin, 14.8.2018. Available online: https://www.bmwi.de/Redaktion/DE/Downloads/A/aktionsplan-stromnetz.pdf?__blob=publicationFile&v=10.

BMWi — Federal Ministry for Economic Affairs and Energy, 2020. Smart Meter und digitale Stromzähler: Eine sichere, digitale Infrastruktur für die Energiewende. Berlin, March 2020. Available online: https://www.bmwi.de/Redaktion/DE/Publikationen/Energie/smart-meter-und-digitale-stromzaehler.pdf?__blob=publicationFile&v=12.

BMWi and BMU — Federal Ministry of Economics and Technology (BMWi) and Federal Ministry for the Environment, Nature Conservation and Nuclear Safety (BMU), 2010. Energy Concept for an Environmentally Sound, Reliable and Affordable Energy Supply. Berlin, 28 September 2010. Available online: https://www.osce.org/files/f/documents/4/6/101047.pdf.

BNetzA — Federal Network Agency, 2017. Strommarkt im Wandel — So funktioniert der Strommarkt 2.0. Available online: https://www.smard.de/home/topic-article/426/486.

Gassmann, O., Frankenberger, K., Csik, M., 2013. Geschäftsmodelle entwickeln: 55 innovative Konzepte mit dem St. Galler Business Model Navigator. Carl Hanser Verlag, Munich, Germany.

Herrmann, A., Börries, S., Ott, R., Steiner, S., Höckner, J., 2018. enera: Flexibilitätsmärkte für die netzdienliche Nutzung. Netzpraxis — Magazin für Energieversorgung, 1.12.2018. Available online. https://www.energie.de/netzpraxis/news-detailansicht/nsctrl/detail/News/enera-flexibilitaetsmaerkte-fuer-die-netzdienliche-nutzung-2018233/.

Loßner, M., Böttger, D., Bruckner, T., 2015. Wirtschaftliches Potential virtueller Kraftwerke im zukünftigen Energiemarkt — Eine szenariobasierte und modellgestützte Analyse. Z. Energiewirtschaft 39 (2), 115—132.

Poplavskaya, K., de Vries, L., 2020. Aggregators today and tomorrow: from intermediaries to local orchestrators? In: Sioshansi, F.P. (Ed.), Behind and beyond the Meter: Digitalization, Aggregation, Optimization, Monetization. Academic Press, London, Oxford, Cambridge, UK.

Wang, X., Liu, Z., Zhang, H., Zhao, Y., Shi, J., Ding, H., 2019. A Review on Virtual Power Plant Concept, Application and Challenges. IEEE Innovative Smart Grid Technologies - Asia (ISGT Asia), Chengdu, China, pp. 4328—4333.

PART 4

Implementation, business models, enabling technologies, policies, regulation

Market design and regulation to encourage demand aggregation and participation in European energy markets

Juan José Alba, Carolina Vereda, Julián Barquín, Eduardo Moreda
Endesa (Enel Group) — Regulatory Affairs, Madrid, Spain

1. Introduction

Traditionally, electricity systems have featured rigid demand which generally does not respond to short-term price signals. Today, with an increasing share of variable generation, demand-side flexibility provided by demand response (DR) is key and will help the system to be more efficient. DR provides flexibility to the system, and its core function is to improve the matching of demand and supply in electricity markets that will be more unpredictable in the near future. Storage and generation will still provide flexibility to the system, but the empowerment of customers will allow the demand side to be much more active than in the past.

Given the distributed nature of demand and the increasing role of digitalization to empower customers, independent aggregators may play an important role to unlock the great potential of demand to contribute to the energy transition.

Aggregation of demand may, indeed, overcome several market failures which are currently preventing demand to deliver greater value to the energy system. However, it is important to properly regulate the aggregator role to avoid system inefficiencies.

This chapter provides an overview of the aggregator role in the European Union (EU) and identifies the benefits they would bring to the electricity system if their introduction to markets is properly implemented.

The chapter's main contribution is to encourage demand aggregation by assessing the role of independent aggregator models for energy markets (e.g., day-ahead, intraday, balancing and reserves markets), and the distributional effects on different stakeholders. This is done by analyzing different

Variable Generation, Flexible Demand
ISBN 978-0-12-823810-3
https://doi.org/10.1016/B978-0-12-823810-3.00006-6

independent aggregator models in terms of their financial impact on final customers, suppliers and balancing responsible parties (BRPs), aggregators, and other balancing services providers.

The chapter is organized as follows:
- Section 2 provides a general overview of how the EU is moving toward DR aggregation;
- Section 3 describes the role of the aggregator in providing DR;
- Section 4 describes a model where the aggregator is also a retail supplier and an alternative model where the aggregator is an independent actor; and
- Section 5 examines three distinct models for independent aggregators; followed by the chapter's conclusions.

2. European vision of the future role of aggregators

The European Commission has introduced the concept of aggregators, DR, and independent aggregators in Directive (EU) 2019/944[1] on common rules for the internal market for electricity. The aim of aggregators is to enable the power market to make better use of the existing potential of DR in a decarbonized system with an increasing share of variable renewables generation, the topic of this volume.

By increasing flexibility in the power system, independent aggregators could play a key role in allowing more intermittent generation onto the system, lowering costs, optimizing market positions, and increasing security of supply. However, the regulatory framework for the participation of market players in DR is still quite different across European countries.[2] According to the analysis performed by the Smart Energy Demand Coalition (SEDC 2017), they fall into three groups:
- Advanced countries with an active DR market, such as France, Belgium, the United Kingdom, and Ireland. Beyond Europe, the United States serves as a role model for the activation of DR through appropriate conditions further described in Chapters 9 and 10;
- Intermediate countries with a partially open DR market including, inter alia, Austria, Germany, and most Scandinavian countries; and

---

[1] DIRECTIVE (EU) 2019/944 OF THE EUROPEAN PARLIAMENT AND OF THE COUNCIL of 5 June 2019 on common rules for the internal market for electricity and amending Directive 2012/27/EU.

[2] Explicit demand response development in Europe (SEDC 2017).

- Countries with closed DR markets including, inter alia, Greece, Spain, Portugal, and Cyprus.

However, the regulatory frameworks in many countries are improving in the sense that they are more supportive of DR. Regulators are interested in facilitating the expansion of DR in balancing markets to enhance competition among market players by building a demand-side counterweight to the industry's current supply-side mentality, as further described in Chapter 5.

In Spain, for example, the regulators are moving in that direction by allowing demand-side aggregation to participate in balancing markets under the same conditions that apply to generation and storage. The Spanish Government has also introduced legislation that recognizes the independent aggregators as new actors in the electricity system. However, the rules, the rights, and the obligations of the independent aggregators must still be defined. In Section 4 of this chapter, models for introducing independent aggregators into electricity markets are examined. The new EU Directive gives freedom to Member States to choose their preferred model for independent aggregators and there are different opinions on how exactly what is the best policy and what is the best form of regulation to enable aggregators to play a more active role in balancing supply and demand.

A regulatory framework should be put in place that is proportionate to the challenges faced by aggregators and ensures that they can access the market without depending on the agreement of the consumer's retailer.

3. The role of aggregators in providing demand response

At the outset, one should distinguish between *explicit* and *implicit* DR. Explicit DR is actively sold into one of the markets of the power system, as distinct from implicit DR, which is primarily concerned with consumers reacting to wholesale prices. This chapter is focused on the former.

When consumers are active in providing explicit DR, they benefit by receiving direct payments or bill reductions in exchange for altering their consumption and/or generation patterns on request, as explained in Chapter 7 for a DR aggregator such as Enel X. The latter may be triggered by the need for balancing energy, by differences in electricity prices, or by a constraint on the network. In the past, most traditional DR programs were designed to curtail load in a small number of hours per year that were likely to coincide with the system peak demand as explained in Chapter 10. Examples of these programs include interruptible tariffs for commercial and industrial customers and direct load control of residential air conditioners. However, today, with increased uncertainty of generation output and significant uncertainty about network conditions, flexibility may be needed at any time, often on very short notice.

This offers an opportunity for aggregators active in explicit DR to combine multiple flexible consumer loads, generation or storage for sale, purchase, or auction in electricity markets. The distinction with the independent aggregator is that the latter has no affiliation to the customer's retail supplier. Aggregators—whether independent or affiliated with a retailer—act as facilitators by bundling flexibility from different customers and generators and offering it to the different actors that need it.

In the current power system, the supplier acquires the amount of energy the consumer is expected to consume and informs the TSO about the consumer's projected consumption on a 15-min to 60-min basis[3] for the next day or hours. This is usually referred to as "scheduled demand" or the "firm program." Moreover, the supplier takes responsibility for imbalances between the supplier's forecast of the flexible consumer's demand and the actual consumption that occurs in real time. All market participants bear responsibility for imbalances they cause, thus all market participants must either be BRPs[4] or delegate this responsibility to a BRP. In Europe, according to EU Regulation (slightly abbreviated),

All market participants shall be responsible for the imbalances they cause in the system ('balance responsibility'). To that end, market participants shall either be BRPs or shall contractually delegate their responsibility to a BRP of their choice. Each BRP shall be financially responsible for its imbalances and shall strive to be balanced or shall help the electricity system to be balanced.

Fig. 17.1 illustrates a case where the supplier is the only actor involved and the only contact point with the consumer. In this example, the consumer (or prosumer) consumes 100 units supplied by its supplier (SUP) nominated by the BRP sup. In this case, all BRPs are in balance, including the system (BRP syst).

Independent aggregators, on the other hand, do not have an energy supply contract with their customers and typically intervene as a third-party service provider next to the supplier. An independent aggregator has the right to offer DR flexibility to the market to the extent agreed with the

---

[3] Depending on the resolution of the power market rules of country in question.
[4] Balancing Responsible Party (BRP) is defined to settle differences between the scheduled and actual values of consumption, generation, and trade.

Figure 17.1 Supplier is the only contact point with the customer who supplies 100 units and its customer consumes 100 units. The position of the supplier is balanced.

Figure 17.2 The consumer has two contact points, aggregator and supplier. The aggregator will activate the prosumer while the supplier will supply 100 units of energy to its customer.

flexible consumer. In this case, the aggregator is entitled to activate the DR flexibility when required and technically feasible. Independent aggregators can have their own BRP, or they can contract with the supplier's BRP[5] as illustrated in Fig. 17.2.

In practice, different models are conceivable to handle the relationship between the independent aggregator, the supplier, and the BRP.

4. Business models for aggregators[6]

According to the new EU directive, the aggregator, as a new energy player, needs to be introduced into the existing organization of the power market in a way that does not jeopardize the existing protocols and respects the

---

[5] Directive (EU) 2019/944 states that independent aggregators shall have the right to enter electricity markets without the consent of other market participants. Contracting with the supplier may be considered as a barrier if no additional model is foreseen in the Member State.

[6] Chapter 6 of this volume evaluates 12 DR and load flexibility programs.

existing legislation while enabling the aggregator to provide services efficiently. It is critical that the roles and responsibilities of these new entrants are clarified. In particular, it is important that the relationships between retailers, BRPs, and independent aggregators are clear, fair, and allow for fair competition.

Since this can be done in many ways, different models are conceivable by considering such relationships and by describing how balance responsibility, transfer of energy, and information exchange are organized.

Before giving models examples on how integrating the independent aggregator figure, this section describes an integrated model where the supplier is also the aggregator as well as a separated model where the aggregator is independent of customer's supplier in order to understand what the differences are from today.

4.1 Integrated model: the supplier is also the aggregator

In the easiest scenario, the energy supplier acts as an aggregator and monetizes the flexibility of its customers in the energy markets. In this case, the customer has a contract with his supplier, bundling energy and flexibility services, i.e., aggregator and supplier are the same agent who also has balancing responsibility. The supplier can perform the "aggregation" function on his own or use a third party to do so.

As shown in Fig. 17.3, the supplier forecasts demand for his consumer of 100 units. The supplier detects an opportunity in the balancing markets and reduces that consumer's consumption from 100 to 80 units and sells a reduction of 20 units to the TSO. The supplier interacts with the TSO through his own BRP.

Figure 17.3 Integrated model where the supplier offers 80 units to its customers, while, at the same time, the supplier is able to sell part of his customer's flexibility (20 units) into the balancing markets to help the system.

4.2 Separated model: the aggregator is independent of customer's supplier

Section 2 explained that, according to the EU directive, the independent aggregator has the right to offer DR flexibility to the market to the extent agreed with the flexible consumer while the supplier is responsible for delivering energy to the consumer.

In this situation, the customer has two contracts, one with his supplier for supplying energy, and one with his aggregator for flexibility services. Aggregator and supplier are different companies, each with different balancing responsibilities.

If at any point (either close-to-real time or in real time) the aggregator activates DR, the consumer's electricity demand is likely to deviate from the forecast consumption pattern for that period. This difference in volume is settled financially between the TSO and the BRP, to which the supplier adheres.

The imbalance provoked in the supplier's portfolio will first appear in the balancing account of the BRP to which the supplier adheres. It may be expected that the result of the financial settlement between the TSO and the supplier's BRP will subsequently be settled between the supplier and the supplier's BRP. The supplier will finally bear the full risk of any imbalances in its supply portfolio as shown in Fig. 17.4.

In addition to the imbalance issue, the consumer's consumption falls short of the amount of energy procured by the supplier. In the absence of further arrangement, the supplier may not invoice this difference and thus, may not recover the full electricity procurement cost. This is best known as

Figure 17.4 Separated model: the aggregator is independent of customer's supplier. The supplier estimated 100 units consumption for its customer and the aggregator sold part of it (20 units) to the balancing markets creating an imbalance of 20 units in the supplier's portfolio.

"bulk energy" and it imposes a second problem for the supplier whose customer is activated by an independent aggregator.

DR can also be activated to provide flexibility to the system in the form of a load increase, i.e., an increase in demand. This is not the most common practice of DR flexibility since consumers tend to adapt their consumption to reduce electricity, by stopping or shifting energy-intensive industrial processes, for example.

Although historically aggregators did not provide DR by asking consumers to consume more than its regular schedule, this situation will change in the short period. Reasons for this are prices volatility where customers will be more incentivized to shift consumption. Price for the energy additionally consumed during load increase may be negative and the aggregator would make monetary gains. Consumers may also have batteries that enable them to absorb energy for later use. If this situation occurs, the supplier would then face an increase in consumption and would be able to invoice a higher amount than without such activation. In this case, the supplier would not face the bulk energy issue.

The imbalance and the bulk energy issues refer to the same time period and amount of energy. Depending on the sourcing cost and the imbalance settlement price, the combination of the two issues provides complex and undesired outcomes for suppliers. Some arrangements between the independent aggregator and the supplier need to be defined to address these issues, which will be analyzed in the following section.

5. Models for independent aggregators

This section introduces a classification of models for independent aggregators that may be implemented in Europe according to the Directive (EU) 2019/944[7] on common rules for the internal market for electricity. It should be noted that other models may exist depending on country preferences or the level of liquidity. There exist different opinions on volume and financial settlement between the different market parties, with a view to avoiding any significant distortive impact on the retailers/BRP.

---

[7] DIRECTIVE (EU) 2019/944 OF THE EUROPEAN PARLIAMENT AND OF THE COUNCIL of 5 June 2019 on common rules for the internal market for electricity and amending Directive 2012/27/EU.

This section describes three possible models for independent aggregators:
- A corrected model;
- A central settlement model; and
- A win-win model.

The model where the supplier acts as an aggregator has been previously defined in Section 3.1 and is not further explained here. Rather, all references to "aggregators" in this section refer to independent aggregators.

5.1 Corrected model

In the corrected model, the issues of the imbalance and bulk energy are both solved by correcting the meter readings. This involves a correction in the consumer's consumption profile based on the amount of flexibility that has been activated by the independent aggregator. The sourcing cost for the DR-affected energy is passed from the supplier to the aggregator through the consumer, based on retail prices. This means that the supplier will supply the total amount of energy to the customer as if there had been no activation by the aggregator. This energy will be based on retail prices as usual. In this situation the supplier does not face the bulk energy issue since it is able to supply the total amount of energy and it does not incur any imbalances.

As illustrated in Fig. 17.5, in the corrected model, the aggregator is delivering 20 to the TSO restoring the balance in its portfolio. To do so, the independent aggregator has reduced the consumption of the consumer from 100 units to 80 in order to sell 20 units to the TSO. Without a correction, the supplier (having sourced 100) would only be able to bill 80,

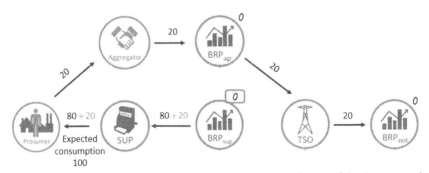

Figure 17.5 Corrected model. The imbalance in the supplier portfolio is corrected based on the aggregator activation (by 20 units) allowing the supplier to bill 100 units as previously foreseen without being "affected" by DR activation.

rendering an open supply position and an imbalance of $+20$ for the BRPs. In this model, the imbalance is corrected during the times DR is activated by restoring the energy balance of the BRP.

Without this correction, the consumer buys 100 units from his supplier and "resells" the remaining 20 to the aggregator which will sell these 20 to the TSO. In this model the "bulk energy" issue is solved through the consumer, based on retail prices. Thus, the retailer can bill the same energy volume as if no activation had occurred. Since energy is transferred through the consumer to the aggregator, the aggregator will compensate the consumer for the energy that has been billed, but not consumed, or vice versa in case of increased load, depending on contract conditions. The grid tariff will still be based on the uncorrected values. Taxation, however, becomes more complex.

This situation may be different in cases where the aggregator tries to reduce 20 units, but only achieves 18 units. This partly contributes to the restoration of the system balance but an imbalance on 2 units remains. This situation is quite common, since aggregators are not always able to activate all customer DR at the same time or some consumers may not respond to the aggregator's attempt to activate DR. Fig. 17.6 shows this example. The supplier is not affected by aggregator activation since there is a correction in the consumer's consumption profile based on the amount of flexibility that has been activated by the independent aggregator and the sourcing cost is passed from the supplier to the aggregator through the consumer, based on retail prices. However, the aggregator who was not able to achieve the 100% of the system operator (SO) order has an imbalance for those units

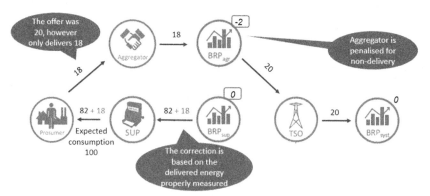

Figure 17.6 Corrected model: the aggregator offered 20 units to the balancing markets; however, it was only able to deliver 18 units and an imbalance is created in its portfolio.

not delivered. In Spain, under current regulations, the aggregator would have to pay an imbalance cost and a penalty as would other parties in the market.

In both situations, either if the aggregator delivers 20 or 18, we assume the measurement or the telemetry requirements are accurate enough so that the TSO is 100% certain of the energy delivered. However, this is a strong assumption. If the aggregators deliver 18 and telemetry requirements "check" that the aggregator delivered 20, the correction leaves the supplier with a new imbalance of +2, equivalent to 102 units in the supplier portfolio. This outcome is portrayed in Fig. 17.7.

Additionally, in this scenario, the independent aggregator is not penalized for nondelivery because the TSO was not able to verify the exact amount of energy delivered.

The imbalance issue may arise for reasons other than the activation of DR by the independent aggregator. For example, the phenomenon known as consumer demand rebound, where consumers offset their response to DR activation by increasing or reducing consumption afterward. Since the independent aggregator is only responsible when DR activation is invoked, the rebound effect creates another imbalance issue for the supplier.

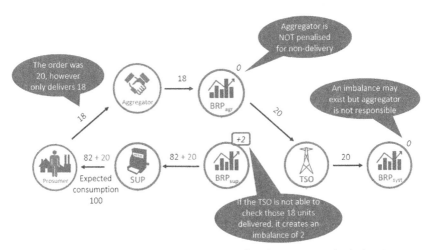

Figure 17.7 Corrected model: the aggregator offered 20 units to the balancing markets; however, it was only able to deliver 18 units from its customers. Since the baseline was not appropriate, the TSO couldn't check the proper delivery; an inappropriate measurement of the activation creates an imbalance in the supplier's portfolio of 2 units.

5.2 Central settlement model

In the central settlement model, the imbalance and bulk energy issues are both solved by a central entity, usually through the TSO, without the need to correct the meter readings, but instead by correcting or adjusting the balancing perimeter. The perimeter is the position any agent has in the market and correcting such position when offering, for example, balancing services means that the position is balanced.

Usually the TSO is responsible for correcting the balancing perimeter following a DR activation, as well as determining the energy compensation to the supplier based on a predefined price formula that is usually based on the market price.

As observed in Fig. 17.8, the process is similar to the corrected model (Fig. 17.5) where the aggregator is delivering 20 units to the TSO restoring the balance in its portfolio. To do so, the independent aggregator has reduced the consumption of the consumer from 100 to 80 in order to sell 20 units to the TSO and imbalance is created in the supplier portfolio. The aggregator, on the other hand, has delivered 20, yet did not source this energy.

In Fig. 17.8, a central entity, usually a role taken by the TSO, corrects the perimeter by transferring energy from the supplier BRP to the aggregator BRP based on a predefined price formula that is applied to the transferred energy. Specifically, in this case, the independent aggregator pays 20*(€/MWh) to the supplier.

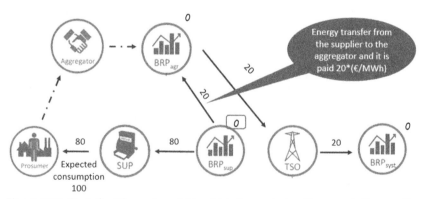

Figure 17.8 Central settlement model in which the central entity corrects the supplier perimeter of 20 units based on the 20 units activated by the aggregator and sold to the OS.

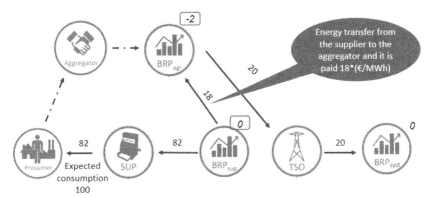

Figure 17.9 Central settlement model in which the central entity corrects the supplier perimeter of 18 units based on the 18 units truly activated by the aggregator. The imbalance of 2 units is held by the aggregator, the agent who didn't deliver 100% of the DR ordered.

The situation may be different in cases where the aggregator tries to reduce 20 units, but only achieves 18 (as previously examined in Fig. 17.6). This partly contributes to the restoration of the system balance but an imbalance of 2 remains as shown in Fig. 17.9.

In both situations, regardless of whether the aggregator delivers 20 or 18 units, we assume that measurement is accurate enough that the TSO is 100% certain of the energy delivered. However, as explained in the corrected model, this is a strong assumption. If the aggregators deliver 18 and telemetry requirements "check" that the aggregator delivered 20, the perimeter correction for the supplier creates a new imbalance of $+2$, equivalent to 102 units in the supplier portfolio as shown in Fig. 17.10.

Finally, as for the corrected model (Section 5.1), where the independent aggregator is only responsible for imbalances at times where DR activation occurs, the rebound effect creates another imbalance issue for the supplier.

5.3 Win-win model

The win-win model addresses, at once, all of the issues: imbalances, including the rebound effect, bulk energy, and problems arising from telemetry requirements. It may be considered a complete model.

In this model, the whole responsibility is transferred from the supplier to the independent aggregator before day-ahead gate closure. Thus, the independent aggregator takes responsibility for its selected consumers and for

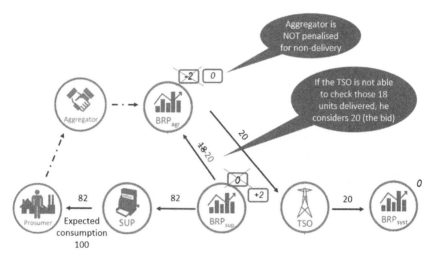

Figure 17.10 Central settlement model where the aggregator didn't deliver 100% of the order and the telemetry requirements fail, creating an imbalance of 2 units in the supplier portfolio.

their imbalances over the next 24 h, allowing it to adjust the portfolio in all energy markets such as intraday market, balancing services, etc.

This model is simple since instead of the supplier taking responsibility for imbalances between the flexible consumer's forecast and actual consumption in real time, the independent aggregator takes this responsibility for those days and consumers for which it sees an opportunity.

A step-by-step description of the process is presented below.

1. Before day-ahead gate closure, the independent aggregator communicates to the SO[8] or to the market operator (NEMOs), the aggregated load curve of its customers who at the same time have an energy contract with any supplier. The independent aggregator has its own BRP and is responsible for the imbalance of the program communicated to the SO. The SO communicates to the affected suppliers the program that the aggregator has defined for its (the suppliers') customers to which it must supply energy as shown in Fig. 17.11.

2. The supplier purchases the energy of all its customers in the day-ahead market (sum of the aggregator's consumption forecast plus the

---

[8] We consider the system operator (SO) is the TSO in the whole chapter and the examples are done with the communication via TSO. However, the same model works with communication done via NEMOs.

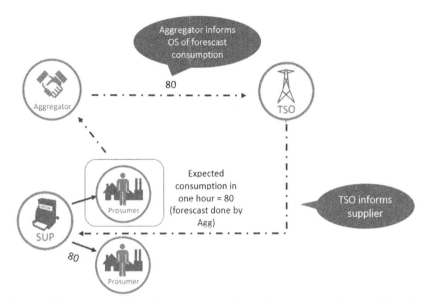

Figure 17.11 Step 1 of the win-win model where the aggregator forecast 80 units consumption for customers who at the same time have a supplier contract. The aggregator informs the OS of its forecast. Later, and before day-ahead date closure, the OS informs the supplier of the customers who are going to be "managed" by the aggregator.

consumption forecast of the rest of its customers). The supplier remains responsible for procuring the energy for all customers, including those that will be managed by an independent aggregator.

3. The OS transfers the aggregated load profile previously defined by the aggregator, from the supplier to the aggregator, at the day-ahead market price.[9] This transfer is not a real one, but an accounting one for the upcoming settlement through which the independent aggregator pays for the energy at the market price before selling it in the markets. The supplier, therefore, is already compensated for such energy.

4. The aggregator offers demand services in the other markets where it detects an opportunity.

---

[9] Transfer price cannot be other than the price of the energy transferred in the last negotiated market before the transfer is formalized. In the proposal presented, this price is the daily market price, since the transfer is made at the close of the daily market. Any other transfer price leads to arbitrage opportunities that will generate inefficient behavior.

5. The aggregator receives the income for the services provided to the system. Measurement deviations from its program are settled by the aggregator BRP. This is an important point and differs from the previous models where the supplier BRP was responsible for such imbalances. As illustrated in Fig. 17.12, the aggregator forecasts 80 units for its customers (100 units of consumption + 20 units reduction) and the supplier is not affected by any imbalance caused by the aggregator. The supplier's customers consumed 60 units and its portfolio is balanced.

6. Since the supplier has to bill consumers, the revenue for the energy consumed, as measured by smart meter each hour,[10] is transferred by the aggregator to the supplier at the day-head market price (as was done previously) in addition to the average cost of imbalances incurred by the supplier for the energy sold to customers for which the aggregator is responsible. This addition is important since suppliers usually internalize in their offers any imbalances their customers may have. Supplier's customers, therefore, pay an amount of money to their supplier for system imbalances. In this model, the aggregator is responsible for customer's imbalances and therefore he must pay the system for such imbalances. However, since the supplier is billing such customers, suppliers (instead of aggregators who bear the cost) are compensated for customer's imbalances. These benefits received by the supplier should

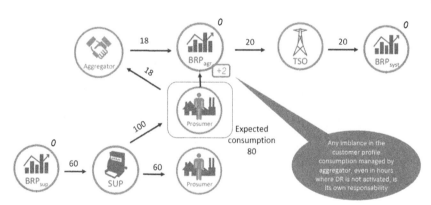

Figure 17.12 Step 5 of the win-win model where the aggregator is responsible for its own imbalance of 2 units. The aggregator forecast 80 units consumption for its customers, and they consumed 82, in the end causing an imbalance of 2 for the aggregator BRP.

[10] Or every 15 min depending on the country.

somehow be transferred to the aggregator since they are the agents assuming the whole imbalances of these customers.

7. The aggregator only buys the energy in those hours where it activates the demand management service. However, it is responsible for the imbalances for the DR customers in other hours, solving the rebound effect and, thus, ensuring that imbalances do not penalize suppliers.

In this model, each agent is responsible for his own participation in the market. On the one hand, the supplier takes responsibility for the imbalances between the flexible consumer's forecast and actual consumption that may occur in real time only for those customers who are not also managed by an independent aggregator. On the other hand, the independent aggregator is responsible for the imbalances between the flexible consumer's forecast and actual consumption only for those customers previously communicated to the SO. The aggregator is free to go to the market only on those days where he foresees an opportunity.

6. Conclusions

Independent aggregators are expected to play a key role in allowing more intermittent generation onto the system, lowering costs, optimizing market positions, and increasing security of supply.

The aggregator, as a new energy player, needs to be introduced in a way that respects the existing legislation while enabling the aggregator to provide services efficiently. As explained in this chapter, this integration can be done in many ways, but it is critical that roles and responsibilities for independent aggregators are clarified. In particular, the relationships between retailers, BRPs, and independent aggregators that should be clear, fair, and allow for fair competition.

European countries are now defining the regulatory framework that will enable DR through aggregators to participate in all electricity markets—and may end up with different regulations.

Although each Member State has the freedom to choose the model they prefer, it is important to look at the system as a whole. In Spain, the model is not defined yet but the most appropriate model should be one in which efficient electricity supply and system stability are guaranteed, while allowing aggregators to obtain benefits for the services they provide without inflicting financial damage on other agents.

There is no doubt that activating demand management without compensation to the supplier makes the business model of the independent aggregate more attractive, but it is still a model based on a subsidy at the supplier's expense and it will lead to greater costs for the system.

The win-win model described in Section 5.3 solves all the key issues: imbalances, including the rebound effect, bulk energy, and the problems arising from telemetry requirements. For this reason, we consider this to be the most complete model. However, the central settlement model described in Section 5.2 is also interesting. It solves two important issues that need to be regulated—imbalances and bulk energy—through a central entity.

Bibliography

Benedettini, Simona, et al., 2019. Assessment and Roadmap for the Digital Transformation of the Energy Sector towards an Innovative Internal Energy Market. European Commission. European Commission.

de Heer, Hans, et al., 2017. Workstream on aggregator implementation models. USEF.

SEDC, 2016. Assessing the Potential and Market Uptake of Demand-Side Flexibility in the European Energy System. SEDC.

Ziegler, Holger, et al., 2017. Demand Response Activation by Independent Aggregators As Proposed in the Draft Electricity Directive.

CHAPTER 18

Do Time-Of-Use tariffs make residential demand more flexible? Evidence from Victoria, Australia

Kelly Burns, Bruce Mountain
Victoria Energy Policy Centre, Victoria University, Melbourne, VIC, Australia

1. Introduction

Time-Of-Use (TOU) tariffs can influence how much electricity is consumed at different times of the day. In Victoria, TOU tariffs (unlike the traditional flat rate tariff) impose prices that vary by time of day and by day of week.[1] Higher prices during peak times are intended to encourage consumers to shift consumption to low price (and demand) periods. Electricity prices that reflect temporal variation in the cost of supply may also reduce costs if consumers respond to time-varying prices by reducing their consumption when prices (and costs) are higher. For example, a more constant demand reduces the need to start and stop more expensive production to meet short-lived demand peaks. More constant demand improves the utilization of production and distribution infrastructure and so reduces average costs and defers or avoids capacity expansion. More constant electricity demand may also reduce greenhouse gas emissions, although this depends on the greenhouse gas intensity of the production that would replace otherwise infrequently used peaking capacity. In these ways, a more constant demand can reduce investment and operating

---

[1] Two-rate time-varying (i.e., TOU) tariffs charge a different tariff for peak (7a.m.–11p.m.) weekdays, off-peak all other times. In addition to peak and off-peak consumption charge, consumers on TOU tariffs also pay a daily charge (cents per day). This daily charge accounts for around 30% of the median annual bill for households in Victoria (Carbon and Energy Markets, 2017).
In Victoria, there are also three-rate time-varying (i.e., Flexible) tariffs. These tariffs charge a different tariff for peak (3p.m.–9p.m. weekdays), off-peak (10p.m.–7a.m.) and shoulder (all other times) periods. Flexible tariffs account for only 0.05% of households in Victoria and are excluded from the analysis (Carbon and Energy Markets, 2017).

Variable Generation, Flexible Demand
ISBN 978-0-12-823810-3
https://doi.org/10.1016/B978-0-12-823810-3.00024-8
411

expenditure and thus help to ensure lower and less volatile electricity prices (Ericson, 2011; Gyamfi et al., 2013). In addition, in the context of increasing variable renewable electricity generation and uptake of Electric Vehicles (EVs), demand that is better able to follow supply is increasingly valued (see Chapters 14 and 15, and Nicholson et al., 2018). Finally TOU tariffs also offer private benefits to consumers by providing the opportunity to reduce their bills if they shift consumption from high-priced periods to lower priced periods, although such shifts typically come at the expense of convenience and utility.[2] However, the success of TOU is predicated on the concept of flexible demand.[3] Only if demand is flexible (as is contemplated throughout this book) can TOU encourage households to vary consumption to match the available supply and thus unlocking these welfare benefits.[4]

Do TOU tariffs make residential demand more flexible and how is this affected by various factors? In this study, a sample of 6957 household electricity bills obtained from consumers in Victoria, Australia, are used to assess whether there is a shift in consumption from peak to off-peak periods in response to the difference in peak and off-peak prices. The elasticity of substitution (the measure traditionally used to assess load shifting behavior under TOU tariffs) is estimated for the full sample and then separately for households that have installed rooftop photovoltaics (PVs) and households that have separately metered (controlled) loads. The sample is also aggregated into three socioeconomic tiers (low, medium, high) based on the socioeconomic ranking of the postcode of each household, to measure how the elasticity of substitution varies in these tiers.

---

[2] Chapter 6 describes how shifting flexible demand can entail a loss of utility for households. For instance, "changing the temperature indoor in order to reduce energy needs, advancing or postponing an activity (such as a productive step within a business or a washing cycle inside home), or shutting down a machinery, all require individuals to give up part of their utility, at the very least because they need to change their programs."

[3] Flexible demand refers to any customer load that need not be on or totally served at all times (refer Chapter 5 for a more detailed discussion). Flexible demand is synonymous with other terms commonly used to describe loads that can be modified or may be flexible, including price responsive demand or load; responsive demand, responsive load; flexible load, shiftable load, elastic demand; and demand response.

[4] In Chapter 5, Sioshansi concludes that virtually all customer demand has some flexibility (as not every electricity using device needs to be on at all times, nor need all devices run during the peak demand period) with only few exceptions.

This study contributes to the literature as follows. First, a large sample of actual household bills is used to assess the effectiveness of TOU tariffs in Victoria as measured by the elasticity of substitution. As far as we are aware, such studies do not yet exist for TOU tariffs in Victoria. Furthermore, globally there have been few such empirical studies since the 1980s and thus there is limited contemporary evidence on how responsiveness has changed following developments in market structure, technology, and regulation.

Second, by drawing on a large sample of household bills, actual consumer responsiveness in a real market is analyzed rather than the more commonly studied small pilots and experiments. As highlighted throughout this book, the key to harnessing flexible demand—especially in the residential sector—is scale. Small residential customers have flexible demand that comes in small increments, and this only brings about changes in the market if flexible demand can be aggregated across a very large number customers. There is evidence of success in harnessing flexible demand in small-scale pilots and experimental studies (see, for instance, Chapter 5). This study examines the application of TOU tariffs in practice in Victoria's deregulated retail electricity market.

The chapter is organized as follows:

- Section 2 discusses the background and literature relevant to this study.
- Section 3 describes the study and presents a preliminary analysis of the data.
- Section 4 describes how consumer responsiveness to TOU tariffs is measured and modeled.
- Section 5 presents the main findings.
- Section 6 concludes.

2. The background and historical performance of TOU tariffs

Early studies of TOU tariffs were of pilot experiments applied to households supplied by monopolies in Wisconsin and six other US states in the 1970s. These studies conclude that households have historically been weakly responsive to TOU tariffs (Fig. 18.1). Despite differences in location, pilot program design, sample data and methodology, these studies report weak responsiveness overall and little variation in results (estimates of the elasticity of substitution are in the range of 0.12—0.24). The main results of this study (the last bar in the figure) are included as a point of comparison to the earlier studies.

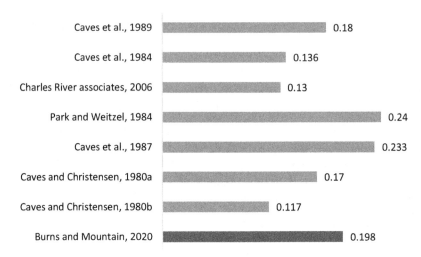

Figure 18.1 *Summary of estimated percentage change in peak to off-peak usage in response to 1% point change in peak to off-peak price.*

A review of the literature identifies various factors that affect responsiveness, summarized as follows:

Major appliances: Households with major appliances are more responsive to TOU tariffs (Caves and Christensen, 1980).

Technology[5]: Technology can improve peak load reductions (Faruqui and Sergici, 2010).

Ownership: Homeowners are more responsive than renters to TOU tariffs (Bartusch and Alvehag, 2014).

Price block duration: Households are more responsive to TOU tariffs when peak price blocks are shorter (Faruqui and Sergici, 2010; Ham et al., 1997).

Seasonality: Households are less price responsive in winter than summer (Bartusch and Alvehag, 2014).

3. A study into the effectiveness of TOU tariffs

The data used in this study are obtained from 6957 household electricity bills for households on TOU tariffs in Victoria. These bills were part of a larger sample of 47,114 bills that were voluntarily uploaded to the Victorian

---

[5] Technology in this context refers to digital technology in meters (commonly referred to as "Smart Meters") and Advanced Metering Infrastructure (AMI).

Government's price comparison website (https://compare.energy.vic.gov.au/) over the period from July 2018 to December 2018.[6] Information from each bill is extracted including consumption in the peak and off-peak period, price charged in peak and off-peak period, solar PV exported to the grid, solar PV produced on the premises,[7] controlled load usage and postcode. Unlike previous studies into flexible demand and TOU tariffs, how consumers respond to peak and off-peak prices is assessed based on objective bill data where prices and usage are known with certainty.

Of the 47,114 bills, 15% (6957) have TOU tariffs. This compares to 13% of households in the population of households in the state of Victoria (Mountain and Burns, 2020). Of the 6957 households on TOU tariffs:

- 55% are in high socioeconomic status post codes, 30% in the middle, and 15% in the low socioeconomic post codes. Although comparative data on the socioeconomic distribution of all households on TOU tariffs in Victoria are not known, the distribution in the sample is similar to the distribution of the population of Victoria (46%, 35%, and 20% of Victorian households are in the high, medium, and low socioeconomic tiers, respectively: Australian Bureau of Statistics, 2011).
- 455 (6%) have separately metered controlled loads (typically electric hot water or underfloor heating). Controlled loads are typically charged at a lower rate than the main loads that operate during off-peak hours (e.g., overnight) and the controlled load rate does not vary by time of day.
- 3914 homes have rooftop PV (56% of the 6957 households in the sample). By comparison, 16% of households in Victoria have access to rooftop PV and only 13% have TOU tariffs (Carbon and Energy Markets, 2017; Climate Council, 2018).[8]

Consumers with rooftop PV produce electricity mostly during hours that are classified as peak (7a.m.−11p.m. from Monday to Friday). This reduces their consumption of electricity from the grid during the peak period. Households with rooftop PV can consume the electricity generated

---

[6] For further details on the data extraction, processing, and measurement methods applied to individual household bills, please refer to Mountain and Burns (2020) and Burns and Mountain (2020).

[7] Data for the amount of solar PV that is self-consumed are not directly reported on household bills. Solar PV that is self-consumed is estimated using the methodology described in Mountain et al. (2020).

[8] Most households are automatically placed on TOU tariffs upon the installation of PV. Thus, households with PV are overrepresented in the sample of households on TOU tariffs.

from their PV system or be paid to feed the electricity generated into the grid. The choice to maximize their own solar consumption or to maximize solar exports is influenced by consumers' access to a premium solar feed-in tariff (a tariff which involves much higher payments for exports to the grid, but which only applies to around one in three households). Since the premium feed-in tariff is higher than the variable retail purchase price, households with access to a premium feed-in tariff have stronger incentives to export electricity to the grid than households that only have access to the conventional feed-in rate (which is lower than the retail purchase prices).

Fig. 18.2 plots the ratio of peak to off-peak prices against the ratio of peak to off-peak consumption for the 6957 households on TOU tariffs in our sample.[9] It is clear from this scatter that there is a wide range of peak and off-peak price combinations under TOU tariffs in Victoria. It is also evident from the dispersion in the scatter plot that there is no obvious relationship between relative prices and relative consumption. The same observation applies to households with TOU tariffs and controlled load or rooftop PV, and for households with TOU tariffs by socioeconomic tier.[10]

Fig. 18.3 reports the median ratio of peak to off-peak consumption for the full sample of households on TOU tariffs, distinguishing households with rooftop PV or controlled load tariffs and by socioeconomic tier.

Measured over the full sample, households on TOU tariffs consume almost as much electricity in the peak period as they do in the off-peak period. Households with TOU tariffs and controlled loads use relatively more in peak

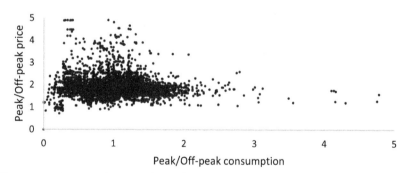

Figure 18.2 *Scatter plot of peak to off-peak consumption and peak to off-peak price.*

[9] Outliers where the ratio of peak to off-peak price or consumption exceeds 5:1 are excluded.
[10] These scatter plots are available from the authors upon request.

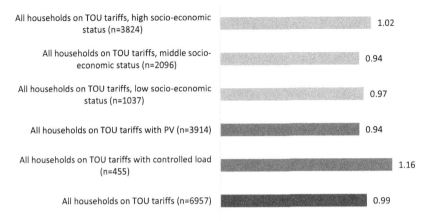

Figure 18.3 *Median ratio peak to off-peak consumption.*

periods because their controlled load consumption occurs in the off-peak period (and so the residual amount of their consumption in the TOU off-peak period is lower). Conversely households with TOU tariffs and rooftop PV purchase less electricity from the grid in the peak period because the PV electricity produced and consumed on the premises occurs during the day time (i.e., the peak period for weekdays—5 out of 7 days each week).

As expected, median annual grid electricity purchases are noticeably lower for households with PV (Fig. 18.4). Median annual grid purchases are higher for households with controlled load and those living in low socio-economic postcodes.

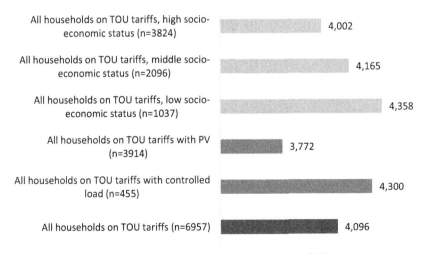

Figure 18.4 *Median annual grid purchases (kWh).*

These descriptive statistics are informative but do not establish evidence on the magnitude and direction of causal relations (if any) or provide a measure of load shifting under TOU tariffs. This is explored in the next section.

4. Measuring consumer responsiveness to TOU tariffs

4.1 A model to measure household responsiveness

The elasticity of substitution measures the percentage change in relative demand to relative prices (relative being peak to off-peak). The elasticity of substitution captures load shifting behavior under TOU tariffs because it measures how relative consumption is influenced by relative prices. Mathematically, the elasticity of substitution is defined as:

$$E_{sub} = \frac{\%\Delta\left(\dfrac{Q_P}{Q_{OP}}\right)}{\%\Delta\left(\dfrac{P_P}{P_{OP}}\right)} \tag{18.1}$$

where: Q_p and Q_{Op} is usage in peak and off-peak period, respectively, and P_p and P_{Op} is peak and off-peak price, respectively.

Load shifting (i.e., the elasticity of substitution as measured by β) under TOU tariffs is estimated using the following model (using Ordinary Least Squares):

$$ln\frac{Q_{i,P}}{Q_{i,OP}} = \alpha + \beta ln(P_{i,P} / P_{i,OP}) + \gamma_1 lnSolar_i + \gamma_2 Decile_i + \varepsilon_i \tag{18.2}$$

where: $lnSolar$ is (log) of the amount of rooftop solar that is self-consumed,
   $Decile$ is the socioeconomic decile (takes values 1 to 10), and
   i is the i-th household in the sample.

All data available from household bills were considered and tested for inclusion in the econometric analysis. However, some variables are excluded because they introduce multicollinearity into the model (e.g., annual grid purchases) or fail the redundant variable test (e.g., controlled load) (Burns and Mountain, 2020). Standard hypothesis testing is then applied to assess the strength and magnitude of load shifting under TOU tariffs in Victoria, for households with rooftop PV or controlled load, and across socioeconomic cohorts.

4.2 Estimated responsiveness of households to TOU tariffs in Victoria

Fig. 18.5 presents the estimation results of Eq. (18.2) for each of the six cohorts. There is a statistically significant shift in consumption from peak times to off-peak times in response to the difference in peak and off-peak prices when measured across the full sample. However, the response is weak: a 1% increase in peak prices relative to off-peak prices only results in a 0.2% shift in consumption from peak times to off-peak times. This means that if the median peak to off-peak price ratio doubled from 1.83 to 3.76, the ratio of peak to off-peak consumption would fall by only 14%.

However, consumers in the low socioeconomic cohort or those that have controlled loads do not respond to the ratio of peak to off-peak prices. Perhaps households with controlled load have shifted their flexible consumption and thus do not respond to TOU tariffs. Explanations as to why low socioeconomic households do not respond to TOU tariffs are likely to be more complex.

Access to rooftop PV does not affect the responsiveness of households to the difference in peak and off-peak prices, compared to those households without rooftop PV. However, the amount of rooftop PV that is produced and consumed on the premises affects the amount of electricity purchased from the grid by that household during the peak and off-peak times. The impact is the highest for the low socioeconomic cohort—a 1% increase in

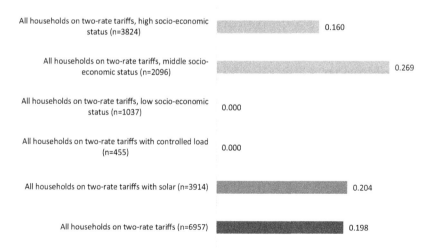

Figure 18.5 *Percentage change in peak to off-peak consumption in response to 1% change in peak to off-peak price.*

rooftop PV consumed on the premises reduces the relative consumption in peak and off-peak periods by 0.12%—and is lowest for the high socio-economic cohort. It may be that peak period grid purchases by low socioeconomic households are more sensitive to rooftop PV because these households are more likely to suffer energy cost pressures and therefore have a greater incentive to maximize their own solar consumption. Conversely, high socioeconomic households may be less sensitive to energy cost pressures and thus the amount they consume during the peak period is less sensitive to how much rooftop PV is produced and consumed on premises.

5. Findings

The results of this study into the effectiveness of TOU tariffs in Victoria provide several important findings in relation to the future of flexible demand and variable electricity generation. The essential findings are summarized as follows:

5.1 Household demand in Victoria is unresponsive to TOU tariffs

The study into the effectiveness of TOU tariffs in Victoria finds that across the full sample of 6957 households on TOU tariffs, a 1% increase in the ratio of peak to off-peak prices results in a 0.2% shift in consumption from peak hours to off-peak hours. In the conventional definition of the elasticity of substitution, this qualifies as inelastic or weak responsiveness. There may be many reasons for this and they merit further examination.

5.2 Rooftop solar has no effect on responsiveness to TOU tariffs in Victoria

Consumers with rooftop PV have incentives to self-consume that depend on their feed-in tariffs in comparison to their purchase prices. This finding undermines the case for discrimination in policy support for TOU tariffs on the basis of whether households have installed rooftop PV.

5.3 Households in the poorest socioeconomic regions do not respond to TOU tariffs

Households in the lowest socioeconomic regions do not respond to TOU tariffs. It may be these households have less efficient appliances and lower rates of employment, and hence spend more time at home and suffer greater inconvenience by shifting load from peak to off-peak hours, than consumers in higher socioeconomic cohorts (as suggested by European Energy Agency (2013); Gyamfi et al. (2013) and Simmons and Rowlands (2007)). Perhaps a lack of awareness of time-of-use differences may help explain these results.

5.4 So much has changed, yet so little

Considering substantial technology, regulatory, and market developments over the past 50 years, it is reasonable to expect that responsiveness to TOU tariffs would be higher than reported in earlier studies. In particular, two barriers to demand management (smart meters and smart prices) have already been addressed in Victoria.[11] Further, this study is of a market deregulated almost 20 years ago, where consumers choose freely offered peak and off-peak prices, where consumption is metered half-hourly by remotely read "smart meters," where competing suppliers have long offered close to real-time consumption data to consumers on easy to use hand-held devices, where technology to automatically control household electricity devices is widely available and inexpensive and where policy and regulation has enthusiastically supported TOU tariffs. Notwithstanding these changes the results indicate that consumers in Victoria in 2019 are similarly weakly responsive to TOU tariffs as reported in the studies in the United States in the 1980s (Fig. 18.1).

5.5 Long peak periods may explain consumers' low responsiveness

The length of the peak price block in Victoria may partly explain the low responsiveness of consumers to TOU tariffs in Victoria. Consumers suffer a larger loss of utility if they are required to defer consumption for longer periods. Perhaps consumers do not shift load because the cost to them is higher than the financial benefit of lower prices in off-peak periods. As pointed out in Sioshansi (Chapter 7) the expected savings must exceed the effort, investments, and/or any inconvenience that consumers are asked to bear.

---

[11] In Chapter 5, Sioshansi argues that the main barriers to engaging residential flexible demand to a meaningful degree are smart meters, smart prices, and smart appliances (the former two are essential).

It may also be the case that consumers are not aware of price differences and the times at which they apply.[12] Education and awareness campaigns may improve responsiveness (see, for instance, Chapter 7 for evidence of how electronic messages prompting customers to respond to price movements improves responsiveness).

6. Conclusion

The findings suggest that households in Victoria respond weakly to the existing TOU tariffs. We consider all households on TOU tariffs including those with rooftop PV or controlled load and we segment households by socioeconomic cohort. We find weak responsiveness at best across the sample. This does not suggest that residential electricity demand is necessarily inflexible (it is argued in Chapter 19 that Australian households are well placed to provide demand flexibility). However, these results need to be explained. Perhaps a 16 hour peak period imposes greater inconvenience to consumers that might otherwise be inclined to shift load to cheaper times if they did not need to defer consumption for such a long period. Australia's regulators are currently reviewing proposals to reduce the peak period from 7a.m. to 11p.m., to 3p.m. to 9p.m. This may increase responsiveness. Evidence that this is actually the case would be valuable before imposing such tariff structures as the default.

It may also be that consumers on TOU tariffs are not aware of the peak and off-peak times and price differences. This might be particularly true for those consumers that did not explicitly choose to be on such tariffs. It would be valuable to understand why responsiveness has been so poor, particularly for those consumers in the lowest socioeconomic areas.

There has been a great deal of investigation into TOU tariffs based on small trials and pilots in retail electricity markets across the world. This is a study of a real market in operation characterized by full penetration of smart meters, long-standing policy support for TOU pricing, and readily available access to affordable smart appliances. Despite these supportive conditions, the large-scale application of TOU tariffs to achieve load shifting has some way to go. These findings are relevant to other countries developing TOU tariffs to harness flexible demand.

---

[12] Chapter 6 presents policy recommendations to promote customer engagement among households to unlock the benefits of demand flexibility, including addressing information asymmetry. They argue that households must be aware of the financial benefits and cannot be expected to seek out this information.

Funding

This work was supported by a multi-year funding grant from the Government of Victoria for the establishment of the Victoria Energy Policy Centre.

Bibliography

Australian Bureau of Statistics, 2011. Census of Population and Housing: Socio-Economic Indexes for Areas (SEIFA), Australia, 2011. Retrieved from: https://www.abs.gov.au/AUSSTATS/abs@.nsf/allprimarymainfeatures/8C5F5BB699A0921CCA258259000BA619?opendocument.

Bartusch, C., Alvehag, K., 2014. Further exploring the potential of residential demand response programs in electricity distribution. Appl. Energy 125, 39—59. https://doi.org/10.1016/j.apenergy.2014.03.054.

Burns, K., Mountain, B., 2020. Do Households Respond to Time-Of-Use tariffs? Evidence From Australia.

Carbon and Energy Markets, 2017. The Retail Electricity Market for Households and Small Businesses in Victoria Analysis of Offers and Bills. Retrieved from: https://s3.ap-southeast-2.amazonaws.com/hdp.au.prod.app.vic-engage.files/4315/0252/0560/CME_-_electricity_-_analysis_of_offers_and_bills.pdf.

Caves, D., Christensen, L.R., 1980. Residential substitution of off-peak for peak electricity usage under Time-Of-Use pricing. Energy J. 1 (2), 85—142. https://doi.org/10.5547/issn0195-6574-ej-vol1-no2-4.

Climate Council, 2018. Powering Progress: State Renewable Energy Race. Retrieved from: https://www.climatecouncil.org.au/wp-content/uploads/2018/10/States-Renewable-Energy-Report.pdf.

Ericson, T., 2011. Households' self-selection of dynamic electricity tariffs. Appl. Energy 88 (7), 2541—2547. https://doi.org/10.1016/j.apenergy.2011.01.024.

European Energy Agency, 2013. Achieving Energy Efficiency Through Behaviour Change: What Does it Take?.

Faruqui, A., Sergici, S., 2010. Household response to dynamic pricing of electricity: a survey of 15 experiments. J. Regul. Econ. 38 (2), 193—225. https://doi.org/10.1007/s11149-010-9127-y.

Gyamfi, S., Krumdieck, S., Urmee, T., 2013. Residential peak electricity demand response — highlights of some behavioural issues. Renew. Sustain. Energy Rev. 25, 71—77. https://doi.org/10.1016/j.rser.2013.04.006.

Ham, J., Mountain, D., Chan, L., 1997. Time-Of-Use prices and electricity demand: allowing for selection bias in experimental data. Rand J. Econ. 28, S113—S141.

Mountain, B., Burns, K., 2020. Loyalty Taxes and Search Costs in Retail Electricity Markets: Have We Got it Wrong?.

Mountain, B., Gassem, A., Burns, K., Percy, S., 2020. A Model for the Estimation of Residential Rooftop PV Capacity. https://doi.org/10.26196/5ebca99c43e1a.

Nicholson, M., Fell, M.J., Huebner, G.M., 2018. Consumer demand for time of use electricity tariffs: a systematized review of the empirical evidence. Renew. Sustain. Energy Rev. 97 (September 2017), 276—289. https://doi.org/10.1016/j.rser.2018.08.040.

Simmons, S.I., Rowlands, I.H., 2007. TOU rates and vulnerable households: Electricity consumption behavior in a Canadian case study. Behavior, Energy and Climate Change Conference Nov 7–9 2007. Retrieved from https://www.google.com/url? sa=t&rct=j&q=&esrc=s&source=web&cd=&cad=rja&uact=8&ved=2ahUKEwj4j_2r 84LsAhX5zzgGHaMvDicQFjAAegQIBhAB&url=https%3A%2F%2Fciteseerx.ist.psu.edu %2Fviewdoc%2Fdownload%3Fdoi%3D10.1.1.500.7577%26rep%3Drep1%26type%3Dpd f&usg=AOvVaw3DU-fF73z8MoClVIh3AKb3.

CHAPTER 19

Empowering consumers to deliver flexible demand

Lynne Gallagher[1,1], Elisabeth Ross[2]

[1]Energy Consumers Australia, Sydney, NSW, Australia; [2]Independent Consultant, Sydney, NSW, Australia

1. Introduction

The flexible use of load and other behind-the-meter energy assets is becoming increasingly important as countries seek to decarbonize their economies, topics covered in accompanying chapters of this book. In Australia's National Electricity Market (NEM),[2] as elsewhere, this means the retirement of fossil fuel generators and a shift toward a generation mix dominated by variable renewable energy. Flexible demand provides a lower cost means to facilitate this transition by limiting the need for investment in expensive, fast response capacity and network infrastructure.

While the potential value of various forms of flexible demand has long been recognized in electricity, little has been done to cultivate this fledgling area in Australia—and elsewhere. Many of the regulatory and other barriers to achieving flexible demand identified a decade ago remain.

There has been some success in enabling flexible demand to provide ancillary and emergency services at the commercial and industrial (C&I) level.

However, few opportunities exist for households and small business consumers to offer and be rewarded for flexible demand beyond the use of off-peak controlled load (using ripple control of hot water systems, air conditioners, and pool pumps) that is a feature in Queensland and to a lesser extent in New South Wales, a few spot price pass-through retail offers and very early "virtual power plant" (VPP) trials that aggregate flexible demand

---

[1] The authors would like to thank colleagues at Energy Consumers Australia for their valuable input and ideas. However, the views expressed here are the views of the authors and do not necessarily reflect the views of Energy Consumers Australia.

[2] The NEM includes the electrically interconnected states of Queensland, New South Wales (including the Australian Capital Territory), Victoria, South Australia, and Tasmania.

Variable Generation, Flexible Demand
ISBN 978-0-12-823810-3
https://doi.org/10.1016/B978-0-12-823810-3.00008-X

from households with batteries (AEMO, 2020a).[3] As discussed in Chapter 5, the means to cost-effectively aggregate small changes in load by many homes and businesses were not available until now.

Smart meters and smart pricing are critical enablers for unlocking flexible demand. As smart connectivity and devices become more ubiquitous households and small business will be enabled to more easily monitor and control their use and, with the greater availability of home energy storage options, manage their output of energy to the grid.[4]

But, as this chapter explores, these enablers are not sufficient to realize the potential for large numbers of households and small businesses to shift their consumption in response to incentives. The focus needs to move beyond a technocratic view of price signals and mechanisms as triggers for behavior change, which tend to ignore the questions of how price signals are received and understood. Policy and market design must take account of other triggers than simply price or reward, and recognize the value that consumers place on control and at the same time their "preference for preserving scarce cognitive and emotional bandwidth."(Samson, 2017). Where these come together is in a market where trusted intermediaries play a role in empowering consumers to make choices, for their own benefit and to benefit the system as a whole.

The balance of the chapter is organized as follows:

- Section 2 describes Australia's experience to date with flexible demand and why it is an increasingly important part of the energy mix;
- Section 3 explores the critical importance of independent intermediaries to the expansion of flexible demand, including the participation of households and small business;
- Section 4 examines how the future energy system could be designed to respond to the expectations and needs of households and small business, and empower consumers to deliver flexible demand; and
- Section 5 summarizes the chapter's conclusions.

---

[3] AEMO does not separately report demand response by residential and small business consumers, with its methodology differentiating between load shifting that it captured in its forecasts as reduced maximum electricity demand, and demand response that is market driven, or deemed price sensitive (AEMO, 2020b). There would be value in an equivalent study in Australia of the potential for load flexibility that is reported for the United States in Chapter 9 (Hledik, 2020).

[4] It is common in Australia to use distributed energy resources (DER) as a term to include all forms of demand-side participation, distributed generation, and storage. As it is not always used consistently, the preferred term of flexible demand is used here, as it is discussed in this volume.

2. Australia's experience to date with flexible demand

The earliest plan to support flexible demand in the NEM was developed by the Australian Energy Market Commission (AEMC). In October 2007 the AEMC initiated a review of *whether the demand-side of the National Electricity Market (NEM) is participating effectively and efficiently in the market* (AEMC, 2009).

The AEMC defined the "demand side" as all small electricity consumers in the NEM; that is, households and small businesses. The AEMC released its final "Power of Choice" report in 2012 (AEMC, 2012). The report outlined a plan that would "put consumers in the driving seat," including:

- **expanding access to markets to reward flexible demand**, through a mechanism to allow consumers to access the wholesale spot price for electricity, including through a third-party aggregator;
- **enabling the widespread rollout of smart meters**, through competition in metering services to encourage commercial investment in smart meters which could then provide consumers with the choice to take up time-varying retail pricing[5]; and
- **more efficient and flexible retail pricing structures**, to be achieved through requiring cost-reflective electricity distribution network tariffs.[6]

The reforms have met with mixed success in achieving greater demand-side participation in the subsequent 8 years, with the participation of households and small business remaining minimal.

There are four markets where consumers with flexible demand can potentially be rewarded for providing capacity to the energy system: frequency control; wholesale electricity market; emergency services; and network support services.

1. **Frequency control**. Since 2012, the NEM has seen the most success in incorporating and rewarding flexible demand at the C&I level in frequency control ancillary services (FCAS).

---

[5] The competition in metering reforms were proposed after all other jurisdictional governments in the NEM declined to mandate the rollout of digital meters, as the Victorian Government had initiated in 2006 and implemented by end-2013.

[6] The AEMC proposed other reforms to support increased activity on the demand side such as allowing consumers to sell their flexible demand to intermediaries instead of their retailer, improved access to their consumption data and incentives for electricity network distribution businesses to incorporate demand-side options into their planning and investment processes.

A new type of market participant was introduced in 2017 that allows intermediaries to aggregate demand response and provide FCAS services. Demand response now makes up about 15% of the raise FCAS markets (Fig. 19.1), demonstrating that C&I customers are willing and able to be flexible in their demand.

Trials are currently under way, testing the ability of virtual power plants (VPP) to provide FCAS. VPPs aggregate and coordinate flexible demand, including rooftop PV, battery storage, and controllable loads such as air conditioners and pool pumps. The first VPP was enabled to provide 1 MW of FCAS in September 2019. While limited data are available to date on the success of the trial, AEMO considers that the available evidence suggests "VPPs can effectively respond to power system events and price signals" (AEMO, 2020a).

2. **Wholesale electricity market**. There have been few opportunities for flexible demand to be rewarded in the wholesale market. There are some examples where customers can choose a retail product that passes through the spot price and so benefit from demand reductions at peak times, but these examples are limited, particularly in the residential sector.

The AEMC has recently implemented a new wholesale demand response mechanism which will take effect from October 2021 (AEMC,

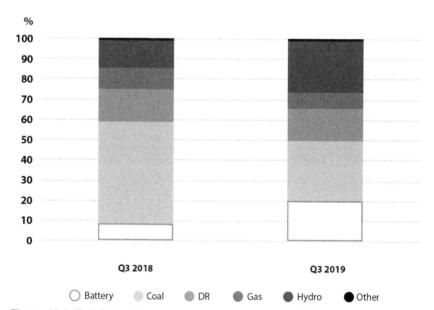

Figure 19.1 (Raise) Frequency control ancillary services (FCAS) supply mix by fuel type. (*Source: AEMO, Quarterly Energy Dynamics. Q3 2019 workbook, available at www. aemo.com.au/.*)

2020a). The mechanism introduces a new market participant, a "demand response service provider" (DRSP). The DRSP will be able to engage directly with a customer without the involvement of that customer's retailer and bid demand response directly into the wholesale market.

While an important development, the mechanism will not be available to small business and residential customers. The AEMC considered the final design of the mechanism could not easily be applied to these customers. Rather, the ability of small customers to participate in the wholesale market will be considered as part of the proposed "two-sided market", which is part of a broader NEM reform process being led by the Energy Security Board (ESB) (ESB, 2020).

3. **Emergency services**. There has also been some success in trialing aggregated demand flexibility to provide emergency services under the NEM's Reliability and Emergency Reserve Trader (RERT) mechanism. In 2017 ARENA and AEMO agreed to conduct a 3-year trial with the objective of demonstrating that flexible demand can provide assistance to maintain reliability during emergencies. Participants recruited both residential and C&I customers.

Two tests of the ability of the portfolio to respond to emergency events have been reported on. While individual participant performance varied, the aggregate delivery exceeded the aggregate contracted amount in both cases (ARENA, 2019). While there is some way to go to roll these trials out on a mass scale, these initial successes have been positive and again demonstrate the willingness of consumers to participate in flexible demand schemes.

4. **Network services**.[7] There has been limited development of a market in demand flexibility to address emerging and future network constraints. Since 2018, network businesses have been able to apply to the Australian Energy Regulator (AER) under the Demand Management Incentive Scheme to invest in demand response and distributed generation (including rooftop solar PV and batteries) where these can address system constraints at the least cost to consumers (AER, 2017a,b).

The AER has subsequently approved a number of demand response proposals most of which are from businesses that have previously had a history of engaging in demand response with residential consumers, including programs such as Summer Savers (United Energy), Power

---

[7] Note that in Australia, there is a vertical separation between generation and retail from the natural monopoly elements of transmission and distribution (see AEMC, 2013).

Changers (Jemena), Peaksmart (Energex), and Ausgrid's Hot Water Rebate.[8] These programs have focused on changing electricity network peak demand (usually in summer) and now also target minimum demand in networks with high rooftop solar PV system penetration using a mix of behavioral rewards, incentives, or control over appliances to shift consumption. While households have responded positively to participating in these trials and programs, it is not clear how these could be rolled out by network businesses on a wider scale, as their core business does not provide them with a direct customer relationship.

Flexible demand is increasingly important because of the urgent need to address the affordability, security, and reliability challenges arising as the NEM transitions to accommodate commitments to lower emissions by 2050 and rapid technological change.

The cost of electricity in Australia is already at an historic high (Fig. 19.2), which impacts negatively on living standards and the contribution of small businesses to growth in the economy.

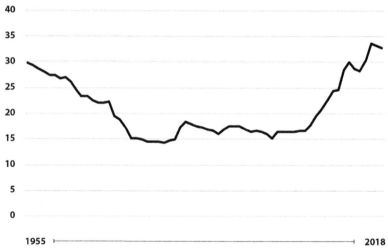

Figure 19.2 Real electricity prices, Australia, national average ($A, c/kWh). *(Source: CSIRO, Australian Bureau of Statistics.)*

[8] See information online on Summer Saver (https://reneweconomy.com.au/united-energy-home-60445/); Power Changers (https://fbe.unimelb.edu.au/cmd/projects/power-changers); PeakSmart (https://www.abc.net.au/news/2019-12-05/electricity-smart-meters-offer-hope-for-reliable-clean-energy/11766766) and Ausgrid Hot Water Rebate (https://www.ausgrid.com.au/-/media/Documents/Demand-Mgmt/DMIA-research/Ausgrid-Hot-Water-DMIA-Projects-Final-Report.pdf).

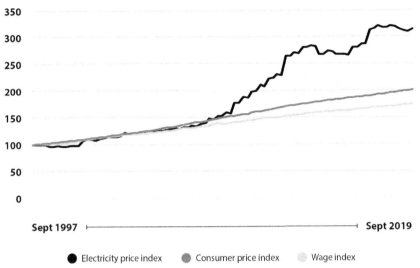

Figure 19.3 Rate of growth electricity prices, general prices, and wages (1997 = 100). *(Source: Australian Bureau of Statistics.)*

In the most recent decade growth in electricity prices outpaced both general prices and wages (Fig. 19.3).

In the absence of flexible demand, the costs of the additional investment required to reconfigure the NEM around renewable generation could be much higher and lead to worse affordability.

Changing patterns of energy consumption could also impact on future investment requirements, and consequently affordability. The shift towards the electrification of transport and heating (where some households currently use gas), and the increased demand for home cooling in hotter summers and home heating in colder winters could all increase the need for investment in NEM system capacity in the absence of flexible demand.

In the other direction, if the changing patterns of work—such as is being experienced in the COVID-19 pandemic—become more permanent, electricity consumption may increase but peak demand may not.

These patterns of changing demand by households are being explored in a 3-year research project, *Digital Energy Futures*, being led by Monash University.[9]

[9] This interdisciplinary project is addressing the knowledge gap in how digital technologies could be an enabler of future demand flexibility of households. The outcome of the project will be critical insights about how households understand and anticipate their futures in relation to emerging digital technologies, and how this knowledge can be used to manage, anticipate, and change their future energy demand behavior. More information is available at https://www.monash.edu/emerging-tech-research-lab/research/energy-futures/digital-energy-futures.

Alongside changing trends in consumption, the continued growth in households and potentially small business uptake in generation assets provides both an opportunity and a challenge.

Australian households have invested in rooftop solar PV at a rate and scale that was not anticipated at the time of the AEMC's Power of Choice Review.[10] From around 340,000 rooftop solar PV systems installed in 2012, there are now 2.4 million systems with 9 GW of capacity. AEMO expects this to increase to reach 12 GW by 2025 (AEMO, 2020c).

The widespread adoption of solar PV, combined with a demonstrated ability and willingness by consumers to use energy flexibly, provides an enormous potential energy resource than could be deployed more cheaply than many alternatives. However, these increasing levels of household rooftop solar PV are impacting AEMO's ability to securely operate the system. This has led to a focus, through the work of the ESB, on the technical and regulatory requirements for "integrating" rooftop solar PV and storage assets into the energy system, including being able to curtail these systems from operating if needed to maintain system security.

A more palatable alternative to consumers than curtailing their excess generation is to incentivize households to use generation and the increasing options for affordable home energy storage to smooth their electricity consumption, such as electric vehicles with vehicle-to-grid capability and modular plug and play systems.[11] The opportunities for the potential for electric vehicles to provide demand flexibility are explored earlier in Chapter 8.

Australia's Chief Scientist, Dr. Alan Finkel AO, recognized the substantial role that consumers could play in the 2017 *Independent Review into the Future Security of the National Electricity Market* (Finkel, 2017).

However, with scarce time and resources the Finkel Review's blueprint focused on the energy system security challenges and the market mechanisms needed to encourage investment in the electricity sector, while lowering emissions. This included the planning and orchestration required for a future energy system. This means that there remains a gap in the planning for a pathway forward to develop flexible demand at scale in the

---

[10] See (Futura, 2011) Final report for the AEMC.

[11] Home energy storage technologies are rapidly changing. The Australian National University's, *Realizing electric vehicle-to-grid services* project was recently initiated https://bsgip.com/research/realising-electric-vehicles-to-grid-services. It was also recently announced that Origin Energy (one of Australia's largest retailers) is supporting the first trial of the Orison plug and play home battery. See https://onestepoffthegrid.com.au/worlds-first-plug-in-home-battery-set-to-be-tested-in-australia/.

NEM, which addresses how flexible demand can be achieved and why it has been so difficult to achieve to date.

Consumers are at the heart of the transition. More thought should be given to how we can best reward consumers for demand management and the power they generate through distributed energy resources like rooftop solar PV. When combined with improved energy efficiency, this will help reduce consumers' electricity bills. The future grid will be more distributed, but its security and affordability will be strengthened through smarter grids, more granular meter data information, and clear data ownership rules to promote new ways of trading, including via a demand response mechanism.[12]

3. The role of intermediaries in flexible demand services

The AEMC in its Power of Choice report saw third-party intermediaries playing a critical role in enabling greater demand-side participation, with growth in these intermediaries, and a diversity of business models, emerging from their reforms. The AEMC also considered retailers would provide demand flexibility as a service offering in a competitive retail market.

Since 2012, it has become apparent that there are limited incentives for retailers to offer innovative flexible demand services. Further, there are constraints on the growth in intermediaries, with retailers continuing to have the predominant relationship with customers. Where intermediaries have emerged is in managing consumers' generation and storage assets, but they do not currently have access to the full range of value streams across the FCAS, wholesale, emergency, and network services markets that would help drive meaningful participation of consumers in flexible demand.

Flexible demand allows retailers to hedge against spot price exposure, while these services also provide a point of differentiation from competing retailers. Despite these opportunities, there are limited examples in the NEM to date of retailers offering innovative services that provide a signal for, or reward, flexible demand.[13]

---

[12] Independent Review into the Future Security of the National Electricity Market, Blueprint for the Future, June 2017, page 6.

[13] There are few examples internationally where retailers have offered flexible demand services at scale or as a standard component of their offerings. However, innovation in pricing is emerging in the United Kingdom such as Octopus Energy, with their Agile and Go tariffs, and Bulb with their smart tariff. Octopus Energy, in particular, has demonstrated rapid growth in market share through innovative tariffs and services. In the United States, innovation in pricing is appearing with the uptake in electric vehicles (see Smart Electric Power Alliance (SEPA) (SEPA, 2019a,b).

In this context it is relevant that Australia's electricity market is characterized by a few large integrated retailers and generators that between them control significant generation capacity as well as retail market share in the NEM.

In 2012, when the Power of Choice report was published, the three largest "gentailers"—AGL, Origin Energy and EnergyAustralia—together supplied 76% of small electricity customers and controlled 35% of generation capacity in the NEM (AER, 2012). While they have surrendered retail market share in recent years, in 2019 they still supplied 63% of small consumers and have increased their generation capacity to 46% (AER, 2020).

This has important implications for the wider participation of consumers in delivering flexible demand.

With vertical integration providing a physical hedge against spot price risk, there is less incentive for gentailers to use flexible demand for this purpose.

Further, while the largest three gentailers have lost retail market share, this has not been accompanied by any significant change in the predominant retailer business model. Competition was introduced in an environment with a regulated "price to beat." Retailers responded by competing on discounts from this price rather than other forms of pricing innovation. Consequently, most consumers are on flat or block rate pricing for consumption which provides no signal or reward for flexible demand. This form of pricing also limits retailers' incentives to curb demand, since they earn a margin on each unit of consumption.

However, in recent years, smaller retailers have emerged that offer spot market exposure or offer to manage part of consumers' energy consumption or generation using smart connected devices. These nascent services include those offered by Amber Electric,[14] which offers access to the spot price for a monthly subscription fee, Energy Locals[15] which offers plans linked to home energy storage and Pooled Energy,[16] which offers pool pump automation, optimizing operation. The small business sector has few opportunities, though in recent years Flow Power[17] and ERM Power (which is a dedicated business only retailer)[18] are investigating extending their C&I business model

---

[14] See Amber Electric https://www.amberelectric.com.au/.
[15] See Energy Locals https://energylocals.com.au/about-us/.
[16] See Pooled Energy https://pooledenergy.com.au/.
[17] See Flow Power https://flowpower.com.au/.
[18] See ERM Power https://ermpower.com.au/?
gclid=Cj0KCQjwu8r4BRCzARIsAA21i_Acq3nd_oM0eCrAwEcJcGSs8k4zcBQbt-44AGWLbnxeFTQy1zxKnFgaAp72EALw_wcB.

into this sector, which engages with consumers to reduce consumption and maximum demand using demand response, energy efficiency, and installation of solar systems. To date these retailers' market shares of the residential market are very low, estimated at most 1%–2%.

Further, innovative new offerings are starting to emerge as the number of households with batteries increases following a number of state subsidy schemes (AEMC, 2020b). A variety of retailers, including the large gentailers, have recently developed VPP offerings, available to customers with batteries. Unlike other types of flexible demand which reduces or shifts load to hedge against high spot prices, batteries provide opportunities for retailers and aggregators to generate revenue.

Independent third party intermediaries, or demand aggregators, offer a very different service to retailers. Their core business is to identify the value of demand flexibility and extract that value across multiple markets, on behalf of their customers. This requires the development of specialist technology and skills to facilitate flexibility and demonstrate its value to consumers. A number of these types of business models are mentioned in Chapter 7.

Intermediaries are not conflicted in managing consumers' flexible demand in the way that gentailers are. Further, intermediaries do not have to tie their flexibility services to a retail product. They can focus on developing their flexibility offerings as a new service in a new market, rather than having to capture a large share of a mature market at the same time.

Intermediaries can also provide a complementary service to network businesses. While network-led flexible demand programs have been successful, network businesses are limited in what they can offer at scale because they do not have a direct relationship with the customer. Further, regulated network businesses are limited to providing network services and are therefore not able to unlock additional value streams that a single asset could provide in other markets.

For these reasons, intermediaries are likely to be better placed than retailers or network businesses to achieve demand flexibility at scale and across multiple value streams, working with households and small business.

While small retailers and network businesses have shown what is possible, new players are required to build on and deliver the value for consumers.

Until recently there have been limited opportunities for intermediaries to enter the market and offer flexible demand services, particularly at the household level, primarily because the retailer has controlled the relationship with the customer.

There are examples of entrepreneurs that have emerged in recent years. It has been announced that OhmConnect is set to enter the Australian market (in partnership with Origin Energy)[19] having demonstrated the value of its gamification platform in the United States where it has engaged over 100,000 households in reducing demand by 500 MW in a summer campaign, turning flexible demand into gift cards, points, and financial rewards.[20] While CitySmart has demonstrated the power of gamification in low-income households managing their energy use, in the Australian context in their *Reduce your Juice* program, it has not been able to attract commercial interest in a partnership.[21] In the technology space, now that solar systems and home energy storage are cheaper, companies such as Reposit Power, Evergen, and Rheem's Home Energy Management system (*EcoNet*)[22], have entered the market to help consumers optimize their energy generation and use through home management systems.

Currently, the value that intermediaries can offer consumers to reward flexible demand is limited by the ability of intermediaries to access markets, particularly the wholesale market. The more that consumers can be rewarded for the value of their flexible demand in multiple markets—frequency, wholesale, emergency, and network services—the greater is the incentive for their participation and investment by intermediaries. Indeed, the ability to "value stack" across markets is a critical factor in the development of flexible demand, particularly at the small consumer level.

The AEMC has recognized the importance of intermediaries, most recently in allowing DRSPs to aggregate demand response in the wholesale market without having to have a relationship with a customer's retailer. This progress must continue to the household and small business level, which has been proposed to be addressed in a future two-sided market (ESB, 2020). In the ESB's framework existing multiple market participant categories would be replaced with two simple categories:

---

[19] https://www.afr.com/companies/energy/game-thinking-to-help-origin-fend-off-disruption-20200528-p54xgp.

[20] See California Energy Commission, http://calenergycommission.blogspot.com/2019/03/ohmconnect-gamifies-energy-use-to-%20shift.html?utm_source=feedburner&utm_medium=email&utm_campaign=Feed%3A+CaliforniaEnergyCommissionBlog+%28California+Energy+Commission+BLOG%29.

[21] For more information about CitySmart see https://www.citysmart.com.au/why/ and see https://reduceyourjuice.com.au/.

[22] See Reposit Power https://repositpower.com/, Evergen https://www.evergen.com.au/ and Rheem https://www.rheem.com/econet/.

- those who use electricity (end users) and
- those who sell it on behalf of end users (traders), which the ESB notes could include both retailers and intermediaries.

While this may simplify the framework, it is worth noting that some existing market participant categories, including the DRSP, were created as a means to allow intermediaries to access markets without being, or having a relationship with, a retailer. It is important this is not lost, whether it be through allowing multiple traders at a connection point or by some other mechanism.[23]

Finally, there are limited examples where intermediaries have negotiated to provide demand flexibility services to network businesses as an alternative to investing in network infrastructure, for example the Mooroolbark Mini-Grid project[24] and the Salisbury Battery Trial (Vincent, 2017).

Intermediaries' access to this market will need to be addressed as future reforms consider the ongoing role of network businesses in the context of the greater penetration and orchestration of DER, including how network businesses can be incentivized to utilize nonnetwork options (KPMG, 2018) and how the regulatory framework can better support information exchange between network businesses and intermediaries.

As this section has explored, intermediaries need access to multiple markets that allow intermediaries to offer new services and differentiate business models in order to develop flexible demand at scale. The market frameworks must support, and not create barriers to, the entry of intermediaries and new business models.

4. Empowering consumers to deliver flexible demand

This chapter has established that there is the potential for households and small businesses to be flexible in their demand, and that it would be beneficial to consumers themselves and for the energy system overall if

---

[23] An alternative to maintaining separate market participant categories could be to allow multiple traders to operate at a single connection point. This model was first proposed by the AEMC (AEMC, 2012). In 2016 the AEMC decided the costs of allowing multiple traders at a connection point outweighed the benefits (AEMC, 2016). However, the concept was raised again by the AEMC in 2018 (AEMC, 2018). Most recently, the AEMC has suggested this issue should be addressed through the ESB's existing reform processes (AEMC, 2020b).

[24] See Ausnet Services https://www.ausnetservices.com.au/Community/Mooroolbark-Mini-Grid-Project.

there was more flexible demand in the future. Further, as flexible demand needs to be aggregated, new services and new business models are needed to facilitate the transacting of flexible demand in multiple markets—whether by retailers or other intermediaries.

With this in mind, this chapter now turns to the final piece of the puzzle: empowering consumers to deliver flexible demand.[25]

Industry and governments envision that when consumers have access to smart meters, smart pricing, and smart connected devices, this is sufficient for autonomous energy management to do all the heavy lifting of identifying flexible demand, and releasing it into the market through intermediaries and through a platform.

But as the designers of autonomous vehicles have come to realize, consumer willingness and ability to adopt this technology goes beyond optimizing the features and functions of the car. Rather, it requires drivers being willing to give up control in stages, from self-managed driving to full automation. Drivers will choose which stage of driving control they are comfortable with, and not all drivers will make the same choice.[26]

The lessons learned from the autonomous vehicles can be applied to the task of designing a future energy system in which flexible demand is a feature.[27]

There is the additional issue of trust, as without it large numbers of consumers will not be willing to make incremental changes in how they use and generate energy to deliver flexible demand. The evidence is that trust is low in the current energy market, with around two-thirds of households in the ECA's regular surveys saying that they do not trust the market to work in their interests (ECA, 2020a).

To build trust, there is a need to respond to consumers' needs for autonomy, control, transparency, and assurance on security and privacy, in the

---

[25] ECA actively engages with new energy technology players, to understand, engage, and support new business models that align with consumers' needs. For example, working with Accurassi (https://www.accurassi.com/), Reposit Power, and as a member of the advisory groups on projects funded through the Australian Renewable Energy Agency (ARENA) including the Distributed Energy Exchange (DeX) project https://arena.gov.au/projects/decentralised-energy-exchange/and the My Energy Marketplace project https://arena.gov.au/projects/wattwatchers-my-energy-marketplace/.

[26] See Society of Automotive Engineers (SAE), https://www.synopsys.com/automotive/autonomous-driving-levels.html.

[27] See the report of the National Science and Technology Council and the US Department of Transportation (NSETC, 2020).

design of the future energy system, and our approach to flexible demand in particular. As the social researchers found in the CONSORT Bruny Island Battery Trial (CONSORT, 2019):

> *A common assumption among industry and government reports on DER is that householders are likely to be willing and unproblematic participants in DER sharing with networks. Our CONSORT social research challenges this assumption: households are diverse, and greater awareness and appreciation of the context in which households make decisions about their energy is crucial to understanding their receptiveness to DER, and their DER preferences.[28]*

Consumers' need for control and choice on how appliances and technology are used in their home and business makes designing a market and incentives for flexibility that much more complex. Some of the limitations or constraints on design are discussed in Chapter 14 on the possibility of a two-sided market and Chapter 18 on the experience of time-of-use pricing in Victoria. The critical importance of designing for how consumers actually behave, and not how they could be assumed to behave has been explored by Dr. Ron Ben-David (Ben-David, 2020) and by Dr. Cameron Tonkinwise,[29] in the context of a consumer vision for the future energy market.

Empowering consumers to deliver flexible demand needs to start with incorporating their goals and values into the energy system, which is also discussed in Chapter 22.

Research in the United Kingdom (UKERC, 2013) explored the universal values and principles that consumers would draw on to guide their energy decisions. These include:

- efficiency and not wasting energy;
- protection of the environment and nature;
- ensuring the energy system is safe, reliable, personally affordable, and widely accessible;
- autonomy and control;
- social justice and fairness (attentive to the effects on people's abilities to lead healthy lives); and
- improvement and quality in terms of long-term trajectories, ensuring changes represent improvement and considering their implications for quality of life.

---

[28] (CONSORT, 2019) page 3.
[29] See ECA Foresighting Forum 2020 video https://energyconsumersaustralia.com.au/ projects/foresighting-forum and presentation https://energyconsumersaustralia.com.au/ wp-content/uploads/1H-Transition-Design-Dr-Cameron-Tonkinwise.pdf.

Similarly, in Australia there has been engagement around the need for a compact around the energy transition (ACOSS, 2020).

In research by the ECA (ECA, 2020a,b) consumers acknowledged that they have a role to play and they want to make a difference in the transition to a lower emissions energy system. However, it was clear that their day-to-day priorities for a comfortable home and family life, or a profitable business, is their prime concern and it can be too difficult, or is too time consuming, to take action.

For consumers to change their behavior, they want to know how changing their energy use could improve their own life, or business, and how it could benefit the community. Further, they want to believe that there is reciprocity—that if they play their part then the benefits will be shared fairly, and not just captured by other participants in the energy system.

ECA's research (ECA, 2020b) identified consumers' attitudes toward changing behavior and consumer expectations of a better energy market in the future.

The attitude of households to changing behavior was summarized as "what's the point"? For the future, their expectations are that:

- **energy will be clean, cheaper, and more affordable**—consumers dislike rising energy prices and felt they are being overcharged by energy companies. A better energy future means lower prices;
- **decisions about energy will be simpler**—energy bills and plans consistently confuse and overwhelm consumers who struggle to understand the breakdown of costs and found comparing providers near impossible. A better energy future means simplified, more comprehensible information.
- **energy will be easier to manage, and increasingly controlled by digital technologies**—apps, real-time information, and smart homes could improve the lives of individuals in the future. This does not mean technology that takes control of everything but gives consumers options and automated energy saving behavior.

Small businesses feel they are already doing everything they can to control costs and changing energy use is either not possible or will not make a difference. For them, a better energy future means energy will be clean, cheaper, and more reliable.

To realize consumers' expectations for a future energy market, three actions are needed in Australia:

- **changes to policy frameworks** that focus on consumer outcome measures;

- **designing for the diversity of consumers** and how they use energy in their homes and businesses; and
- **a national flexibility plan** that sets out an ambition for the scale of flexible demand in the NEM, and in particular for households and small business consumers, describing the elements and stages needed to meet the target over the next decade.

Changes to policy frameworks are required in five areas to be reoriented to consumer outcomes to enable more flexible demand.

1. Regulating for consumer control, not controlling consumers. Consumers are looking to technology to provide them with control, including over their energy use and bills (ECA, 2020a). Regulation that is seen to take away their control, or consumer protection frameworks that do not support their control, are likely to undermine consumers' willingness to be flexible in their demand. There is a need to create a social licence for control.

Within the current framework, consumers are not aware of the potential for the use of their assets to impact on other consumers, or the ability for the energy system to be operated securely. The framework needs to be broadened to incorporate opportunities for consumers that are willing to curtail their generation or exports, to be rewarded for doing so. One potential approach was explored recently in South Australia, where consumers were asked to consider different options for pricing their connection of rooftop solar PV to the network. The options were for consumers willing to have their generation be interruptible to pay a lower price, while consumers who preferred to have a guarantee of not being interrupted would pay a premium. This concept should be further explored, and extended beyond just the interruptions that might be initiated by the network, to operations initiated by the system operator.

The current consumer protection frameworks need to be reviewed against the background of the increased blurring of the dividing line between energy specific protections and economy wide protections, and the complexity of the new energy technology environment. Until new energy technology becomes "plug and play" consumers will need to coordinate across installers, electricians, and equipment providers which provides multiple opportunities for things to go wrong. ECA has been at the forefront of the development of a New Energy Technology Consumer Code, together with industry partners, that can adapt to new requirements for consumer protection as they arise.

2. Rewards and incentives that consumers can understand and respond to. Without (retail) pricing and incentives that reward flexibility, consumers will not change their energy use behavior. There cannot be a one-size-fits-all approach to pricing and incentives, given the diversity of how households and small businesses use energy. Further pricing and incentives must be designed to take into account consumers' trade-offs between reward and effort.

There is a widespread view that consumers find retail pricing that charges more for using energy at peak times unappealing. This is borne out by the evidence in Australia that more than 95% of consumers in the NEM remain on flat rate or block pricing (with the slow take-up of smart meters, on a voluntary basis, also contributing).[30]

Time-of-use (retail) pricing, of which there are various forms, exposes consumers to "risk." Consumers lack the information to be able to make a judgment about the trade-off between inconvenience and reward. Without simple tools, it is a complex and time-consuming decision, and consumers bear the risk of getting it wrong. Consumers are also unclear if there is a pay-off from peak pricing. If sufficient energy use cannot be shifted to off-peak (and most off-peak hours are still overnight), then bills could be higher.

Digital technology and smart connected devices create the opportunity to close this gap, so that consumers can be more flexible in their use of specific appliances, at minimal inconvenience and with an understanding of the pay-off.

Pool pumps, hot water systems, and electric vehicles all have the characteristics that their charging can in large part be separated from their use (as discussed in Chapters 7 and 8). Consumers are likely to be more willing and able to program the charging of these devices in off-peak periods on a regular basis, including through automation. By separately charging for these devices, the price signal to avoid peak and reward off-peak use can be a stronger incentive than if all of the consumer's load was charged in this way.[31] In this context:

---

[30] Energy Consumers Australia analysis of retail market offers, and AER data on incidence of cost-reflective network tariffs charged to retailers, see https://www.aer.gov.au/networks-pipelines/network-tariff-reform.

[31] This is currently being explored in work by the California Energy Commission, including the work led by Dr. Karen Herter considering new tariffs, technologies, and other measures to expand energy efficiency and demand management in California through the 2020 Load Management Rulemaking project (CEC, 2020a) and the separate research project, Complete and Low-Cost Retail Automated Transactive Energy System (RATES) (CEC, 2020b).

- South Australia Power Networks has proposed a solar sponge tariff which, if adopted by retailers, could incentivize load such as hot water being shifted to the middle of the day, enabled by smart control of hot water systems[32,33] and

- ECA has proposed an electric vehicle charging tariff, to be adopted in Victoria, that could have a retail off-peak rate of 10 cents/kilowatt hour, with peak charging only applying for 2% of the time (ECA, 2020d).

Consumers find being more flexible in their use of home cooling and heating more difficult on a regular basis, as it depends on social practices and perceptions of comfort among all members of the household, combined with the low energy efficiency of Australia's housing stock.[34] The loss of amenity or inconvenience can simply outweigh paying a higher price. However, while consumers may not respond to pricing or financial incentives, there is evidence that many consumers are often responsive to nonfinancial incentives when asked to change their energy use on the hottest days in summer, to keep the power on. ECA research found that 75% of consumers say they would reduce their use on extreme days for little or no reward (ECA, 2020a). The Power Changers program demonstrated the value of nonfinancial rewards on critical peak days with consumers earning points and donating them to community organizations (Jemena, 2019).[35]

A future energy system in which there is greater demand flexibility can be achieved, but only if consumers finding it rewarding and without too much inconvenience. This means that we can expect consumers to

---

[32] Further information on the consumer data right for energy can be found at https://energyconsumersaustralia.com.au/projects/consumer-data-right.

[33] See https://reneweconomy.com.au/networks-offer-super-cheap-pricing-to-soak-up-australias-solar-sponge-10879/. The value of this load should not be underestimated, for example Hawaii is using smart connected devices with smart pricing to aggregate hot water load, as a virtual power plant. https://www.globenewswire.com/news-release/2019/10/01/1923461/0/en/Shifted-Energy-to-Equip-2-400-Water-Heaters-in-Hawaii-with-Grid-Interactive-Technology-to-Create-Virtual-Power-Plant.html.

[34] The quality of Australia's housing stock is poor, and there is a limited history of what is known as weatherization in the United States. Through the COAG Energy Council, work is underway through the Trajectory for Low Energy homes to begin to tackle this problem. In recent years, in winter there have been reports in the media for the first time of people experiencing hypothermia from a lack of heating in their homes. The choice can be to "heat or eat."

[35] The trial, led by the Center for Market Design Associate Professor David Byrne, showed average peak electricity consumption on hot days down between 23% and 35%.

continue to have simple pricing plans for all or some of their consumption load. But these simple pricing plans may innovate away from traditional pricing models, including subscription pricing (Brattle, 2020).[36]

3. **Real-time information available to support decision-making**. There is a significant shift underway in allowing consumers to access data that they need to inform their real-time decision-making. Further, from 2022 consumers will be given greater control to securely share their energy use data with a trusted third party through the consumer data right.[32] Australia could look to the United States to develop a similar mechanism as Green Button Connect, which has demonstrated that data portability and interoperability need to be brought together in a format that is accessible to consumers (Mission Data, 2019).

Today most consumers are relying on their bill for information on energy use. Consumers with smart meters (100% of customers in Victoria, and around 20% elsewhere in the NEM) can access their historical data via a website, with few having real-time monitoring with an in-home net-worked device. For a very small proportion of consumers with smaller retailers on "pass through" retail offers, access to their data is available through an app on their smartphone.

As smart connected devices become more widely available, it will support the development of information and tools for consumers to monitor and control their energy use in real time. Consumers will need to be given the functional capability in smart controlled devices to choose their level of control, how much they wish to automate, and how much control they are willing to outsource to others.

4. **Trusted and independent sources of advice**. With the shift to greater demand flexibility, and low levels of trust in the energy market, consumers will continue to rely on independent sources of advice. These could include member or subscriber organizations for consumers and their communities, for example Renew[37] and Choice.[38]

A recent review of coordination and funding for financial counsellors (Sylvan, 2019) recommended a national strategy to address the challenges facing this sector, which plays a significant role in supporting consumers.

---

[36] Horizon Power, which is an integrated utility in regional Western Australia, trialled a subscription plan, and the results reported on by the Bankwest Curtin Economics Center (BCEC).

[37] Information on Renew can be found at https://renew.org.au/.

[38] Information on Choice can be found at https://www.choice.com.au/.

Digital services such as price comparison websites and switching services have not developed in Australia to the same extent as they have elsewhere. More could be done to work with innovators and entrepreneurs in this space, to support and engage consumers in their decisions around flexible demand.

5. **Resilience and inclusion integrated into safety nets**. Work done for the AER by the Consumer Policy Research Center (CPRC, 2020) has shown that there are opportunities for fairer and more inclusive markets for all consumers. Traditional approaches have set a bright line that divides those considered vulnerable—and therefore are given assistance—and those that are not. The CPRC report identifies how regulators could play a more significant role in ensuring that an essential service energy is accessible to consumers who are in difficult circumstances or where the complexity of the market creates barriers.

In addition to designing a market and its features for consumer outcomes, there is a need to design for the diversity of consumers. At the outset, the ambition should be to provide the opportunity for all households and small businesses to participate in delivering demand flexibility, if they choose to, by whatever means is simplest and convenient to them.

In designing an energy system that integrates flexible demand in a significant way, the architects need to test and refine possible measures and mechanisms, drawing on behavioral insights. Traditional approaches to market design do not typically account for the wide variety of consumer behaviors.

The ECA has developed a segmentation framework for exploring the decision-making behaviors of households, which is now being extended to small businesses (ACIL Allen, 2018). Households can be categorized within nine segments (Fig. 19.4), where they are differentiated by their rankings on motivation, ability, and opportunity. This framework could be used as a starting point for exploring design approaches to measures and mechanisms that could be appropriate for engaging consumers on flexible demand choices.

Testing and refining could mean adequately resourcing innovation and living labs such as the Energy Systems Catapult Living Lab,[39] or the Green Village in the Netherlands.[40] We also need to explore practical and effective regulatory sandbox approaches, noting the AEMC's advice in March 2020 to the Energy Council on this could be implemented within Australia's rules framework (AEMC, 2020c).

---

[39] See www.es.catapult.org.uk.
[40] See https://www.thegreenvillage.org/projects/living-lab.

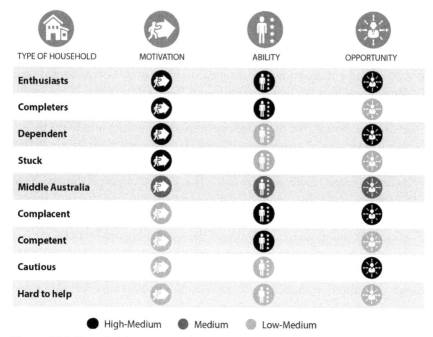

Figure 19.4 Household segments. *(Source: ACIL Allen, 2018. Supporting Households Framework, December 2018 for ECA's Power Shift research program.)*

The evidence internationally is that to achieve a future energy market that meets consumers' expectations, and that benefits the energy system overall, requires a plan with a specific focus on flexibility, which should be co-designed and implemented by decision-makers, stakeholders, and consumer representatives. A flexibility plan for the NEM would need to go beyond statements of ambition and step out the elements and stages necessary to realize target outcomes. In the context of Australia's federated governance model, it would need to describe the different accountabilities at national and state level, and address the issues that have been highlighted in this chapter.[41]

---

[41] In the United States see, for example (FERC, 2010) and (CAISO, 2013). In the United Kingdom see, for example (Ofgem, 2017). In July 2020, The Energy Systems Catapult released its *Europe smart energy and flexibility market study* (see Energy Systems Catapult, 2020). In Australia outside of the NEM, the Western Australian Government's Energy Transformation Taskforce recently released its Distributed Energy Resources Roadmap, which could incorporate flexible demand within its 36 actions.

5. Conclusions

There is huge potential for households and small businesses to offer additional capacity to the market by providing flexible demand. Consumer investment in behind-the-meter energy technology has far exceeded expectations over the last decade, and consumers have demonstrated the ability and a willingness to change their demand, provided it suits them. The question for policy makers is how to develop a framework that allows consumers to easily and conveniently offer demand flexibility into the market and be appropriately rewarded for doing so.

While reforms were proposed in 2012 to address this question, in 2020 the NEM has seen little progress to date in encouraging greater demand-side participation, including through intermediaries. Yet it is becoming increasingly critical to tap into this latent flexibility to help keep energy bills down as the market shifts toward greater penetration of variable energy resources.

This chapter argues there are two key issues that must be addressed in order to realize the potential demand flexibility that households and small businesses have to offer.

First, intermediaries have an important role to play in extracting the value of demand flexibility, and the role of intermediaries must be explicitly considered in current reform processes. While retailers and network businesses have successfully trialed demand response, to promote demand flexibility at scale requires both the incentive and the ability to value stack across multiple markets. Intermediaries are likely to be better placed to do this, including having the incentives, necessary skills and technology.

Second, to empower consumers to provide flexible demand requires rereframing the issues and challenges facing the NEM to focus on the consumer perspective. Consumers must be rewarded and incentivized to provide flexibility in a way that they understand and can respond to. This requires more than standard financial incentives and price signals. Rather, a deeper understanding of what drives customer decision-making, including their diverse values, goals, and expectations for the future is needed.

If the opportunities provided by digital technology can be brought together with trusted intermediaries, in a national plan, the potential for Australia's 10 million homes and small businesses to be flexible in their demand can be realized—delivering a more affordable future energy system that is truly more distributed and more democratized.

Bibliography

ACIL Allen, 2018. Supporting Households Framework. December 2018.

ACOSS, 2020. New Energy Compact, People Centred Vision for the Australian Energy System. Retrieved from: www.acoss.org.au/new-energy-compact/#: ~ :text=The%20New%20En ergy%20Compact%2C%20is.for%20the%20Australian%20energy%20system.

AEMC, 2009. Review of Demand-Side Participation in the National Electricity Market. Final Report, 27 November 2009.

AEMC, 2012. Power of Choice Review — Giving Consumers Options in the Way They Use Electricity. Final Report, 30 November 2012.

AEMC and KPMG, 2013. The National Electricity Market — A Case Study in Microeconomic Reform, 13 December 2013.

AEMC, 2016. Multiple Trading Relationships, Final Rule Determination, 25 February 2016.

AEMC, 2018. Economic Regulatory Framework Review, Promoting Efficient Investment in the Grid of the Future. July 2018.

AEMC, 2020a. Wholesale Demand Response Mechanism, Rule Determination. June 2020.

AEMC, 2020b. Retail Competition Review 2020. June 2020.

AEMC, 2020c. Regulatory Sandboxes — Advice to COAG Energy Council on Rule Drafting.

AEMO, 2020a. Virtual Power Plant Demonstration. Knowledge Sharing Report #1, March 2020.

AEMO, 2020b. Demand Side Participation Forecast Methodology. February 2020.

AEMO, 2020c. Renewable Integration Study: Stage 1 Report. April 2020.

AER, 2012. State of the Energy Market 2012.

AER, 2017a. Demand Management Incentive Scheme.

AER, 2017b. Demand Management Allowance Mechanism.

AER, 2020. State of the Energy Market 2020.

ARENA, 2019. Demand Response RERT Trial Year 1 Report. March 2019.

BCEC, 2018. Power plans for Electricity. BCEC Research Report No. 16/18, September 2018.

Ben-David, Ron, 2020. Response to Consultation on Two-Sided Markets, 17 May 2020. Retrieved from: http://www.coagenergycouncil.gov.au/publications/two-sided-markets.

Brattle Group, 2020. Fixed Bill, Making Rate Design Innovation Work for Consumers, Electricity Providers, and the Environment. June 2020.

CAISO, 2013. Maximising Preferred Resources: Demand Response and Energy Efficiency Roadmap. December 2013.

CEC, 2020a. Load Management Rulemaking (Docket #19-OIR-01). Retrieved from: https://efiling.energy.ca.gov/Lists/DocketLog.aspx?docketnumber=19-OIR-01.

CEC, 2020b. Complete and Low-Cost Retail Automated Transactive Energy System (RATES).

CONSORT, 2019. The Bruny Island Battery Trial. Project Final Report-Social Science, April 2019.

ECA, 2020a. Energy Consumer Sentiment Survey. Final Report, June 2020.

ECA, 2020b. Future Energy Vision Research. Household Findings, February 2020.

ECA, 2020c. Future Energy Vision Research, SME Findings. February 2020.

ECA, 2020d. Prices to Devices Tariffs: Developing a More Cost Reflective EV Tariff for Victoria. June 2020.

Energy Transformation Taskforce, 2020. Distributed Energy Resources Roadmap. April 2020.

Energy Security Board, 2020. Moving to a Two-Sided Market.

Energy Systems Catapult, 2020. Towards a Smarter and More Flexible European Energy System. July 2020.

FERC, 2010. National Action Plan for Demand Response. June 2010.

Finkel, 2017. Independent Review into the Future Security of the National Electricity Market: Blueprint for the Future. June 2017.

Futura, 2011. Investigation of Existing and Plausible Future Demand Side Participation in the Electricity Market, 16 December 2011.

Jemena, 2019. Jemena Power Changers. Final Evaluation Report, 2019.

KPMG, 2018. Optimising Network Incentives: Alternative Approaches to Promoting Efficient Network Investment.

Mission Data, 2019. Energy Data Portability. January 2019. Retrieved from. http://www.missiondata.io/reports.

NSETC, 2020. Ensuring American Leadership in Automated Vehicle Technologies, Automated Vehicles 4.0. January 2020.

Ofgem, 2017. Smart Systems and Flexibility Plan.

Samson, A. (Ed.), 2017. The Behavioural Economics Guide 2017, with an Introduction by Cass Sunstein. Retrieved from LSE Research online: http://eprints.lse.ac.uk/84059/1/The%20behavioral%20economics%20guide%202017.pdf.

Smart Electric Power Alliance, 2019a. A Comprehensive Guide to Electric Vehicle Managed Charging.

Smart Electric Power Alliance, 2019b. Residential Electric Vehicle Rates That Work.

Sylvan, L., 2019. The Countervailing Power: Review of the Coordination and Funding for Financial Counselling Services Across Australia, 3 October 2019. Retrieved from: https://www.dss.gov.au/communities-and-vulnerable-people-programs-services-financial-wellbeing-and-capability/review-of-the-coordination-and-funding-for-financial-counselling-services-across-australia.

UKERC, 2013. Transforming the UK Energy System: Public Values, Attitudes and Acceptability. Synthesis Report, July 2013.

Vincent, M., Colebourn, H., 2017. Salisbury residential battery trial, chapter 22. In: Jones, L.E. (Ed.), Renewable Energy Integration: Practical Management of Variability, Uncertainty, and Flexibility in Power Grids, second ed. 2017. Retrieved from: https://www.elsevier.com/books/renewable-energy-integration/jones/978-0-12-809592-8

CHAPTER 20

Markets for flexibility: Product definition, market design, and regulation

Richard L. Hochstetler
Instituto Acende Brasil, São Paulo, Brazil

1. Introduction

As discussed in Part One of this book, achieving supply and demand equilibrium will become more challenging as the share of the electric power supply derived from variable generation increases.

In the past, the required instantaneous equilibrium of power supply and demand was achieved almost solely by supply-side adjustments provided by dispatchable power plants, as discussed in Chapter 5. In the coming years, this will become increasingly difficult as a larger share of power generation will be derived from renewable power sources, which are not controllable by the system operator, but rather by the supply of their underlying natural energy resources (i.e., solar radiation, wind, hydro inflows …).

In this context, system operators will face a triple challenge to maintain system balance:

- a decreasing share of controllable power plants that can be dispatched to adjust supply;
- an increasing share of supply from uncontrollable generation whose supply patterns most often do not match load patterns; and
- an increasing degree of stochasticity, due to supply-side random variation of renewable generation (in addition to the random load fluctuations).

Fortunately, as explored in Parts Two and Three, technological innovations are enabling increased flexibility and control from the demand side of the market. Advances derived from digitalization, telecommunications, and electric-power storage technologies are making it possible to coordinate demand in response to systemic conditions.

These technological advances open a wide range of demand response possibilities that can greatly contribute to deal with the triple challenge that arises from variable renewable generation. However, for these demand

Variable Generation, Flexible Demand
ISBN 978-0-12-823810-3
https://doi.org/10.1016/B978-0-12-823810-3.00003-0

response possibilities to materialize, it will be necessary to adapt markets to provide proper price signals and incentives.

This chapter will focus on the market adjustments that are needed to foster the supply of flexibility from consumers. The chapter is organized as follows:

- Section 2 identifies the different forms by which demand response can provide flexibility;
- Section 3 introduces the different strategies that can be deployed to foster demand response;
- Section 4 discusses how improved product definition can provide better price signaling that incentivize consumers to adjust demand to current market conditions;
- Section 5 explores market mechanisms that can be adopted to foster further demand responses, which would not be feasible otherwise and that are more economical than supply responses; and
- Section 6 explores regulatory issues that must be addressed to enable certain types demand responses; followed by the chapter's conclusions.

2. How consumers can provide flexibility

Although, historically electricity price elasticities have been very low, consumers do have great potential to provide flexibility to electric power systems. Many customers could alter the time of usage of particular electric appliances, if provided with proper information and incentives. Some consumers may even be willing to occasionally curtail demand, if offered adequate remuneration. Most of the consumers' electric appliances have a fast activation time, which suggests that, with centralized and dynamic coordination, it would be viable to provide short-term load balancing services. As households and businesses invest in electricity storage and electric vehicles, they also become capable of injecting energy in the power system, amplifying the spectrum of demand response services they could provide.

The prime difficulty in obtaining demand response is customer engagement. Not only must customers be willing and able to adjust their consumption behavior, but they must also be aware of system conditions in order to provide the proper response at the right time.

While electricity customers do have electric equipment that could be managed to provide demand response, tapping into these resources is a challenge. They are typically small-scale and dispersed among millions of consumers, which, in the past, made the deployment of these resources infeasible. But technological innovations now make the coordination of these dispersed resources a possibility.

Various surveys lay out the realm of demand response possibilities that will be increasingly explored in the coming years (such as Feuerriegel and Neumann, 2016; and EPE, 2019). The different types of demand response services can be summarized by the four S's (Alstone et al., 2017):

- **Shape**, which refers to long-term changes in the pattern of the daily load curve to help accommodate the increased penetration of variable generation with a particular daily supply pattern;
- **Shift**, which refers to ability to dynamically adjust demand from period to another in response to system conditions;
- **Shed**, which refers to curtailment of consumption in critical periods of peak demand or of scarce supply; and
- **Shimmy**, which refers to rapid adjustments to meet the power system's short-term needs.

The ability to provide the different types of demand response depends not only on the consumers' disposition to adjust their demand to current electricity market conditions but also on the consumers' electricity usage portfolio.

Electricity consumption behavior is determined by what economists call a "derived demand" function. Consumers don't consume electricity directly, rather it is an input for the production. Thus, the demand for electricity is "derived" from the demand for these other goods and services.

This fact helps explain why the short-term price elasticity is typically very low. First, because the cost of the final goods and services produced with electricity typically includes many other inputs, the effect of electricity prices variations on the final cost to consumer tends to be diluted. Secondly, electricity usage usually involves capital investment in durable goods, such as appliances, that have relatively long life span. Thus, responses to changes in relative prices often are delayed because consumers wait for their investment in these durable goods to depreciate before substituting them for another appliance that is better suited to current relative prices.

Some of the goods and services for which electricity is used are more conducive to demand response than others. Electricity is used for a wide range of services: lighting, heating and cooling, running motors, telecommunications, digital computations, just to name a few. Some of these uses are highly inflexible, but others can be modified with relative ease, as explored in Chapters 5 and 7.

Recent technological innovations are bringing about changes that make demand response more viable. The emergence of "smart' appliances has made it possible to automate or to remotely control their operation. This has come to be known as the Internet of Energy and could substantially increase the demand response potential.

Empirical studies demonstrate that the adoption of enabling technologies substantially enhances the effectiveness of demand response initiatives. For example, the adoption of time-of-use rates has, on average, resulted in a peak-demand drop of 4%, but when implemented together with enabling technologies the peak-demand drop was boosted to 26%. Similarly, critical-peak pricing programs alone obtained a peak-demand drop of 17%, but together with enabling technologies they attained a 36% drop (Faruqui and Sergici, 2010).

The emergence of new types of electric equipment is also amplifying the type of demand responses that can be offered. Chapter 8 explains how the acquisition of electric vehicles and batteries enables consumers to provide a wider range of demand responses: not only permitting consumers to reduce demand but also to increase the supply of electricity.

The demand response capability of consumers' energy resources can be evaluated based on five major attributes (Eid et al., 2016):

- **Direction**, which refers to the resource's ability to adjust demand (increase or decrease load), supply (injection of energy into the system), or both;
- **Electrical Composition**, which refers to the extent to which the resource meets capacity (kW) and energy (kWh) needs;
- **Temporal Characteristics**, which refers to the speed by which resources can react to system needs;
- **Duration**, which refers to how long the resource can provide a particular service at a time;
- **Location**, which refers to where the resource is connected on the grid.

Considering these five attributes, one can identify various power system needs that can be met by demand response (ordered by their respective temporal characteristics and duration):

- **Ancillary Services** aimed at meeting short-term capacity needs of the power system, which encompasses:
  - Primary Reserves (also known as Frequency Containment Reserves) that are automatically provided in matter of seconds and
  - Secondary Reserves (or Frequency Restoration Reserves) that can be provided within a few minutes;

- **System Balancing** (also referred to as Tertiary Reserves or Replacement Reserves) to meet the power system's capacity and energy needs in the time range of a few minutes to a few hours;
- **Network Constraints**, which consists of medium-term services (in the 2-h duration range), aimed at alleviating problems derived from transmission or distribution congestion;
- **Economic Load Response**, which seeks to minimize system operating costs by incorporating load response in the spot market (both in day-ahead and intraday transactions);
- **Avert Capacity Expansion**, which consists of the provision of demand response services during critical periods of peak demand or scarce supply to avert costly investments in capacity expansion.

The advantage of demand response is that it can provide flexibility without incurring costly investments in additional capacity that are needed when resorting to supply-side solutions. However, demand response is not entirely free of investment costs. Investments in smart meters, in appliances that can be managed automatically or remotely, and in centralized management systems are needed to enable many demand-response services.

Furthermore, the deployment of consumers' demand response capability will require changes in the way electricity is marketed, in the way electricity markets are structured, and in the way electricity provision is regulated.

3. Strategies to foster demand response

Given the current trend of an increased share of electricity supplied by variable generation, flexibility will become increasingly scarce in electricity markets. Demand response is a potential source of flexibility that should be sought to meet the triple challenge brought forth by the increased penetration of variable generation.

Three major strategies to foster demand response from consumers are explored in this chapter:

- Product definition;
- Market design; and
- Regulation of demand-response programs.

A brief introduction to each of these three strategies is presented in this section. A more in-depth discussion of each strategy is provided in the next three sections, addressing:

- Their respective value propositions;
- The challenges associated with each approach; and

- Some practical examples of their implementation in various electric power markets.

One strategy to promote demand response is to better define market products so as to elicit proper price signals. The resulting market-price equilibrium for each of the resulting products would provide the information and the incentives for consumers to adjust their consumption based on power system conditions: reducing demand when the price surpasses their marginal utility of electricity consumption and increasing demand when the price is below their marginal utility. Typically, electricity is differentiated by time and location. The main challenge in product delineation is to determine the degree of differentiation that is pertinent and useful. Consumers can be overwhelmed by high frequency price variations and complex rate structures. New technologies that allow automatic adjustments and the emergence of service providers that actively manage demand on behalf of consumers now potentialize this strategy's ability to elicit demand response.

A second strategy consists of enabling consumers to actively participate in electricity wholesale markets, either directly (such as large industrial customers) or indirectly by means of service providers that aggregate demand response from a group of consumers. Market design changes to allow consumers to submit demand-response bids in wholesale markets would further unleash the demand response potential. This market mechanism would not only save consumers from incurring the costs of purchasing electricity during periods of higher marginal costs (as in the first strategy) but also would provide consumers a payment for the demand reduction. This additional compensation has the potential of significantly amplifying the magnitude of demand response.

Finally, a third strategy that can be employed to harness flexibility from consumers is to adopt demand-response programs specifically designed to mitigate system capacity constraints. These capacity constraints could be due to limited generation capacity or from transmission or distribution constraints. In contrast to the previous two strategies—that employ general, market-oriented approaches to minimize operational costs—this strategy focuses on obtaining specific demand responses to avoid the cost of investments in capacity expansion. Because of their specificity, they are designed case by case and typically require regulatory approval, as discussed in Chapters 12 and 13.

4. Product definition

A natural first step in stimulating demand response is usually to improve product definition because it basically consists of providing better information to consumers regarding system costs, as well as providing consumers better incentives to react to system conditions. This approach involves relatively simple and low-cost implementation.

The strategy in product delineation is to provide proper price signals that reflect current supply costs, so as to enable consumers to spontaneously adjust their consumption.

Electricity markets are often segmented based on daily demand variation (peak and off-peak demand), but segmentation based on supply variability is a novelty for most electric markets.

The benefits of more precise product definition can be illustrated by the comparison of two product delineations for a power system with a large share of variable generation.

In the first case, electricity is defined as a single product. Consumers are offered electricity at a flat rate equivalent to the average cost per kilowatt-hour of energy delivered over the span of a month.

In the second case, electricity is marketed as two different products: one being electricity during periods of scarce supply and another in periods of abundant supply.

Fig. 20.1 illustrates supply and demand in periods of scarce and abundant variable generation and the resulting market equilibria in each case.

When electricity is marketed as a single product, at the flat rate (p_{single}), consumers choose to consume electricity uniformly in both periods (q_{single}). Thus, the market equilibrium remains fixed (E_{single}) regardless of the supply conditions. Because consumers are oblivious to current marginal costs, their consumption behavior is misaligned, resulting in large deadweight welfare losses (equivalent to the shaded areas). During periods of scarce supply, the marginal cost of generation raises to c_{scarce}, much above the consumers' marginal utility that is equivalent to p_{single}. Conversely, in periods of abundant supply, the marginal cost of generation drops to $c_{abundant}$, while consumers' marginal utility remains unaltered. In both cases consumers could be better off if they were aware of the current costs and could adjust their consumption behavior accordingly.

When electricity is marketed as two distinct products during periods of abundant and scarce supply, two market equilibriums emerge: consumers opt

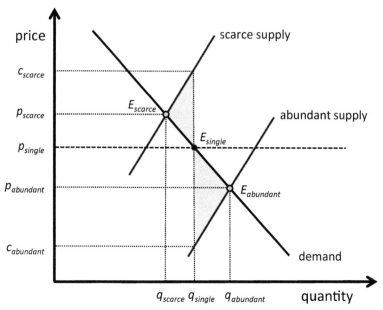

Figure 20.1 Market equilibria for different product definitions based on supply variability.

to consume a greater quantity ($q_{abundant}$) during periods of abundant generation and the market price is low ($p_{abundant}$); and choose to reduce consumption (q_{scarce}) during periods of scarce supply when the market price is high (p_{scarce}). In both cases, the resulting equilibria ($E_{abundant}$ and E_{scarce}) occur at the point where marginal utility equals marginal costs, maximizing social welfare.

Simply making it possible for consumers to adjust consumption to current market prices may incite significant demand response, but is no guarantee, as in the case of time-of-use tariffs in Victoria, Australia, reported in Chapter 18. It is important to remember that the demand response may take some time to be observed, but regular pricing patterns, can instill long-term change in consumers' consumption behavior.

The key to this strategy is providing more detailed and timely information to consumers. Much can be improved in this respect. For many consumers the main source of information of electricity market conditions is their monthly bill. So, electricity consumption behavior tends to react to monthly averages, rather than to current marginal costs. Furthermore, retail rates often have complex rate structures which are difficult for consumers to understand and interpret. This gives rise to slow and muffled demand response (Bushnell and Mansur, 2005; Ito, 2014).

Furthermore, many customers are serviced by standard electromechanical meters that do not distinguish demand in different days of the week, nor at different times of the day, which makes it impossible to charge consumers based on their individual load curve. Thus, a prerequisite to adopting more precise pricing and billing of electricity is massive investment in the replacement of the electricity meters.

This can be a significant hurdle in emerging economies, but even in industrialized economies it is a relevant issue. As pointed out in Chapter 5, the latest data indicate that only 56% of electricity customers in the United States have advanced metering infrastructure (EIA, 2019), although the trend has been of a steady substitution for smart meters.

It may seem that the adoption of real-time pricing for all customers would be the ultimate solution in this strategy. Real-time pricing enables more timely demand response to system conditions, which is likely to increase market efficiency. Not only does it benefit the customers responding to real-time prices but also tends to lower spot market prices in times of scarcity, which benefits all customers. However, the welfare benefits obtained from customers switching to real-time pricing tend to present diminishing returns. In fact, as more customers switch to real-time pricing, those who switched previously are likely to be harmed, as the savings derived from their demand response is decreased (Borenstein, 2005; Borenstein and Holland, 2005).

It is also important to evaluate to what extent consumers can effectively monitor and react to such detailed information. Dynamic, real-time pricing is certainly beneficial to consumers who have automated equipment that can be programmed to adjust operation based on current market prices or for customers who contract demand-management services from an aggregator. But for many customers high frequency price data just becomes noise, which results in more confusion than enlightenment. Thus, policy makers must evaluate what information is most pertinent and effective in their particular context, since consumers typically prefer simple, transparent, and predictable pricing together with auxiliary signaling (Sioshansi, 2008; Wolak, 2011; Dütschke and Paetz, 2013).

For example, a power system with a high share of generation derived from wind power, whose production pattern is irregular, dynamic pricing is most likely the best way to promote demand response. On the other hand, for a power system with high share of solar generation, whose daily production pattern is very consistent from day to day, the main concern is the signaling of the daily load pattern, which can be adequately addressed by fixed time-of-use rates, since this provides clear and predictable pricing to orient consumers' consumption behavior.

In closing, a move toward a better delineation of products, so as to become more adherent to the power system's marginal costs, is desirable to improve demand response. However, such changes must be tempered by practical considerations which often limit the effectiveness of better price signaling.

4.1 Example: Brazil's tariff flag mechanism

Brazil provides an interesting case study. Most of the electricity in Brazil is derived from hydroelectric power plants, whose production capacity can vary substantially due to volatile rainfall patterns. Reservoir storage capacity is more than sufficient to smooth-out seasonal variations, so the relevant variability occurs not in the time frame of hours or days, but rather over the span of months or even years.

In this context dynamic pricing is necessary to adjust to the irregular hydro conditions, but there is no need for high frequency (daily or hourly) price changes.

To meet this need, Brazil adopted the Tariff Flag Mechanism for retail customers, which consists of a monthly price adjustment based on projected supply and demand conditions for the following month (Lima et al., 2017).

The Tariff Flag for the coming month is announced during the last week of each month and signaled by colored flags:

- "Green" in periods of normal supply conditions;
- "Yellow" in periods of medium supply scarcity;
- "Red 1 in periods of high supply scarcity; and
- "Red 2" in period of extremely high scarcity.

Although the signaling is simple, the methodology to determine the Tariff Flag each month is actually quite complex. It is based on an algorithm that takes into account:

- Projected hydro generation relative to their contractual supply obligations and
- Projected average spot market price for the coming month.

Hydropower generators have no control over their supply in Brazil, since dispatch is based on a centralized-mathematical model that determines the current opportunity cost of water stored in the hydro reservoirs. Thus, it is easy to understand why hydro generators revenues are very susceptible to changing hydro conditions.

Hydro conditions also are of concern to consumers because many of the existing long-term contracts pass the hydro risk on to consumers. The indicator utilized for hydro generation relative to their contractual obligations is the hydro Generation Scaling Factor (GSF), which is the ratio of hydro generation relative to its "firm energy." The firm energy (*Garantia Física*) is the amount of energy that the hydropower plants are deemed capable of supplying at a given reliably level. The firm energy attributed to each hydropower plant is determined by the regulatory agency and serves as the limit for long-term contracting.

A GSF above 100% indicates hydropower plants are producing a surplus; and a GSF below 100% indicates hydropower plants are possibly producing less than the amounts committed in long-term contracts. Thus, when GSF levels are low, hydropower plants need to acquire additional energy in the spot market to honor their long-term contractual obligations.

The Tariff Flag Mechanism has an algorithm that defines price bands at which each Tariff Flag is to be triggered. The bands are automatically adjusted each month based on the projected hydro GSF. The System Operator then simulates the operation of the coming month to obtain the expected average spot market price. The Tariff Flag is then determined based on the price band in which the average spot market price falls into (ANEEL, 1999a).

Figs. 20.2 and 20.3 show the GSF, the corresponding price bands of each Tariff Flag, and the expected average spot market price for each month in the period spanning from November 2017 to December 2019.

Notice that the Tariff Flag is "Green"—regardless of the projected spot price—in periods in which GSF is high (Jan/18–Apr/18 and Jan/19–Apr/19). Conversely, when GSF is low (Nov/17, Jun/18-Oct/18 and Jul/19–Sep/19), the Tariff Flag price bands all shift downward, resulting in a situation in which the "Red 2" price band dominates the spectrum.

This system is relatively simple for consumers to understand, providing timely warnings for consumers to adjust their consumption to current market conditions. It also helps load-serving companies' cash flow, by better synchronizing costs and revenues, since tariffs are only readjusted once per year.

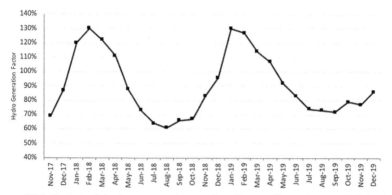

Figure 20.2 Projected hydro generation scaling factor. *(Data Source: ANEEL, 2020. Bandeiras Tarifárias: Relatório do Acionamento (Superintendência de Gestão Tarifária). Brasília, Agência Nacional de Energia Elétrica (ANEEL, 2020). Elaborated by the author.)*

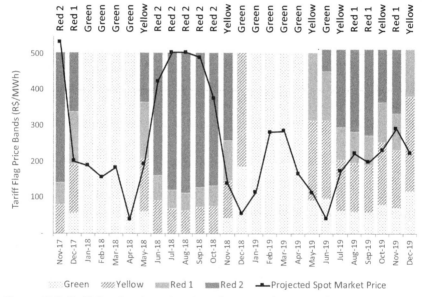

Figure 20.3 Tariff flag bands and projected spot market price. *(Data Source: ANEEL, 2020. Bandeiras Tarifárias: Relatório do Acionamento (Superintendência de Gestão Tarifária). Brasília, Agência Nacional de Energia Elétrica (ANEEL, 2020). Elaborated by the author.)*

5. Market design

To induce further demand response—beyond that which can be achieved by consumers' spontaneous reactions to market prices—one must resort to market design changes that offer additional compensation to consumers that provide the desired demand response.

From a systemic perspective, it does not matter if supply and demand balance is achieved by a voluntary adjustment of generation or consumption. If consumers have well-defined consumption patterns, one could consider a reduction in 1 kW-hour of the consumer's usual demand as equivalent to the supply of an additional kilowatt-hour from a generator. Thus, one could adopt a market mechanism based not only on supply bids from generators but also on demand-response bids from consumers.

The United States' Federal Energy Regulatory Commission (FERC) adopted this idea in Order 745 that established a "compensation approach" for demand response resources (FERC, 2011). The Commission argued that enabling demand-response bids in wholesale energy markets provides three major benefits:

- Minimizes costs;
- Mitigates generators market power; and
- Supports system reliability and resource adequacy.

FERC Order 745 paved the way for the implementation of demand-response bids in wholesale markets across the United States, such as the Economic Load Response in PJM's Open Access Transmission Tariff, and CAISO's Demand Response and Load Participation. Thus, today consumers can submit demand-reduction bids for the day-ahead, real-time, and ancillary services markets (PJM, 2020; CAISO, 2020). Chapter 10 by Helman provides an overview.

The determination of the amount of compensation that should be paid to customers who provide demand-response bids has been a contentious issue. FERC Order 745 determines that "demand response resource must be compensated for the service it provides to the energy market at the spot market price for energy, referred to as the locational marginal price (LMP)."

Many objections to this policy have been raised by specialists (Ruff, 2002a,b; Bushnell et al., 2009; Hogan, 2010; Brown and Sappington, 2016). Treating demand-response bids as resources akin to supply bids can be misleading, since it involves "paying anybody for something they might have bought but didn't," which Ruff (2002a) refers to as a "double-payment," since the customer not only obtains the savings related to the demand reduction but also the revenue from the sale of the energy to another consumer.

Such a policy can lead to overcompensation of consumers, since it is difficult to evaluate what the consumers' normal consumption would be if they were not providing demand-response services. A matter that becomes even more complex to evaluate as consumers opt to invest in their own generation.

This policy can also lead to a "crowding out" of other demand response approaches, such as demand response that would spontaneously occur due to market price variations.

The introduction of this policy also introduces a transient problem, since it tends to reduce spot market prices, frustrating incumbent generators' expectations. Baseload power plants rely on higher spot market pricing during a certain percentage of the time to recover their sunk costs in capacity investments. Thus, changes introduced by demand response can threaten their return on investments.

Paying consumers to reduce their regular consumption at the current spot market price can be more cost-effective than increasing generation in many cases, but not always. It depends on the supply and demand characteristics at the margin.

The demand-response bid will provide savings when supply is price-inelastic. A demand-response bid results in a shift to the left of the demand curve. Typically, there will be many demand-response bids at different prices, which have the effect of making the demand curve flatter. To facilitate the analysis, Fig. 20.4 shows the effect of the introduction of a large number of demand-response bids at single price equal to p_1. The incorporation of demand-response bids results in a reduction of the equilibrium price from p_0 (if there were no demand-response bids) to p_1. The consumers who submitted the winning demand-response bids are paid

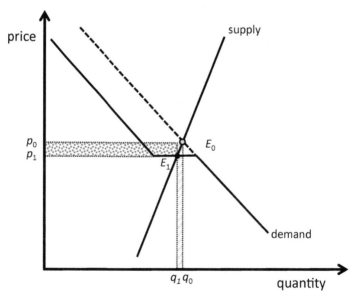

Figure 20.4 Effect of a demand-response bid with low price-elasticity of supply.

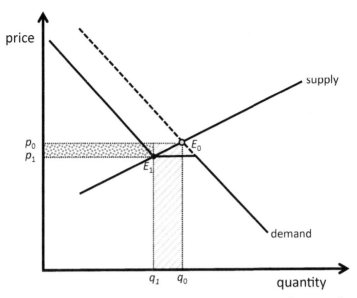

Figure 20.5 Effect of a demand-response bid with high price elasticity of supply.

p_1 for the demand reduction from q_0 to q_1. The cost of this compensation must be paid by the remaining customers. In this case, the savings derived from the reduction of the spot market price (equivalent to the dotted area) are greater than the cost of the compensation paid for the demand–response bids (equivalent to the hatched area), thus consumers obtain a net gain.

However, as shown in Fig. 20.5, this would not be the case if supply curve were more elastic (flatter). In this case, the savings obtained from the lower market price (equal to the dotted area) would be smaller than the demand response compensation (equal to the hatched area). Since this cost has to be picked up by a smaller customer base, FERC refers to this as the "billing unit effect."

One way to avoid the risk of promoting inefficient demand response would be to limit the consumers' compensation to the net welfare gain from the transaction. However, FERC opted not to reduce the compensation offered to consumers, but rather to restrict the incorporation of demand-response bids in wholesale markets to situations in which demand response is expected to provide net benefits.

FERC Order 745 ensures this by introducing a "net benefit test." The test identifies the conditions in which demand-response bids become cost-effective. Each month, the Regional Transmission Operators must apply the net benefit test, based on expected supply and demand conditions, to identify the specific conditions at which demand-response bids are expected to provide net benefits.

This entails identifying situations in which market equilibrium is expected to occur in the inelastic portion of the supply curve, so the cost of demand reduction is smaller than the alternative of acquiring more generation supply.[1]

The net benefit test is burdensome, but the biggest hurdle in implementing demand-response bids is the determination of the consumer's "baseline load." Unlike generators which must produce the energy they bid in wholesale markets, consumers that submit demand-response bids only need to avoid consuming the energy they supposedly were expected to consume. This gives rise to a moral hazard problem: what is to stop a consumer from claiming they would consume more, so as to submit even more demand-response bids?

To avoid this problem, wholesale markets have developed intricate rules to determine the consumers' baseline load. Several approaches can be used (Goldberg and Agnew, 2013):

- Historical load record based on "day matching";
- Load projections based on statistical analysis (regression models);
- Control group comparison (clustering approach); and
- Self-reported baselines combined with an incentive scheme.

The most common approach is based on the consumer's historical load record. This is usually based on a rolling average of the consumer's load, adjusted for weekends, holidays, and other extraordinary conditions. The main drawback of this approach is that it is backward looking and will result in a biased baseline whenever there are changes in the underlying conditions.

Projected loads, based on statistical analysis, seek to provide more precise estimates considering expected conditions such as weather. The fragility of this method is that it requires a relatively long set of historical data, which entails: (i) using data that may be very outdated and (ii) a model specification which may not adequately capture all the relevant variables.

Control group comparisons seeks to avert the drawbacks of the two previous approaches by defining the baseload *ex post*, based on the

---

[1] Another factor that could also hinder demand-response bids from being cost effective is the "rebound effect." When the system faces extended periods of high prices, demand-response bids provided by shifting demand from one-time interval to another may not be helpful because they do not result in a net demand reduction in the relevant time frame. Such demand shifting is only beneficial when consumption shifts from a period of higher-marginal cost to a period of lower-marginal cost.

consumption of comparable consumers during the relevant period in which the transaction took place. The weakness of this approach is how to determine who are the comparable consumers.

This approach will be used in Brazil in an "energy efficiency auction" to determine baseline consumption. The auction aims to reduce demand in Boa Vista, the capital of the only State that is still not connected to the national grid and that faces some of the highest generation costs in the country. The control group will be formed of randomly selected consumers from the pool of consumers recruited to participate of the program, classified by customer class and city zone (ANEEL, 2019b).

Finally, there is the possibility of resorting to self-reported baselines from consumers, combined with an incentive scheme to avoid the problem of "baseload inflation." The mechanism consists of recruiting a great number of participants, so that at any one occasion, only a portion of participants submitting demand-response bids are called upon to reduce loads. Those who are called upon to reduce loads receive payment for their load reduction relative to their self-declared baseline. The remaining participants, that are not called upon to reduce loads, must consume their declared baseline loads and are subject to penalties for any deviation. This penalty system is what provides consumers incentives to provide their best baseload projections (Muthirayan et al., 2020).

The search for different approaches to properly determine the baseline load reveals how pertinent and sensitive this particular issue is for the success of this approach.

Since this approach is centered in wholesale market transactions, aggregators are necessary for retail customers to participate. How the market of aggregation services will evolve is an open question (Fox-Penner, 2010). Will utilities continue to provide bundled services including generation and grid services under regulated rates ("energy service utilities") or will they evolve to become neutral platforms, divesting of any generation assets to provide only grid services ("smart integrators")? Will utilities dominate the landscape, or will independent energy service providers become increasingly popular? These are key questions that will depend on the interplay of mutating business models, regulatory initiatives, consumer behavioral trends, and technological innovations.

5.1 Example: Colombia's electricity market

Colombia's electric power dispatch is determined by bids submitted in the wholesale markets (*Mercado Mayorista de Energia*). Since 2015, aggregators (*comercializadores*) have been allowed to participate by submitting demand-response bids in the day-ahead market (CREG, 2015). Demand response participants must be willing to submit bids for all 24 h of every day over the entire year.

Currently, demand-response bids are only considered in periods of "critical conditions," when spot prices surpass the "activation scarcity price" (*precio de escasez de activación*) by at least 8% (CREG, 2017).

The bids specify the price and the amount of energy they are willing to curtail in the next 24-h dispatch cycle. Aggregators who are called upon to curtail demand receive payment equal to the difference between the spot market price and the scarcity price. These rules ensure that demand-response bids will always provide net welfare gains.

To participate of the program, aggregators must:
- Identify the nodes at demand response resources are connected to the grid;
- Meet the required metering requirements;
- Inform consumers of the conditions for participation in the demand-response bidding process;
- Provide direct control to remotely shut down the supply to customers that offer "voluntary demand curtailment" (*Demanda Deconectable Voluntaria*) or provide managed load reduction coordinated by the aggregator; and
- Request the System Administrator (*Administrador del Sistema de Intercambios Comerciales*) to calculate the baseline load based on the information available.

Based on bids submitted by generators and aggregators, the System Operator determines the predispatch schedule for the following day, informing the winning bidders of their commitments and the respective spot market prices for each time interval of the following day.

The computation of baseline load varies depending on how their demand is metered. For consumers with "direct independent metering" that is readily observable by the System Administrator, baseline load is determined by the average consumption during the previous 105 days, differentiating weekdays (Monday through Saturday) from Sundays and holidays. Days in which the consumer was called upon to provide demand response are excluded from the sample.

For consumers whose demand response is managed by an aggregator, the baseline load is determined by projections obtained from statistical models that take into account the long-term demand trend and variation between the different days of the week.

Deviations of up to 5% of the demand-response bid quantities are tolerated, but deviations above this tolerance level are charged the difference. If the demand-response offers included in the predispatch schedule are redispatched in real-time operation due to unforeseen contingencies, they receive the same compensation provided to redispatched power plants.

The incorporation of demand-response bids in the Colombian market have already provided benefits. For example, in March and April 2016, when the system faced severe supply scarcity, demand-response bids reduced demand by around 3%. With more variable generation being aggregated to the system, demand-response bids should become even more relevant. Simulations suggest increased savings in the coming years (Rodas-Gallego and Mejía-Giraldo, 2020).

6. Regulation

Finally, the third strategy to spur demand response looks beyond marginal costs, providing compensation to consumers for demand responses that avoid fixed costs associated to capacity expansion. This strategy involves tailor-made solutions for particular systems, in particular conditions. These demand-response programs may be designed to avoid generation, transmission, or distribution capacity expansions.

These programs are analogous to overbooking programs in the airline industry. Overbooking programs seek to entice ticket-bearing passengers to voluntarily forego their scheduled flight in those infrequent occasions when all passengers of an oversubscribed flight show up at the boarding gate (Sioshansi and Vojdani, 2001).

To illustrate the cost-effectiveness of a particular demand-response program, consider a generic example of program to reduce critical-peak demand, so as to avoid investment in additional power plant capacity.

The two graphs in Fig. 20.6 show how one can determine the optimal installed capacity of three different types power plants (Technologies 1, 2, and 3) to meet a particular load and how much should be met by a demand-response program.

The top graph reflects the total costs of each technology when operating at different capacity factors. The point at which each line intercepts the vertical axis represents the fixed-annualized costs of capacity of the respective technology, while the line's inclination represents their respective marginal costs. The inner contour of the intersecting lines indicates the least cost of provision of electricity at the different capacity factors.

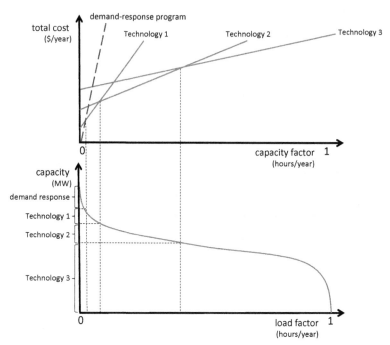

Figure 20.6 Optimal configuration to meet a particular load duration curve.

The bottom graph shows the load duration curve, which represents the percentage of time a particular capacity level is required to meet demand. The thin-dotted vertical lines connecting the two graphs indicate the segments of the load duration curve that are supplied at least cost by each of the three technologies. Technology 1, that has the lowest fixed costs (albeit highest operational costs), is most cost-effective to supply the left portion load duration curve that is required relatively infrequently (peak load). Technology 3, that has the lowest variable costs (albeit highest fixed costs), is the most cost-effective to meet the right portion load duration curve that is required almost constantly (base load). And Technology 2 is most cost-effective at meeting the middle portion of the load curve. The thin-dotted horizontal lines indicate the optimal amount installed capacity of each technology to service the duration curve (as shown on vertical axis of the bottom graph).

Given a particular cost for demand-response program, one can identify the segment of the load curve that is optimally met by consumer curtailment. The total cost of the demand-response program is shown on the top graph (dark-dashed line). Given that this program does not involve any fixed costs it intercepts the vertical axis at zero. Notice that cost per kilowatt-hour of the demand-response program is higher than that of all three power plants (reflected by the line's higher inclination). Nevertheless, the demand-response program is cost-effective because it averts the fixed costs associated to acquiring more capacity of Technology 1 that would be needed to meet that critical peak demand that seldom occurs.

In this example, the cost of the demand-response program is offset by the savings form annualized-fixed costs of additional generation capacity, but a demand-response program could also be justified by the savings achieved from averting annualized-fixed costs of additional transmission or distribution capacity.

Thus, this strategy further expands the potential of demand response by providing compensation to consumers that avert capacity constraints at a cost that is lower than the annualized-fixed cost of such capacity.

Demand-response programs generally involve the targeting of a few critical-peak periods that are expected to occur during the year. During these periods, consumers who opt to participate of the program are incentivized to curtail demand, either by:
- Being charged higher prices (critical-peak pricing) in exchange for a yearly compensation or
- Receiving rebates for demand reductions relative to expected consumption (critical-peak rebates).

The main challenge in the deployment of this strategy is that the cost effectiveness of each program generally has to be evaluated on a case-by-case basis under regulatory oversight, making it impossible to draw up universal policies.

Nevertheless, historically this has been the demand response strategy that has been most explored. Because the value proposition of this strategy is not limited to marginal generation cost savings, but rather, from the savings of fixed costs associated to installed capacity, it can provide the biggest bang for the kilowatt-hour of demand response.

The effectiveness of demand-response programs is observed in empirical studies. Recent estimates indicate that the long-run demand response elasticity achieved with demand-response programs in the United States range from 0.4 to 0.6. These demand-response elasticities are much higher than the typical price elasticities of demand, which are in the range of -0.05 to -0.2 (Faruqui and Sergici, 2010; Stewart, 2020), indicating that demand-response programs can greatly augment the provision of demand-side flexibility.

6.1 Example: Brazil's 2001 electricity demand-response program

This strategy can also be employed to respond to periods of supply scarcity, such as in prolonged droughts in a system that relies primarily on hydropower generation. In May 2001, Brazil's power system faced a dire situation: it was the end of the rainy season (when hydro reservoirs should be near full capacity) but storage levels were below 30%. With the limited resources available, the power system would not be able to meet electricity consumption until the following rainy season. It looked like rolling blackouts would have to be adopted, since new generation capacity could not be built fast enough to meet the system demand.

To mitigate the effects of the energy shortage, the government adopted an aggressive rationing program that may be one of the world's largest demand-response programs ever deployed.

A major campaign was launched to build awareness and help educate consumers of ways they could save energy. Retail customers were attributed quotas roughly equivalent to 80% of their previous year's consumption level, with bonuses given to those who consumed less and penalties to those who consumed more. Large industrial and commercial customers were also attributed quotas, although they varied by sector and economic activity. Large consumers were also given the opportunity to trade quotas, which allowed firms that could curtail electricity consumption at lower costs to sell their surplus quotas to those willing to bid the highest price (Maurer et al., 2005).

The demand-response program lasted from June 2001 to February 2002 and was highly successful. Rolling blackouts were averted and the economic impacts of the energy shortage were minimized.

There was widespread participation, with an average consumption drop of approximately 20% in the initial months, with even higher reductions among residential and commercial consumers, as shown in Fig. 20.7.

Considering that annual consumption, shown in Fig. 20.8, was growing at a rate of 4.8% per year prior to the implementation of the demand-response program, one can infer that consumption was reduced by over 40 Terawatt-hours (TWh) over the 9 months the program was in place (and much more thereafter, since demand did not immediately raise to previous levels after the program ended—indicating that it led to lasting efficiency gains).

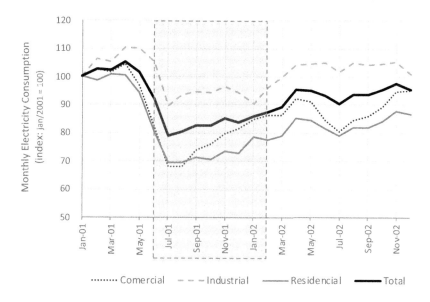

Figure 20.7 Monthly electricity consumption in Brazil by customer class. *(Data Source: ANEEL. Elaborated by author.)*

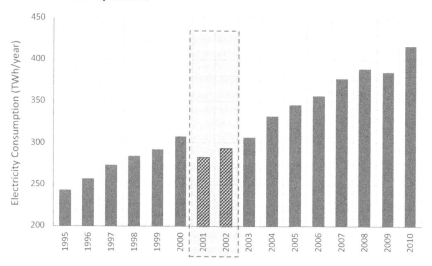

Figure 20.8 Annual electricity consumption in Brazil. *(Data Source: Empresa de Pesquisa Energética. Elaborated by author.)*

7. Conclusion

Flexibility derived from demand response shows great potential to meet power system imbalances. Some of these resources have already been tapped into, but in the coming years much more demand-side response will become viable due to technological innovations that facilitate the coordination of dispersed energy resources.

This chapter explored three major strategies to promote demand response, which are:

- Better delineating market products so as to elicit more precise pricing that reflect power system conditions;
- Adapting markets to enable consumers' resources to actively participate in the wholesale market by means of demand-response bids;
- Designing specific demand-response programs to curtail demand during periods the system approaches capacity constraints.

Each strategy seeks to obtain welfare gains by different means.

The first approach seeks to increase welfare by allowing consumers to adjust their electricity consumption to point where their marginal utility equals the systems' marginal costs.

The second approach seeks Pareto improvements by prompting consumers to adjust their demand in exchange for a compensation that is lower than the cost of resorting to additional supply.

The third approach derives cost savings by compensating consumers for demand curtailment in critical periods of high demand—or of scarce supply from variable generation—so as to avert the cost of capital investments in additional generation, transmission, or distribution capacity.

All strategies can provide sizable savings. While the third strategy— demand-response programs—tends to provide higher returns per kilowatt-hour, it is only deployed a few hours per year. The savings from the other two strategies—better product delineation and market design to incorporate demand response—tend to be smaller, but can be achieved on a daily basis, which can add up to large sums over time.

The three strategies have been pursued to some extent in electricity markets across the globe, but all three can—and should—be explored much more intensely in the coming years.

Bibliography

Alstone, P., et al., 2017. 2025 California Demand Response Potential Study – Charting California's Demand Response Future (Final Report on Phase 2 Results). Lawrence Berkeley National Laboratory for the California Public Utilities Commission, Berkley.

ANEEL, 2019a. Nota Técnica 25/2019-SGT/SRG/SRM/ANEEL. Agência Nacional de Energia Elétrica, Brasília.

ANEEL, 2019b. Nota Técnica no 0074/2019 – SPE/ANEEL. Agência Nacional de Energia Elétrica, Brasília.

ANEEL, 2020. Bandeiras Tarifárias: Relatório do Acionamento (Superintendência de Gestão Tarifária). Agência Nacional de Energia Elétrica, Brasília.

Borenstein, S., 2005. The long-run efficiency of real-time electricity pricing. Energy J. 26 (3), 1—24.

Borenstein, S., Holland, S., 2005. On the efficiency of competitive electricity markets with time-invariant retail prices. Rand J. Econ. 36 (3), 469—493.

Bushnell, J., Mansur, E., 2005. Consumption under noisy price signals: a study of electricity retail rate deregulation in San Diego. J. Ind. Econ. 53 (4), 493—513.

Bushnell, J., Hobbs, B., Wolak, F., 2009. When it comes to demand response, is FERC its own worst enemy? Electr. J. 22 (8), 9—18.

Brown, D., Sappington, D., 2016. On the optimal design of demand response policies. J. Regul. Econ. 49 (3), 265—291.

CAISO, 2020. Demand Response and Load Participation: Demand Response and Load Participation. California Independent System Operator, Folsom. Available at: http://www.caiso.com/participate/Pages/Load/Default.aspx.

CREG, 2015. Resolución 011/2015. Bogotá: Comisión de Regulación de Energia y Gas.

CREG, 2017. Resolución 140/2017. Bogotá: Comisión de Regulación de Energia y Gas.

Dütschke, E., Paetz, A., 2013. Dynamic electricity pricing — which programs do consumers prefer? Energy Pol. 59, 226—234.

Eid, C., Codani, P., Perez, Y., Reneses, J., Hakvoort, R., 2016. Managing electric flexibility from Distributed Energy Resources: a review of incentives for market design. Renew. Sustain. Energy Rev. 64, 237—247.

EIA, 2019. Annual Electric Power Industry Report (Form EIA-861). Energy Information Agency, Washington. Available at: https://www.eia.gov/electricity/data/eia861/.

EPE, 2019. Resposta da Demanda: Conceitos, Aspectos Regulatórios e Planejamento Energético (Nota Técnica EPE-DEE-NT-022/2019-r0. Empresa de Planejamento Energético — Ministério de Minas e Energia, Brasília.

Faruqui, A., Sergici, S., 2010. Household response to dynamic pricing of electricity: a survey of 15 experiments. J. Regul. Econ. 38, 193—225.

FERC, 2011. FERC Order 745. Demand Response Compensation in Organized Wholesale Energy Markets (Docket Number RM10-17-00). Federal Energy Regulatory Commission, Washington, DC.

Feuerriegel, S., Neumann, D., 2016. Integration scenarios of demand response into electricity markets: load shifting, financial savings and policy implications. Energy Pol. 96, 231—240.

Fox-Penner, P., 2010. Smart Power: Climate Change, the Smart Grid, and the Future of the Electric Utility. Island Press, Washington.

Goldberg, M., Agnew, F., 2013. Measurement and Verification for Demand Response (Prepared for the National Forum on the National Action Plan on Demand Response: Measurement and Verification Working Group). Department of Energy/Federal Energy Regulatory Commission, Washington.

Hogan, W., 2010. Demand response compensation, net benefits and cost allocation: comments. Electr. J. 23 (9), 19—24.

Ito, K., 2014. Do consumers respond to marginal or average price? Evidence from nonlinear electricity pricing. Am. Econ. Rev. 104 (2), 537—563.

Lima, D., Perez, R., Clemente, G., 2017. A comprehensive analysis of the demand response program proposed in Brazil based on the Tariff Flags mechanism. Elec. Power Syst. Res. 144, 1—12.

Maurer, L., Pereira, M., Rosenblatt, J., 2005. Implementing Power Rationing in a Sensible Way: Lessons Learned and International Best Practices. The World Bank — Energy Sector Management Assistance Program, Washington.

Muthirayan, D., et al., 2020. A minimal incentive-based demand response program with self-reported baseline mechanism. IEEE Trans. Smart Grid 11 (3), 2195—2207.

PJM, 2020. Open Access Transmission Tariff — Attachment K — Appendix Section 3.3A Economic Load Response Participants. PJM Regional Transmission Organization, Audubon. Available at: https://agreements.pjm.com/oatt/4483.

Rodas-Gallego, E., Mejía-Giraldo, D., 2020. Market-based impact of a demand response program in the Colombian power market. IEEE Latin Am. Trans. 18 (3), 537—544.

Ruff, L., 2002a. Demand response: reality versus "resource". Electr. J. 15 (10), 10—23.

Ruff, L., 2002b. Economic Principles of Demand Response in Electricity. Edison Electric Institute, Washington.

Sioshansi, F., Vojdani, A., 2001. What could possibly be better than real-time pricing? Demand response. Electr. J. 14 (5), 39—50.

Sioshansi, F. (Ed.), 2008. Competitive Electricity Markets: Design, Implementation, Performance. Elsevier, Amsterdam.

Stewart, J., 2020. Utility customer supply of demand response capacity. Energy J. 41 (4), 99—121.

Wolak, F., 2011. Do residential customers respond to hourly prices? Evidence from a dynamic pricing experiment. Am. Econ. Rev. Pap. Proc. 101, 83—87.

CHAPTER 21

Energy communities and flexible demand

David Robinson
Oxford Institute for Energy Studies, Madrid, Spain

1. Introduction

Can energy communities, which are designed to support "demand aggregation," contribute to flexibility in an electricity system that is increasingly challenged by intermittent generation? This chapter argues that they can and should do so, in their own interest and in the wider social interest. It also identifies economic policies, as well as new business and technology platforms, that are necessary.

A chapter from a previous book[1] in this series made the case for providing efficient price signals to consumers so that their decisions "behind the meter" (BTM) would contribute positively to the energy transition. The vision was that consumers would be able to use the BTM assets to determine their own energy mix and their reliance on system, while also enhancing overall system security and lowering system costs, especially by providing flexibility that facilitates integration of intermittent renewables. For that vision to be realized, consumers should receive efficient price signals. However, in most countries, price signals are often inefficient and not aligned with the interest of the system as a whole. Hence the need to reform markets, regulations, and fiscal policy, as explained in the previous book.

This chapter extends the argument in two ways:

- First, it addresses the difficulty of engaging consumers in providing demand flexibility. Efficient price signals are a necessary condition, but often not sufficient to motivate consumer engagement. The chapter discusses and illustrates how new technologies and business models can facilitate consumer engagement through aggregation.

---

[1] David Robinson, "What market design, fiscal policy and network regulations are compatible with efficient BTM investments?", *Behind and Beyond the Meter: Digitalization, Aggregation, Optimization, Monetization*, Edited by Fereidoon P. Sioshansi, Academic Press, February 2020.

Variable Generation, Flexible Demand
ISBN 978-0-12-823810-3
https://doi.org/10.1016/B978-0-12-823810-3.00022-4
477

- Second, drawing on recent European Union legislation and experience, it asks whether and under what conditions energy communities can support the energy transition, in particular by providing demand flexibility that lowers costs for the community and for the rest of society. The chapter is organized as follows:
- Section 2 summarizes the case for providing efficient price signals to guide investment in distributed energy resources behind the consumer's meter, and what reforms to consider, drawing on experience in the European Union (EU), especially Spain. This section summarizes the arguments from the previous book in this series.
- Section 3 explains how technology and business models involving aggregation can facilitate demand flexibility, and are a necessary compliment to efficient price signals.
- Section 4 analyzes the potential for energy communities, as a form of aggregator, to contribute to the energy transition and address the related challenges.
- The concluding section summarizes the argument and draws implications for public policy and community energy strategy.

2. Efficiency and price signals

This section covers three topics: the link between prices and efficiency objectives; the importance of sending efficient prices to consumers in the decarbonized electricity system of the future; and reforms needed to provide efficient signals to consumers.

2.1 Efficiency objectives

A key feature of electricity sector liberalization is that prices should reflect the efficient economic cost of supplying electricity.[2] Liberalization, initially in England and Wales and later throughout the European Union, pursued three specific efficiency objectives.[3]

---

[2] Joskow, P. (2006). Introduction to electricity sector liberalization: lessons learned from cross-country studies. In Sioshansi, F. and W. Pfaffenberger (Eds.). Electricity market reform: An international perspective, pp. 1—32.

[3] In addition to these three efficiency objectives, there are two others which were not the focus of liberalization in the 1990s: social efficiency, when goods and services are optimally distributed within an economy, also taking account of externalities; and X-efficiency, when a firm has an incentive to produce maximum output with a given amount of input. https://quickonomics.com/five-types-of-economic-efficiency/.

- *Allocative efficiency*: Allocative efficiency occurs when all goods and services are distributed according to consumer preferences; prices equal marginal cost of production. During liberalization, governments dropped most price controls on electricity. This reduced subsidies and cross-subsidies as generators and retail suppliers tried to avoid selling at a loss. As a result, prices tend to reflect marginal costs.
- *Productive efficiency*: Productive efficiency occurs when the optimal combination of inputs results in the maximum output at minimal costs. Energy-only markets were designed to provide incentives to replicate the existing cost-based merit order dispatch, using a price-based dispatch to ensure that the cheapest generator available at any time was dispatched.
- *Dynamic efficiency*: Dynamic efficiency occurs over time, as investment in innovation and new technologies reduce production costs. A key aim of liberalization was to transfer investment risk to investors, and thereby sharpen the incentives to make efficient decisions. Previously, investment decisions were typically made by government or regulated companies, with the costs and risks passed to consumers. This often led to expensive and inefficient decisions, the costs of which were passed to consumers. Liberalization aimed to stop this practice.

It was never easy to design electricity markets for competitive activities and regulation for natural monopolies in a way that provided efficient price signals. The record is especially mixed for dynamic efficiency, with governments having largely taken control over the mix and total capacity of generation. They have done this especially to accelerate the process of decarbonization, by promoting investment in renewable energy through Feed-In Tariffs, quotas, and more recently through central government auctions. They have also intervened to provide financial support for firm and flexible plants to back up intermittent renewables and to ensure resource adequacy. Most of these payments occur outside of the energy market. The result is that most investment decisions are driven by governments, and that the energy markets do not provide efficient signals. Nevertheless, European legislation for the sector is based on the liberalization model. National policymakers are expected to follow the core principles of liberalization, in particular the idea that competitive energy markets should be the basis for organizing activities that are not natural monopolies. They should do their best to meet efficiency objectives through the competitive price mechanism and eliminate price distortions. It is particularly important to provide efficient signals to encourage flexibility in the modern electricity system.

As described in Part One of this volume, it is now widely understood that the decarbonized electricity system in most countries will rely primarily on intermittent renewable electricity (wind and solar), that a significant share of the energy resources will be distributed and that digitalization will facilitate coordination of these resources. It is also understood that flexibility is central to managing this system, primarily to facilitate the integration of the intermittent renewables and to help manage congestion on the network. Flexibility can be provided by generation, storage, interconnection, and demand management. The first three offer supply-side flexibility; they have always been available to electricity systems, although storage less so than generation and interconnection. However, until recently, demand was generally considered unpredictable and uncontrollable; supply was planned and operated to follow and adapt to variations in demand, as explained in Chapter 5. Today, with the growth of intermittent renewables, generation is increasingly unpredictable and cannot be dispatched. Largely as a result of ITC developments, demand-side management has become a potentially important source of flexibility for the system, both to integrate renewables and to manage local network congestion.

Demand-side flexibility refers here to all measures taken by consumers or their representatives that affect the timing and the amount of their net demand for electricity from the system.[4] This flexibility can be delivered from management of self-generation, storage, and active demand response, as well as by using smart electrical equipment and optimization software. The consumer may manage all of these resources directly or allow remote management by aggregators that use platforms to integrate and optimize resources owned by many consumers.

2.2 Importance of consumers and demand-side flexibility

When the goal was gradually to reduce emissions, this could be achieved by raising the penetration of renewable power in the system, with little or no active consumer participation. In Spain, for instance, renewable power now accounts for over 40% of electricity generation and almost none of that comes from distributed generation involving rooftop solar PV; consumers have noticed little difference, apart from rising levies (in tariffs) to pay for it. Eventually, energy consumers will have to become more active since electricity only accounts for about 20% of primary energy use and further

---

[4] See, for instance, Chapters 5 and 7 (both by Fereidoon Sioshansi) in this book.

decarbonization will directly affect personal investment decisions over mobility (EVs), heating and cooling (heat pumps, residual water recycling), buildings (insulation), and more. With the EU's adoption of the goal of carbon neutrality by 2050, it is now clear that strong citizen support and participation in the energy transition is essential to achieving that goal. One key role for citizens as consumers is to provide flexibility from their distributed energy resources (DERs).

In this new context, demand flexibility is not just about consuming more or less energy. As explained in the previous volume in this series, it is also about managing a growing number of resources that are physically behind the consumer's meter and of shifting load to manage the variability of renewable energy and congestion on the network. For these resources to be deployed in a way that benefits the consumer and society as a whole, consumers and their representatives (e.g., aggregators, retail suppliers) must receive efficient economic signals that encourage flexibility.

2.3 Reforms to provide efficient price signals

There are a number of policy reforms required to provide efficient price signals to consumers behind the meter. Drawing on European experience and in particular the case of Spain, here is a summary of the main reforms proposed in the previous book.

- *Fiscal reform:* Fiscal reform is needed to eliminate incentives to generate one's own electricity simply to avoid paying taxes and policy-related levies that are collected through the variable component of regulated electricity tariffs. Reform is also required to ensure that electricity and fossil fuels each internalize the environmental externalities they cause; in many cases, electricity does bear these environmental costs but fossil fuels do not. Financing policy costs through variable energy charges can also discourage demand flexibility by distorting marginal cost signals. Failure to reform fiscal policy can potentially raise system costs, enable the wealthy to benefit, leave others to bear the fiscal burden, and slow electrification, which is recognized as being key to decarbonization and improved energy efficiency. Eventually, governments will need to fix these problems. It would be far better to fix them now, rather than later when it will be politically even more difficult. The main proposal is to finance support for renewable energy and other public goods through the government budget, not through levies included in the electricity tariffs. Any remaining policy costs not financed through the government budget should be recovered through a fixed charge on consumers, not through a variable charge.

- *Access tariffs design:* Regulators should design network access tariffs to provide incentives to invest in, and operate, distributed energy resources that contribute to lowering the overall costs of the system as well as the costs of the individual consumer owning the DER. There are two extreme approaches to access tariff design that have recently been debated in Spain: one recovers virtually all network costs through one or more fixed charges—per kW contracted, possibly differentiating between peak and off-peak—and the other through variable charges per kWh consumed, with the variable charge depending on the time of consumption.[5] Both are workable and can offer signals to use the system when it is uncongested, while fully recovering the network costs. The fixed charge approach—with very low or no variable component—is designed primarily to kick-start consumer investment in electrification since it encourages electricity use at all times. The variable charges approach—with little or no fixed charge—is designed to encourage efficient use of distribution networks, especially using them when they are uncongested. A dynamic version of the variable charges approach would involve real-time pricing for network access. In future, the efficiency incentives in the variable charging approach will become increasingly important.

- *Wholesale market access:* Today, wholesale electricity markets reflect the economics and technology of the last century. They were designed for a small number of large, mainly thermal, generation stations with significant variable costs. Now, with the growth of intermittent renewable generation, with zero marginal costs, and the development of distributed energy resources, markets need to be reformed to enable effective competition among all resources. Governments and regulators should ensure that consumers and their representatives can participate in all wholesale markets for energy and flexibility services, especially the flexibility markets where margins are typically higher. The products traded in these markets should reflect the new system needs and recognize what services can be provided by the full range of providers—not just by large generation sets.

---

[5] There are many interesting models being debated. For instance, see Chapter 12 in this book by de Vries and Doorman in this book; they develop the idea of capacity reservation. Also see a recent paper by Ryan Hledik and Peter Fox-Penner of The Brattle Group on "FixedBill+", https://www.brattle.com/news-and-knowledge/news/working-paper-highlights-the-benefits-of-new-approach-to-electricity-fixed-billing.

- *Local markets for congestion.* The development of intermittent generation and distributed energy resources increases the complexity of managing networks, especially at the distribution level. Where flows used to be predictable and unidirectional—from large-scale generation to consumers—they are now multidirectional and much less predictable. Fortunately, distributed energy resources offer demand-side flexibility to assist in the management of distribution networks. Establishing local congestion markets is an important step in managing those networks. These markets should be designed explicitly to encourage participation of distributed energy resources. The distribution system operator (DSO) will typically be the buyer of flexibility services. Energy consumers, either individually or through aggregators and energy communities, will compete with other resources to supply flexibility to the DSO. The DSO must ensure fair access to these markets. In particular, where the DSO is owned by a vertically integrated utility, regulation must monitor and prohibit conflicts of interest of two kinds. The first is the potential for the DSO to favor its own distribution network business as the provider of flexibility when there are better alternatives—such as demand-side flexibility. The second is the potential for the DSO to favor an affiliated company—an aggregator or retail supplier—over unaffiliated providers of flexibility services.
- *Reforming entire electricity systems:* Electricity markets are no longer fit for purpose; they were designed in the last century for a different combination of technologies and economics. Policymakers should recognize the need to think about reforming entire electricity systems, starting with a new energy market design. For example, Keay and Robinson have recommended a two-market approach,[6] rather than partial reforms that address only some of the problems of energy-only markets. Following this approach, consumers are given the choice between

---

[6] Malcolm Keay and David Robinson, The Decarbonized Electricity System of the Future: The "Two Market" Approach, Oxford Institute for Energy Studies, June 2017. https://www.oxfordenergy.org/wpcms/wp-content/uploads/2017/06/The-Decarbonised-Electricity-Sysytem-of-the-Future-The-Two-Market-Approach-OIES-Energy-Insight.pdf There are other interesting ideas for substantially changing markets. For instance, see the Australian Energy Security Board's proposal for "Two-sided market", https://www.aemc.gov.au/news-centre/media-releases/consultation-open-energy-security-boards-two-sided-market-paper. See also, the article by Rahmat Poudineh and Farhad Billimoria, "Market design for resource adequacy: an insurance overlay on energy-only electricity markets", *Utilities Policy*, 59, 100935.

buying intermittent renewable electricity at a stable and relatively low price, or firm—normally fossil-fired—energy at a higher and more volatile price. The proposed system is designed so that consumers have an incentive to shift their consumption to times when renewable energy is operating so as to avoid paying for firm energy at higher and volatile price. This gives consumers an incentive to be flexible, facilitating the penetration of renewables and reducing reliance on firm fossil-based energy. It enables consumers to express their preferences on the energy mix and how much they wish to rely on the system. Furthermore, the aim of this approach is to simplify and encourage the provision of flexibility from consumers who would not need to participate in complex flexibility markets.

3. Beyond prices—aggregation business models and technologies

The economic signals recommended above are designed to provide (more) efficient incentives to everyone, including consumers and their representatives. However, these price signals are usually not enough to elicit active consumer participation, especially in the case of small consumers. This is partly because costs may be greater than the benefits, especially where the consumer has to invest in equipment. But there is also a behavioral challenge associated with consumers knowing how to optimize energy use. This challenge will increase along with the growth of potentially flexible energy resources behind the meter, including EVs,[7] solar panels, batteries, and smart devices, not to mention other nonelectrical devices, such as thermal storage and residual water recycling for heating and cooling.

Furthermore, there is evidence that consumers don't understand or don't want to respond to price signals—for instance, see the chapter in this volume by Burns and Mountain. So what is needed to interest consumers in providing flexibility?[8] There are at least two related ingredients to delivering demand-side flexibility, in addition to efficient economic signals:

- technology—especially automation—that facilitates consumer participation and willingness to be flexible and

---

[7] See Chapter 8 of this volume, by Sioshansi.
[8] See Chapter 7 of this volume, by Sioshansi, for insights on the role of automation, Artificial Intelligence and aggregation to motivate consumers to be flexible.

- business models—involving aggregation—that enable the consumer to benefit from providing flexibility.

The examples below are illustrations of how these ingredients are being put into practice.

Aggregation is present in most business models to encourage demand flexibility.[9] It has at least two dimensions. One has to do with coordinating a large number of consumers. The other has to do with the coordination and optimization of many different kinds of distributed energy resource. For instance, an aggregator may specialize in the optimization of thousands of EV batteries to provide a particular service, such as frequency control, to the grid. Another aggregator may specialize in optimizing all distributed resources behind its consumers' meters. In both cases, the aggregator acts as a Virtual Power Plant (VPP) that draws on distributed energy resources to trade in one or more energy markets and potentially also in other markets (financial services, weather services, communications).

Aggregation requires technology to drive down the cost of providing demand flexibility. Here are some examples that illustrate what is now possible, or will be soon: Sonnen's virtual power plant (VPP) in Germany, which uses distributed batteries to store excess wind energy with settlement using blockchain transactions; EW's project with Austrian Power Grid, which uses a range of distributed energy resources for grid balancing, also using blockchain; Tesla's Autobidder platform, which is an optimizing platform that has been used in Australia and is now available in the United Kingdom; Octopus which developed a platform, called Kracken, which aims to optimize all sources of distributed energy; and Voltalis, whose business proposition is the ability through aggregation of demand to reduce energy consumption from its connected customers, in ways that lower its consumers' electricity bills and system costs.

3.1 Sonnen's VPP in Germany[10]

The Energy Web Foundation (EWF) announced recently that Sonnen Group had unveiled a VPP in northeastern Germany. The VPP comprises a

---

[9] See Chapter 16 in this volume by Löbbe et al. It provides a Germany case study of industrial demand flexibility connecting small and medium sized enterprises to a virtual power plant (VPP).

[10] The ideas on Sonnen's VPP are from: https://medium.com/energy-web-insights/sonnen-leverages-energy-web-chain-ew-origin-for-virtual-power-plant-that-saves-wind-energy-862d54df4bed.

network of distributed residential energy storage systems. It absorbs surplus wind generation, thereby avoiding curtailment of the renewable energy, by charging batteries when the wind is abundant. Avoidance of curtailment is important in Germany, which curtailed 3.2 TWh of wind energy in Q1 2019. The technical backbone of the VPP is an application using EW Origin, which matches forecast wind energy oversupply with available storage in the VPP. The network operator reports the anticipated surplus energy and the VPP makes an offer to absorb some of the surplus. If the offer is accepted, financial settlement occurs instantly and nearly for free on the EW Chain using EW Dai, a digital currency.

The key advance with this new blockchain-based technology and business model is to reduce the cost of transacting with and settling services provided by distributed energy resources (DER), including distributed batteries, electric vehicles, smart thermostats, and other electrical devices. Previously, the cost was prohibitive and hence DERs were unattractive to system operators. This development aims to offer DER services from the smallest devices to the grid, with the potential to scale up the VPP to include millions of distributed devices able to provide flexibility services.

3.2 Austrian Power Grid testing DER for frequency regulation[11]

Early in 2020, The Austrian Power Grid (APG) and EWF announced the "Flex-Hub" proof-of-concept process to enable small-scale DER to offer frequency regulation services to the grid. According to the announcement, this will be based on EW's blockchain platform to streamline the qualification and registration of service providers, as well as bid management and financial settlement.

APG is responsible for real-time balancing of electricity generation and consumption in order to maintain frequency stability and has been pioneering *horizontal integration* of European reserve markets; for instance with Switzerland and Germany. However, the trend toward decentralized generation and smart consumption creates new opportunities to manage the growing challenges facing grids. The Flex-Hub concept aims to develop *vertical integration* of power markets, drawing on the flexibility of distributed energy resources to help integrate renewables.

---

[11] The information included in this section on this APG initiative was reported here: https://medium.com/energy-web-insights/austrian-power-grid-and-energy-web-foundation-launch-proof-of-concept-to-use-distributed-energy-d9a378f5f5ee.

Until now, small-scale distributed energy resources were not directly connected to wholesale electricity markets. The Flex-Hub proof of concept includes a small number of units (for instance batteries) but APG expects about 1 million households to eventually participate in the flexibility market through this innovation. A central feature of the approach is that electricity customers can register their DERs themselves, while accredited third parties such as installers can independently verify the details of any unit. The buyer of the flexibility service, typically the distribution system operator, can trust that the bidders are reliable and once DERs are dispatched, the blockchain system enables easy financial settlement.

3.3 Tesla's Autobidder[12]

Cornwall Insight reported recently that Tesla's Autobidder platform aggregates and optimizes assets down to the BTM domestic scale for wholesale trading and grid services. The platform has been used in South Australia and is now available in Great Britain. Tesla sees this platform as part of their offering across the value chain, allowing them to offer fully integrated energy service to the system, including an aggregation of home batteries, solar panels, and electric vehicles bundled with large-scale generation. In the words of Cornwall Insight, "If Tesla builds on this with an application for a supply license, the company will span the GB value chain as a fully integrated energy technology and flexibility provider."

3.4 Octopus—Kraken platform[13]

Future business models may depend on the effectiveness of a company's customer platforms for integrating all energy sources in real time (just as Uber, which is essentially a platform, has revolutionized the car hire business), and for managing the end-to-end experience for customers. One example is particularly notable due to its recent popularity: the Kraken platform. According to its developer, Octopus Energy,

> We're building "Kraken," a cloud-based energy platform for interacting with both consumers (via the web, mobile and smart-meters) and the industry (eg data flows, consumption forecasting, trading on the wholesale market). Unlike our competitors, we're a tech-led company.[14]

---

[12] The information reported here is from Cornwall Insight, "Tesla builds capacity across the energy value chain", June 2, 2020.

[13] See *EEnergy Informer*, June 2020, for an article on the Kraken platform.

[14] https://octopus.energy/careers/back-end-developer/.

In May, Origin Energy, one of the three big Australian suppliers, acquired a 20% stake in Octopus Energy, reportedly attracted by its Kraken platform. Octopus claim that Kracken "will work to streamline and automate many interactions between Origin and its electricity customers, and lay the foundation for the multiple and varied engagements that will come with the increased adoption of rooftop solar, battery storage and electric vehicle charging." Origin argues that this platform "provides a highly automated, but high-quality, platform, which manages the end-to-end experience for customers, including connection, billing and allows Origin to develop real-time analytics of customer needs."[15] Apart from its agreement with Origin, Octopus has reached agreements with other companies that wish to develop and use the Kraken platform, including EON in the United Kingdom[16] and Good Energy, a UK-based green energy supplier.[17]

3.5 Voltalis[18]

Voltalis calls itself the "First operator of internet of energy".[19] The central business proposition of Voltalis is its ability through aggregation of demand to reduce energy consumption from its connected customers, in ways that lower its consumers' electricity bills and system costs. It lowers its customers' costs by remotely managing their electricity demand, for instance reducing demand from electrical devices when electricity prices are high. This makes commercial sense for Voltalis provided it can sell the demand flexibility (i.e., the ability to increase or reduce demand at very short notice) in a market. In some countries, for instance France, demand-side flexibility can be sold into day-ahead and other energy markets, as an alternative to generation, and into short-term flexibility markets as well. In other countries, demand flexibility has much less market access so far. However, wholesale markets are opening worldwide. This occurred initially in the United States after Order 745 from the Federal Energy Regulatory

---

[15] https://reneweconomy.com.au/release-the-kraken-origin-strikes-500m-deal-to-smarten-customer-systems-in-digital-distributed-world-86305/.

[16] https://www.eon.com/en/about-us/media/press-release/2020/eonnext-eon-and-kraken-technologies-form-strategic-partnership-for-eons-uk-residential-and-commercial-customer-business.html.

[17] https://group.goodenergy.co.uk/reporting-and-news-centre/press-release-news/press-release-details/2019/Good-Energy-to-implement-new-customer-service-platform-in-2020/default.aspx.

[18] Voltalis is also featured in this book in Chapter 7 (Sioshansi).

[19] https://www.voltalis.com.

Commission in 2011 (later backed by the Supreme Court of the United States in 2015); in September 2020, the FERC approved a historic final rule, Order 2222, enabling distributed energy resource (DER) aggregators to compete in all regional organized wholesale electric markets[20]. In Europe, a recent directive on electricity market design requires Member States to "allow and foster participation of demand response through aggregation" and "allow final customers, including those offering demand response through aggregation, to participate alongside producers in a non-discriminatory manner in all electricity markets."[21] There is also evidence of progress in the same direction in Japan and Singapore.

In principle, and in the view of independent aggregators like Voltalis, everyone benefits.[22] The system benefits if demand flexibility is less expensive than the alternative available: for instance, a reduction in demand of 10 MWh is offered at a price below the cost of increasing generation by the same amount, so that the wholesale market will settle at a lower price. This ensures retailers save on their sourcing costs. The benefits in terms of lower energy sourcing costs are usually several times greater than the costs for retailers (in particular the cost of buying demand response volumes and not billing them to consumers whose demand is curtailed, referred to as an imbalance cost). Hence demand response provides retailers with a net benefit, and retailers can pass on these net benefits to their customers, i.e., all consumers. The connected consumer benefits through a lower electricity bill resulting from a more efficient use of energy. And independent aggregators benefit from selling the flexibility.

The Voltalis business model depends on its aggregation technology platform as well as consumer willingness to participate. The platform includes a "Voltalis Box" in each customer's residence, which is connected to a system for optimizing all customers. The box allows Voltalis to adjust demand remotely for all of the devices connected to the box. The customer

[20] https://www.ferc.gov/news-events/news/ferc-opens-wholesale-markets-distributed-resources-landmark-action-breaks-down.

[21] Directive (UE)2019/944 dated 5 June 2019, article 17, a part of the Clean energy package adopted by the European Union.

[22] In some EU countries, there is considerable debate about the distributional impact of "independent aggregators" on other stakeholders, especially retail suppliers. Independent aggregators are defined by EU legislation as market participants engaged in aggregation who are not affiliated to a customer's retail supplier. Chapter 17 in this volume identifies a number of distributional impacts of independent aggregator activity and proposes models to neutralize the impacts, especially on retail suppliers.

receives the box for free. In addition to enabling lower energy consumption, hence savings on the electricity bill, the box provides the customer detailed (5-min) information about the consumption of all connected devices. In return, the consumer agrees to allow Voltalis to remotely manage demand of connected electrical appliances. Two keys to the success of this model are that the box is free and consumers benefit without having to be active in the management of their demand. However, Voltalis can only make a profit with a significant volume of demand to manage and where markets accept demand flexibility as a product.

The examples described above illustrate how companies are using new technology to facilitate the management of aggregated amounts of distributed resources. They suggest that the way forward involves aggregation and optimization platforms developed by companies that are not traditional players in the power sector.[23]

4. Energy communities and demand flexibility

This section considers the role that energy communities could potentially play in supporting the energy transition, specifically in acting as aggregators that could encourage demand flexibility. It focuses on the European Union, which has adopted legislation supporting the growth of these communities, and illustrates the benefits and challenges of integrating these communities within the wider system.

4.1 What are energy communities and why is there interest in them?

The term energy community refers to a wide range of collective agreements or actions involving citizen participation in the energy system. For instance, participation could involve community members sharing of one or more of the following activities: renewable electricity generation; collective self-consumption[24]; the storage, distribution, and sale of electricity, gas, or other energy sources; trading with the outside energy system; as well as investment and financing these activities. Some energy communities are part of a broader

---

[23] See *Power After Carbon; Building a Clean, Resilient Grid*, Peter Fox-Penner, Harvard University Press, 2020. Among other topics, he deals with new business models in the power sector.

[24] The term "collective self-consumption" is a form of energy community that is growing in importance. In this model, several consumers can own the same generation plant, sharing the consumption of the generated electricity.

municipal or regional agreement that includes other activities (water, finance, municipal services). Energy communities may take various legal forms and are frequently cooperatives or municipal corporations. They are usually not-for-profit and have wider social, environment, economic, and political objectives. However, energy communities could be based on private contractual agreements where profit was an explicit objective.

Recently, the interest in energy communities has been growing. There are at least four reasons.

- *First, they appeal to citizens.* Members emphasize the importance of being part of a community that is able to control activities that matter to them. Many consider the communities as a citizen-led alternative to relying on traditional electric utilities whose market power and commercial practices have alienated consumers. This aversion to existing utilities sometimes has a political motivation, notably in countries where regional or local governments aim to have greater autonomy or even to separate from the country of which they are part (e.g., Cataluña and northern Italy).

- *Second, there are economic motivations*, especially where these communities may be formally exempted from paying certain taxes, levies and system costs, or *de facto* able to avoid them. As explained earlier in this chapter, the economic justification for paying less should be related to the ability to lower system costs, not simply as a way to shift costs, levies, and taxes to others.

- *Third, Covid-19 provides an additional reason* for the anticipated growth of energy communities. This crisis has increased the importance of residential energy consumption, as well as highlighting its costs, encouraging consumers to consider new ways of reducing costs. It has also stimulated interest in working and living outside of city centers. Furthermore, the awareness of the potential for consumers to take control has probably been heightened during the Covid-19 crisis, as citizens have become more accustomed to the digitalization of working, shopping, and communications. Certainly, the sense of solidarity among citizens has increased during the Covid-19 crisis and this too may help to explain the growing interest in community-led initiatives.

- *Fourth, and most importantly, the declining cost and increased availability of distributed energy resources, along with digitalization of economic activity*, has made it possible for citizens to generate and store their own electricity and to manage their demand, from behind their own meter or the behind the meter of the Energy Community. The potential for consumers to control energy production and consumption changes everything as far as the sector is concerned. In particular, it supports the development of a distributed (horizontal) energy system model that competes with traditional (vertically integrated) business models.

4.2 Energy communities, EU legislation, and national legislation

Under EU law, energy communities are an alternative model for producing, consuming, or sharing energy.[25] Europe's legislation refers to two main types of energy community and creates a framework within which they can grow.[26] Article 2(16) of the 2018 Renewables Directive (2018/2001) defines a "Renewable Energy Community" and Article 2(11) of the 2019 Electricity Directive (2019/944) defines "Citizen Energy Communities." Both are referred to here as "energy communities," but it is worth noting some differences. A Renewable Energy Communities are restricted to generating renewable energy (electricity and other) and are effectively controlled by shareholders or members that are located in proximity to the renewable energy projects; they are essentially local energy communities. On the other hand, Citizen Energy Communities are restricted to electricity and allow for the possibility of "virtual" communities including members who are distant from any generation facilities. In spite of these differences, both types of community share characteristics. They share a primary purpose, which is to provide environmental, economic, or social community benefits, rather than financial profits. They are also entitled to produce, consume, store, and sell energy within the community and have access to markets outside the community. Neither allows a traditional energy utility to control the community and, in the case of the renewable energy community, these traditional energy companies cannot be shareholders. However, there is no prohibition under EU law to their participation in the provision of certain services to these energy communities. Furthermore, energy companies are free to participate in energy projects that involve private contracting and that are not covered by the EU legislation referring to energy communities.

---

[25] See also Chapter 4 (Löbbe et al.), Chapter 6 (Reijnders et al.) and Chapter 14 (De Clercque et al.) of the previous book in this series.

[26] https://www.compile-project.eu/wp-content/uploads/Explanatory-note-on-energy-community-definitions.pdf.

This EU legislation must be transposed into national legislation, in principle before June 2021, with the following overall vision and objectives as a guide. In the words of Directive 2019/944 (preamble point 43),

Community energy offers an inclusive option for all consumers to have a direct stake in producing, consuming or sharing energy. Community energy initiatives focus primarily on providing affordable energy of a specific kind, such as renewable energy, for their members or shareholders rather than on prioritising profit-making like a traditional electricity undertaking. By directly engaging with consumers, community energy initiatives demonstrate their potential to facilitate the uptake of new technologies and consumption patterns, including smart distribution grids and demand response, in an integrated manner. Community energy can also advance energy efficiency at household level and help fight energy poverty through reduced consumption and lower supply tariffs. Community energy also enables certain groups of household customers to participate in the electricity markets, who otherwise might not have been able to do so.

In short, the legislation recognizes many reasons to support energy communities. The focus here is on two passages in the above text: "to facilitate the uptake of new technologies and consumption patterns, including smart distribution grids and demand response, in an integrated manner"; and enabling "certain groups of household customers to participate in the electricity markets, who otherwise might not have been able to do so." I interpret this to be a clear intention for energy communities to support demand flexibility and enable participation in electricity markets that may well be outside the community. With the support of energy communities, which are a form of aggregator, individual consumers could provide flexibility within their communities and to the wider system. But will they?

While recognizing the potential benefits of ECs, it is important not to ignore the potential downside. A paper by CEER[27] registers the concerns. In their conclusion, in their guidance for adopting national legislation, they summarize their worries.

However, CEER believes that energy communities should not become a vehicle to circumvent existing market principles, such as unbundling, consumer rights or the cost sharing principles applied to energy grids. Energy communities should be able to compete on a level playing field, meaning the regulatory framework

---

[27] Regulatory Aspects of Self-Consumption and Energy Communities, CEER Ref: C18-CRM9_DS7-05-03, 25 June 2019. https://www.ceer.eu/documents/ 104400/-/-/8ee38e61-a802-bd6f-db27-4fb61aa6eb6a.

should be such that they do not face undue barriers nor create undue distortions in existing markets.

The provisions adopted in the CEP remain relatively open to interpretation, and transposition into national law will be critical to the viability and role of energy communities.

In other words, energy communities are all very well as long as there is a level playing field but the encouragement given to them by recent European legislation is liable to encourage governments to tilt the playing field in their favor. This leads to three basic concerns that need to be addressed by national legislation in order for energy communities to be established in a way that is consistent with EU legislation.

- *Competition could make communities unsustainable*—As long as community members are free to defect to outside suppliers, it is difficult to develop a sustainable energy community model. Communities could create barriers to defection (contractual, regulatory, or fiscal) but this undermines competition. National legislation should be firmly against barriers, but also provide the basis for communities to succeed by providing a competitive energy service and a range of services that its members value.

- *Balkanization raises costs*—An even more fundamental issue is balkanization, that is breaking up the Single Market. In 2016, Green and Strbac estimated the benefits of fully integrating the EU energy market; EUR 12—40 billion/year by 2030. These benefits reflect the cost of the national approach rather than a European approach, for instance providing renewables support at national level rather than EU level.[28] It is essentially "community spirit" that leads to this outcome (i.e., people are happy to pay for "our" renewables but not for "theirs"). If this happens at an even more localized basis, via energy communities, the economic losses would be even greater. We are supposed to be moving toward a Single Electricity Market in the interests of keeping costs down, not away from it. National legislation should support integration, not balkanization.

- *Avoidance of system costs and levies is unsustainable*—One of the reasons why energy communities are popular is that they sometimes offer the prospect of avoiding various system costs, taxes, and levies. There is a strong case for reducing payments to the system if the communities are reducing system costs (e.g., for networks). However, if communities

---

[28] Richard Green and Goran Strbac, Reforming European electricity markets for renewables, Oxford Energy Forum 104, February 2016. https://www.oxfordenergy.org/wpcms/wp-content/uploads/2016/03/OEF-104.pdf.

do not lower system costs, they may be shifting costs, taxes, and levies to other consumers and citizens. As the CEER document points out, this undermines existing cost sharing principles. Apart from being unfair to those who must bear a heavier burden, this can lead to increased costs for the system as a whole and—if this happens at scale—pose problems for the financial stability of the system and slow the energy transition. European legislation does provide some safeguards.[29] First, participation in an energy community is open and voluntary; members and shareholders are entitled to leave the community and do not lose their rights and obligations as consumers. So energy communities always face the prospects of losing dissatisfied consumers if a better offer comes along. Second, these communities must contribute in an adequate and balanced way to the overall cost sharing of the system. A reduction in their share of costs and levies is therefore limited. Third, energy communities are financially responsible for imbalances they cause in the electricity system. Fourth, Member States shall ensure that energy communities are able to access all electricity markets in a nondiscriminatory manner, opening the door to an efficient integration of the community in the system, with economic benefits to be shared.

National legislation will need to strike a balance between supporting energy communities and ensuring that they help to lower the costs of the energy transition, not increase them. In particular, it should not undermine the basic idea behind the Single Market, namely integration as opposed to balkanization. The legislation should also ensure that consumers have the freedom to select their retail supplier and that the community does not favor its own service over that of alternative suppliers; especially when the community has control over the distribution network.

4.3 Examples of energy communities that encourage demand flexibility

There are at least three ways that consumers in energy communities may be encouraged to provide demand flexibility.

First, the business models and technologies mentioned in Section 3 of this chapter should be available to all consumers, whether or not they are members of an energy community. The energy community may not wish to adopt these models and technologies but it should not discourage other aggregators from offering them.

---

[29] See, in particular, Article 16 of EU Directive 2019/944 of 5 June.

Second, it is interesting to consider how an energy community could enable integration into the wider system, allowing its distributed energy resources to trade in external markets. Stedin's *Layered Energy System in the Netherlands*[30] is an interesting model to consider. This involves local markets for each energy community. Because each community will normally not be able to supply its demand at all times, the idea is that the local market will have an open connection with traders on a wholesale level who can also participate in the local market. According to this model, the external energy supply will be subject to a premium, which makes it more expensive than the local energy. All trade within the local market has to fit within the physical limits of the grid involved, which is managed by the distribution system operator. According to its proponents, this system has several advantages.

First, the system stimulates local production and shared use of renewable energy. This has helps to reduce unnecessary grid investments, lowers losses and enhances community building. Second, consumers are free to choose their own supplier and are not obliged to join the local market; so it can co-exist with the existing supplier model. Third, anyone can participate in the local market with a service provider of their choice (even without their own flexibility) or produce energy. Any subsidies will be shared by all members. Fourth, it avoids the disappearance of flexibility, which is a risk if the energy community has minimal connection with the system. Finally, the system will lower the cost of energy.

One of the problems with the proposed model is that it prefers local generation to generation from outside the community; seeking to trade only when there is no local generation. This is inconsistent with the goals of the Single Market. Unfortunately, this seems to be a common feature of energy communities and one wonders to what extent the local generation is truly more economic than alternatives.

Third, there are community energy models that are consistent with the principles of the Single Market. These encourage demand flexibility and enable integration with the wider system. Indeed, to avoid loss of members, energy communities should have incentives to adopting technology platforms that facilitate demand flexibility and participation in markets. An example of a community, which embraces this idea is Ripple Energy[31,32] which has launched what it says is the United Kingdom's first community owned wind farm. Ripple is partnering with Co-op Energy and Octopus

---

[30] "Whitepaper: Layered Energy System", Energy 21, Stedin, November 2018.
[31] https://rippleenergy.com.
[32] Cornwall Insight (Nick Palmer), Daily Bulletin, Issue 3712, 11 June 2020.

Energy to supply the homes. Existing Co-op Energy and Octopus Energy customers are invited to take part and new customers can switch to Co-op Energy to join the Ripple community. Customers will own part of the wind farm by buying shares in the cooperative that owns the Graig Fatha wind farm. Around 2000 customers will be able to join this first pilot project. The upfront cost for a typical household would be around £1900 and this share would expect to generate enough electricity to meet the needs of a typical household for 25 years. The wind farm will become operational in early 2021.

4.4 Proposed guidelines for energy communities

Energy communities in the European Union should be aware that national legislation will reflect the EU Directives. Their plans should reflect the wider objectives of that legislation as well as the specific social, political, and environmental goals for communities. In view of these broader goals and while sticking to the requirement that these communities not be driven by the pursuit of private profit, a central feature of any community energy project should be the technical capability and the business or social model that enables integration and optimization of the community resources within the wider electricity system. Being able to tap that flexibility efficiently creates sources of value for the community, including the following:

- *Lower costs.* Flexibility to consume renewable energy when it is operating means relying less on higher-cost fossil energy from the system. The community may decide that it should contract for some of its renewable energy from a regional source (outside the community) in order to reduce costs for community members.
- *New revenue sources.* Flexibility and energy can be sold to the system, in wholesale energy and services markets and to the DSO in local congestion markets. The community should have the means of optimizing the sale of these services by selecting the markets into which to sell and when to trade. This is a key incentive to be integrated with the wider system.
- *Optimization with the wider system.* To lower costs and tap new revenue sources using many different distributed energy resources (solar panels, batteries, many smart devices in the home, EVs) it is increasingly important to have software and hardware that optimizes the use of all these resources within the community and outside it (e.g., wholesale markets, network charges, congestion markets).

- *Self-reliance and resilience.* Greater flexibility, in particular through storage, makes the communities better able to cope with supply problems from the system. It may also provide greater reliability to the system as a whole since a problem in one part of the system need not have a significant cascading effect throughout the entire system.
- *Risk management.* The development of flexibility inside the community helps to manage risk in a number of ways. First, lower costs and ease of participation (through automation) lowers the risk of losing members to alternative suppliers offering aggregation services and smart equipment (for free). Second, the integration of consumer flexibility with the wider system reduces the risk and cost of balkanization, and lowers the probability of regulatory opposition. Third, flexibility raises revenues and lowers cost, thereby reducing reliance on subsidies and cost shifting. This improves the prospects for a sustainable community that does not rely on government support.

To exploit these potential benefits, energy communities need access to suitable technology and services. Under EU legislation, it is not (nor should it be) illegal for third-party companies to help energy communities reach their objectives. There are restrictions. The EU legislation does not permit utilities or other energy companies to own or control energy communities. However, national legislation should not get in the way of communities being permitted to acquire the necessary skills, technologies, and services from whichever company is best able to provide them. The communities may choose not to contract with a utility or large energy company, but they should demand legislation that does not restrict their right to do so, or to contract with other qualified firms; a restriction of that kind would very likely raise prices and lower service quality. To address concerns about energy companies gaining control, legislation should ensure communities have the necessary training and advice to contract third parties through transparent, competitive procedures, for instance auctions. This would justify financial support to the communities, at least initially.

5. Conclusions

This chapter has argued that efficient price signals are necessary but not sufficient to actively engage consumers and to elicit the demand flexibility that is now a central requirement of modern decarbonized electricity systems. In countries where short-term markets work well, aggregation platforms are now being introduced to capture the economic benefits of flexible demand; to be successful, these platforms need to provide economic benefits to the consumers offering flexibility, to the aggregator and to the system as a whole.

Second, most energy communities have a strong interest in promoting sustainable, low carbon energy resources, while also pursuing a range of other political, economic, and social objectives favored by their members. In principle, they are excellent candidates to further the energy transition, not only through local renewable energy generation but also by offering flexibility through demand management and storage. However, there is a risk that these communities will try to separate themselves from the system, discourage competition and rely on subsidies; this would raise the costs of the energy transition. National legislation should discourage communities from going in that direction. Rather, it should provide incentives for communities to develop distributed energy systems that are an integral part of the wider system, thereby lowering the costs of the energy transition for themselves and for society at large.

Third, as a matter of due diligence—to avoid problems later, in particular the loss of members—energy communities should be given the tools to improve their economic performance. For instance, although traditional energy enterprises are restricted from controlling energy communities, the latter should be free to select their providers of services and products in order to provide the most efficient service possible, while also meeting other objectives. They should also be offered training and support - probably financed by the government - to select and manage these providers efficiently.

Fourth, it is important to stress that scale is required to stimulate the development of profitable aggregation services and a supply chain of other services, technologies, and products aimed at facilitating consumer flexibility. This is most likely to occur through system-wide aggregation services and through markets that apply to all consumers. Individual energy communities are relatively small and unlikely to contribute much in this respect. However, they certainly have an interest in benefiting from scale economies. One can imagine very large virtual citizen energy communities, as well as systems (like Stedins Layered Sytem) that would integrate multiple, smaller, renewable energy communities.

Finally, in order for energy communities to reach their potential to support the energy transition, policy needs to encourage experimentation to learn about consumer preferences and about which models (business, technology, regulation) work best at motivating efficient consumer participation. Regulatory sandboxes and regulatory pilots are excellent tools for experimentation of this kind. Energy communities could well be one of the organizational frameworks which enable consumers to begin to drive the energy transition. It is essential that these communities have the opportunity and the incentives to drive it in the right direction.

CHAPTER 22

Flexible demand: What's in it for the consumer?

Mike Swanston
The Customer Advocate, Brisbane, QLD, Australia

1. Introduction

Australia has a remarkable track record in the adoption of renewable energy. Despite a somewhat erratic national energy policy, investment in utility-scale and rooftop and behind-the-meter generation shows no signs of slowing.

Small and medium consumers (and PV sales companies) across the board appreciate the fact that the cost per kilowatt of energy from the roof is now demonstrably cheaper than grid energy, with falling PV system prices and the fact that energy from the rooftop attracts no distribution system charges (AEMC, 2018) and (IRENA, 2020b). The strong community preference is that new energy shall come from renewable sources, mandating that wind generation and solar PV, both utility scale and rooftop, will continue to develop rapidly. Consumers and the wider community embrace clean energy as a necessity, "doing their bit" for the environment whilst developing a level of energy independence, saving money as their investments offer paybacks of less than 5 years, even as generous incentives fade.

But as the penetration of variable renewable energyincreases, especially deep in the network in the form of rooftop solar PV, the pressure to act in response to the daily and seasonal cycles of high levels of variable generation is now moving beyond the highly technical corridors of networks and market operators, and becoming an issue for consumers. Load shifting, demand balancing, hosting capacity, and importantly supply security are all factors—whilst perhaps phrased in more "customer friendly" terminology—that are starting to enter the lexicon of customer connection agreements, retailer product development, and even the mainstream media.

One recent headline, *"The rise of solar power is jeopardising the WA energy grid, and it's a lesson for all of Australia"* (ABC, 2019) typifies the way consumers are starting to be exposed to the fact that the Variable Renewable Energy (VRE) phenomena is not just "business as usual."

Variable Generation, Flexible Demand
ISBN 978-0-12-823810-3
https://doi.org/10.1016/B978-0-12-823810-3.00013-3

Consumers have long been involved with the concept of flexible demand, responding to the "carrot" of lower energy prices. Under the banner of "demand reduction," products such as off-peak controlled load and time-of-use tariffs have been around for a long time, and can take credit for saving many millions of dollars in avoided inefficient network augmentation and reducing the use of low-merit (and often high emitters of CO_2) peak generation (Productivity Commission, 2013).

Flexible demand now means more than just demand reduction, and the benefits to consumers are expanding in scope and importance, to the point of participation being almost nondiscretionary. If we are to *efficiently* embrace the ever-increasing share of variable renewable energy, the role of the energy consumer must become a flexible, responsive, and even dispatchable entity to be coerced, compensated, convinced, co-opted, and controlled. But just how to engage consumers to play their part in committing flexible demand remains a challenge, as the diversity in their needs widen and market arrangements become more complex.

Gallagher and Ross in Chapter 19 highlight the required framework and steps to be taken to empower consumers to deliver flexible demand, and how flexible demand is a logical and important part of our efficient transition away from high carbon emissions. Presenting a compelling case, with the right rewards, to consumers is the next stage of adapting to the new energy environment; with the progress proving highly dependent on how consumers view the risk and assess the benefits of doing so.

This chapter considers the consumer's role in the development of flexible demand capability that has been taking place, focusing on Australia. The chapter is organized as follows:

- Section 2 briefly reviews the context of the Australian energy landscape and considers how the consumers' approach to renewable energy is framing the imperative for flexible demand.
- Section 3 outlines the range of benefits that flexible demand can deliver the consumer and proposes how these benefits serve not only those consumers with DER but also present wider benefits to other consumers and the community generally.
- Section 4 highlights how grid security is becoming a significant issue of concern for all energy consumers.
- Section 5 considers how many existing and familiar demand response tools are being expanded in scope and purpose to meet changing energy needs.
- In Section 6, the newer forms of demand response, particularly in the context of DER hosting capacity, are discussed.

- Finally, in Section 7 the practicalities of implementing for DR in the consumer environment are examined and the issues important to consumers in realizing the tangible benefits from demand response identified.

The chapter's conclusions are presented in Section 8.

2. The consumer context, as the love for renewables continues unabated

The National Energy Market (NEM), the bulk energy supply system that covers Australia's populous eastern and southern states, already has close to 17 GW of generation capacity from wind and solar sources, being 28% of the total generation capacity in the market. Almost three-quarters of the national solar power generation capability is located on over 2.3 million—that is almost one in every four—consumers' rooftops. The penetration of variable renewable energy (VRE) in the state of South Australia is close to 50%. In the Western Australian network, an isolated grid supplying 2 million people, one in three houses has a solar installation, the combined output of which exceeds that of the biggest coal generator (AER, 2020).

In addition, the rate of uptake of renewable energy, particularly at rooftop levels, continues unabated, as shown in Fig. 22.1.

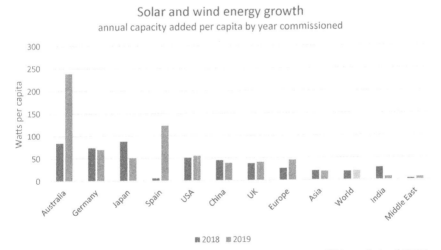

Figure 22.1 Renewable generation growth per capita. (*Source: TCA analysis of IRENA, 2020b, Renewable Power Generation Costs in 2019, Abu Dhabi, International Renewable Energy Agency. Retrieved from: https://www.irena.org/-/media/Files/IRENA/Agency/ Publication/2020/Jun/IRENA_Power_Generation_Costs_2019.pdf data.*)

Australia is by no means alone in this trend. Many regions globally continue to expand their share of renewable energy sources. Australia, Germany, Spain, California, and Japan all continuing to experience rapid growth in the penetration of solar and wind generation on a per capita basis, supported by falling input prices and continued government support for the development of renewable energy (IRENA, 2020a). Whilst Spain and Germany have a higher amount of variable renewable generation per capita that Australia (IRENA, 2020a), the island that is the Australian National Energy market operates under the challenging condition of having no interconnections to other large markets.

This increasing penetration of renewables, along with marked development of energy-related technologies and communication over the past decade has led to a sustained change in the way consumers source and account for their energy. For networks and the energy market, the intraday demand has become more volatile. These changes are, in turn, driving out more traditional high capacity, high inertia fossil-fuelled generators whose technical characteristics are not able to adapt to this increased degree of variability, as shown in Fig. 22.2.

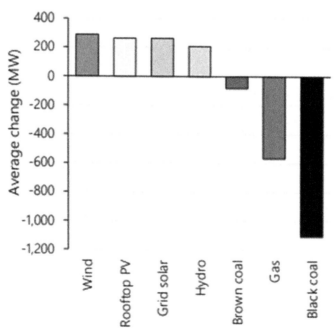

Figure 22.2 Reductions in thermal generation, Australia—Q1202 versus Q1 2019. *(Source: AEMO.)*

Curtailment of renewable generation—a phenomenon that is unfortunately becoming more common—now features at both the utility and consumer levels. With more generation from solar and wind farms that are generally some distance from strong network connections, grid operators are reconsidering transmission loss factors, effectively derating large VRE installations. Grid operators in places like California and Australia are now experiencing a host of issues such as overgeneration, minimum load, and negative prices, as highlighted in Chapter 1. In response, grid level battery storage is becoming more common, encouraged by rising prices for grid stabilization services and an increased volatility in wholesale energy prices.

Local distribution networks are also exposed to increasing levels of DER, advising of the greater likelihood of inverters "winding back" as reaching network voltage limits becoming more common, saying "no" or limiting the capacity of many new PV connections. The situation is also emerging with consumer-owned Virtual Power Plants (VPPs), where simultaneous export called by aggregators at risk of being restricted due to network capacity limitations.

The risk of inefficient new investments is emerging. The term "risk" is appropriate as there is rarely an equivocal or even reliable and repeatable form of cost benefit in the response to increasing VRE, with the term "least regrets" featuring in many analyses.

Large new regional interconnectors and strengthened transmission links are proposed, allowing greater system security by removing bulk energy transfer constraints from a wider diversity of generation sources and regions. System stability measures such as grid storage, synchronous condensers, and advanced protection schemes are becoming commonplace. Locally, distribution network operators are proposing significant investments ranging from network augmentation to intelligent dynamic voltage control systems (ARENA, 2019). Rule changes to implement mandatory control of customers' behind-the-meter generators by the market operator are being drawn up.

Whilst many of these investments, all of which eventually "hit" the consumer's energy bill in the form of depreciation or the utilities return on investment, are valid and important, it is clear that engaging consumers in demand response represents an efficient option in helping to meet these challenges by presenting community-acceptable alternatives that avoid the construction of long-lived, dominant network assets. In times of such rapid change in the way energy is generated and used, avoiding investment in assets that will be depreciated over long times—up to 50 years—is clearly of benefit to energy consumers.

In addition, the lead times to construct large network assets or implement new forms of control can take many years, in some cases "too late" to avoid constraints on the access by consumers to clean, low cost sources of energy. Similar behavior is seen in California and the United Kingdom, where regulators create conditions favorable for the development of so-called "nonwire" solutions as an alternative to the preference of constructing large network assets (Johnston, 2020).

Many opportunities will arise for consumers to realize new streams of benefit should they wish to participate in flexible demand actions. Peak load shifting, a flexible demand option that has been benefiting consumers for some time, remains important in its contribution to reduced network augmentation and more efficient use of efficient generation. However, with so much power from variable, nondispatchable sources being generated on the grid, the market operator is warning the stability of the power system could soon be in jeopardy. Therefore, the conversation now includes a large element of what *must* be done to maintain a stable, reliable, affordable, and sustainable power system for all.

As a relatively large energy market with world-leading levels of rooftop solar PV penetration, Australian consumers are very much the "canaries in the coal mine" as the role of the consumer adjusts to become an integral part of the supply—demand energy balance. A tipping point in the way energy is considered has been reached, with a substantial proportion of Australia's supply network expected to reach reverse power flows significantly earlier than previously thought (Johnson, 2019). A sense of urgency now prevails amongst regulators, generators, market operators, and networks to play their part in addressing the opportunities and challenges of a rapidly increasing proportion of variable generation.

3. How can consumers benefit from flexible demand?

There are several ways of looking at how flexible demand can benefit energy consumers. For instance, Stagnaro & Benedettini in Chapter 6 investigate the potential contribution of small electricity customers to the provision of flexibility services to power systems.

The consumer benefit resulting from the application of flexible demand in a variable-generation environment can also be viewed from the perspective of the consumer, as shown in Fig. 22.3. Benefit falls broadly into four categories, ranging from direct financial payments to meeting the community and societal benefits of a low carbon future.

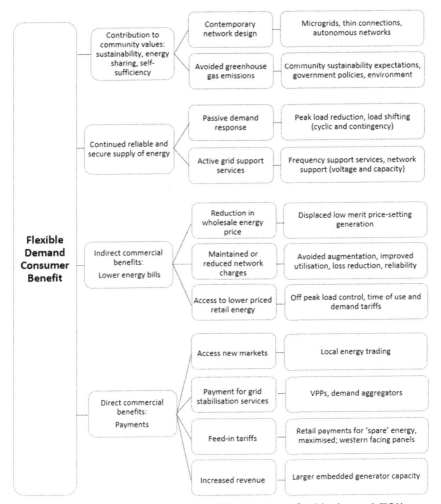

Figure 22.3 Categorizing the form of benefits from flexible demand (TCA).

It is important to note that *all* energy consumers benefit from the application of flexible demand, not just those with active behind–the-meter devices such as PVs, batteries, or EVs. The form of benefit, however, varies. Also, many of the actions and mechanisms to deliver benefit through demand response are already in place for consumers in some jurisdictions, such as passive demand response through off-peak load control and time-varying tariffs to address peak demand.

In addition, the benefits are not as "cut and dried" in distinct categories as the figure may suggest, with many overlapping. For instance, participation in active grid support or 2-sided markets provides direct income for consumers who elect to offer energy in a peer-to-peer trading relationship, whilst also offering reduced wholesale and network costs by supporting the reduced dispatch of lower merit market generation or taking pressure off peak demand that fuels network augmentation needs. All consumers, not just those participating in demand response actions, draw benefit from many of these aspects of benefit, such as reduced carbon emissions, lower network and wholesale prices, and a more secure network.

Several benefits are explored in detail in subsequent sections of this chapter.

As noted in Chapter 7, larger industrial and commercial consumers have for some time engaged with retailers offering curtailable energy capability or demand time-shifting that offers lower energy prices or more flexible operating arrangements. As the impact of utility solar and wind raise new supply security challenges, these contracts are evolving from wholesale price hedging and local network contingency response to join a fast-evolving wider market-based demand response capability. In the near future though, market interventions such as Reserve Trader arrangements are being applied (AEMO, 2020a,b,c).

This shift to a centralized demand response scheme for large consumers raises some changes to the benefit delivery, however. To date, networks and retailers have been able to target individual large consumers with agreements and offers that are very location-specific or relate to a particular generation portfolio that their retailer may own. As the wider Reserve Trader schemes come into play, the priorities of the centralized scheme, being mainly a response to generation adequacy, overshadow the local arrangements. Then again, this tension of needs is also prevalent at the small consumer level, where balancing the varied opportunities on offer becomes a major consideration for consumers with DER.

4. Grid security—a critical consumer benefit

In some jurisdictions, the growth in wind and solar generation is increasing the complexity of the operation of the power system and will continue to do so as further described in Part One of this volume. This presents two significant risks to the secure supply of energy affecting all consumers, from opposite ends of the energy supply spectrum. In times of peak demand, the availability of adequate generation is forecast to tighten, as large dispatchable generators exit the market. On the other hand, when demand is low and rooftop generation is high, the falling grid or "residual" demand challenges the stability of the grid's "safe operating envelope."

Whilst grid-scale battery energy storage is rapidly filling the fast and medium-period response requirements, consumers too are being called on to play a role. Mechanisms to reward consumers for their participation are emerging, albeit slowly and at this stage mainly through trials, the details of which are explored later in this chapter.

4.1 High demand and supply adequacy

Consumers are being called on to offer greater levels of flexible demand to play a larger role in supporting the adequacy of generation.

As large thermal generators exit the utility generation sector to be replaced by wind and solar generation, reserve margins are tightening. Their lower level of "dispatchability" presents greater risks of supply adequacy in times of extreme demand such as heat waves. The role of demand response to reduce demand at a region-wide level has in recent times become more critical in ensuring a reliable and secure supply of energy to all consumers.

Market economic conditions have changed markedly in Australia as the share of renewables increases. The market influence of renewables has forced a change to the economic dispatch conditions and therefore challenges the long-term viability of traditional thermal generators. While 2200 MW of new generation investment was added to the NEM over the 5 years to June 2017, almost 4000 MW of capacity was withdrawn over the same period (AER, 2020).

The Australian Market Operator (AEMO) considers another 30 GW of new grid-scale renewable generation is likely to be developed to replace a large proportion of Australia's aging coal-fired generation that will reach the end of its technical life and retire by 2040 (AEMO, 2019).

Further investment in hydropower, in particular pumped storage, is under consideration. This is a limited option in Australia, however, given the relative lack of favorable geographical opportunities. Dispatchable grid-level storage is being developed at a significant rate, with a number of large grid-connected battery storage installations providing the majority of grid frequency control ancillary services and other balancing functions.

Concerns of generation adequacy to meet seemingly more frequent demand peaks in hot summers are on the increase, as highlighted in below box.

Australia's summer of extremes pushed the grid to the limit, Market operator says.

The summer of 2019/20 was Australia's second hottest on record, trailing only the previous summer, with maximum temperatures 2.11 degrees warmer than the 1961-90 average. December 2019 alone had 11 days when temperatures averaged above 40°C, equaling the number of such days during all previous years since 1910. Bushfires also charred large swathes of the forests of eastern Australia.

Most of the announced new generation projects are variable renewable energy generators, which often do not generate at full capacity during peak demand times or may be positioned in a congested part of the network. As a result, while providing significant additional energy during many hours of the year, these projects are forecast to only make a limited contribution to meeting demand during peak hours.

The Australian Market Operator issued 178 directions to intervene in the market to deal with actual or potential supply shortages or system security issues in 2019—20, a 10-fold increase over the past 3 years.

Number of summer directions issued by the Australian Market Operator

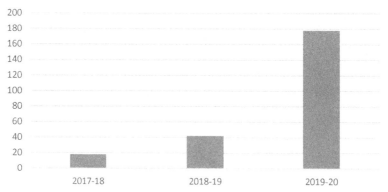

Australian Market Operator, 2019

4.2 Low residual demand and grid stability

At the other end of the demand curve, *minimum* net load is creating challenges. Network operators are now differentiating between customer energy demand and the actual power flow through the grid, using terminology such as "operational demand," "underlying demand," and "residual demand." Defining these terms, where at any point in time:

Total energy demand ('operational' or 'underlying' demand)

equals

behind the meter generation (predo min antly PV)

plus

grid ('scheduled' or 'residual')demand

To illustrate the challenge of falling residual demand, it is useful to consider South Australia and Queensland, where approximately 34% of households and small businesses have rooftop PV systems installed. With the growth in rooftop PV, minimum grid or residual demand is now occurring when high levels of solar PV output coincide with low levels of demand—typically on mild, sunny days in spring and autumn when air conditioners are not in use. On those days, a larger proportion of solar PV energy is exported to the network, and the security and safe operation of networks at both a macro-utility level and locally are taken to "the edge of the operating envelope." Local network voltages rise and the operating conditions for network protection systems are stretched. Traditional network security and stability techniques must adapt, as the risk of cascading failures is high (AEMO, 2019).

Fig. 22.4 (Alan Rai, 2019) illustrates this widening gap between the consumer energy demand and the falling level of energy being sourced and

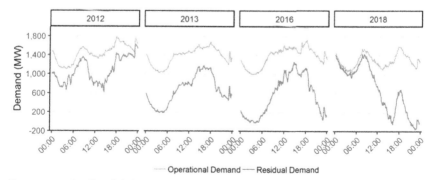

Figure 22.4 Profile of daily operational and residual demand in South Australia, 9 May by year (Alan Rai, 2019).

transmitted through the grid. Considering demand in South Australia in recent years—choosing May 9, a date that represents an innocuous time of year for the power system in the state—the residual grid demand in 2018 at times falls below zero, representing a net regional generation across the state. This issue is explored in more detail in Chapter 14.

This behavior is expected to become much more common. The proportion of energy requirements being provided by wind and solar at low-load times can reach close to 50% in some regions in Australia, California, and Germany, particularly in times of traditionally lower demand on the sunny afternoons in the "mid-seasons" of Spring and Fall (AEMO, 2020a,b,c).

The frequency and magnitude of this phenomenon will only increase. Fig. 22.5 shows the actual wind and solar penetration in the Australian NEM for each half-hour period in 2019, with projections to 2025 indicating the potential instantaneous penetration under midrange and step-change forecasts.

The implication from Fig. 22.5 is that in the near future under even a quite conservative scenario, there will be periods when over 50%—sometimes possibly 100%—of customer energy demand will be met by solar and wind generation.

At a grid level, there are concerns regarding system inertia, fault level, fault ride through, and stability risks from the coordination of inverter operation should the system frequency shift as a result of generation trip or loss of major transmission capability. The challenges in maintaining a stable generation and transmission portfolio are further discussed in the ERCOT context in Chapter 2 by Ross Baldick and in Italy in Chapter 3 by Terenzi.

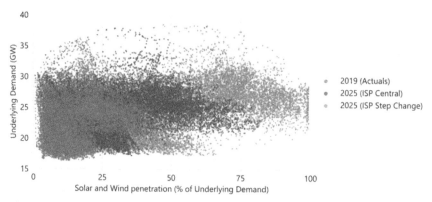

Figure 22.5 Instantaneous penetration of wind and solar, Australia. *(Source: AEMO.)*

Closer to the consumer connection, other issues arise. The rise in local network voltage due as the power flow reduces and changes direction as a result of a high proportion of energy sourced from inverters is generally understood. Underfrequency load shedding schemes inherent in most networks in the world require redesign and retooling so as not to shed generation in times of falling frequency. It is these local issues that highlight the need to encourage demand flexibility "deep" in the distribution network and behind the meter to encourage demand flexibility by consumers—not only how much energy is used, but how much can be shifted by using appliances differently and stored in EVs and batteries.

The role of creating new business models and incentives for aggregators is considerable. Gallagher and Ross in Chapter 19 explore the enablers and opportunities to empower consumers to deliver flexible demand.

Close coordination between load shedding schemes and inverter settings is required. In countries like Australia where a significant proportion of inverters are already deployed, it will take some years for revised operation requirements to filter through the population of consumer-owned generators.

4.3 How can a consumer's flexible demand support system security?

Off-Peak Controlled Load (OPCL) has been the mainstay in clipping evening peak demand in the residential and small commercial sectors for many years. For consumers, network businesses provided attractive controlled-load tariffs for systems like hot water systems, pool pumps, and air conditioners, facilitated by signaling systems such as audio frequency injection or time clocks.

In contrast, wholesale demand response by residential customers was largely nonexistent, as most customers remain on volumetric rate electricity offers. Chapter 5 explains some of the reasons, recognizing that some customers are charged on time-varying time-of-use (ToU) rates.

This voluntary reduction or shift of electricity use by consumers, in return for lower energy tariffs, helped reduce levels of demand as part of peak demand management or, where remote control existed, reducing demand during rare contingency conditions. Demand reduction is commonly used in the USA, Japan, New Zealand, and the UK. In Australia it is used to meet over 10% of peak demand for electricity. New Zealand began using DR in 2007 and now meets over 16% of peak demand through DR programs (UKERC, 2013; SSE, 2020; Lawson, 2011).

As the share of VRE grows, time-based load shifting still has a role on peak demand lopping but is proving inadequate to meet the growing needs of a more dynamic power supply system. A number of jurisdictions have created a wholesale demand response mechanism for large energy users. In Australia, a regulatory rule change is nearing approval, where the agreement will allow large industrial and commercial energy users to trade their demand under market mechanisms, allowing them to be paid to reduce consumption to support a reliable energy system in a manner that is potentially more affordable that other sources of generation (AEMC, 2020).

Hochstetler in Chapter 17 discusses the development of markets for flexibility.

Whilst demand response by large consumers is certainly effective, in many cases the resource of aggregating many small consumers is excluded from the arrangements. The schemes carry a minimum demand threshold, on the assumption that expanding the capability to small energy users would be too complex and cost prohibitive to implement and participate. In addition, small consumers are considered unlikely to provide a sufficient level of certainty over their demand response, as well as driving up costs and increasing complexity in billing (Mazengarb, 2020).

The door, however, is opening on the involvement by small consumers—i.e., those consuming less than 100 MWh per annum. Granted, investment by Australian consumers in battery storage on-premises is presently slow, as high system prices, flat energy tariffs, and continued attractive feed-in tariffs for solar PV dictate a relatively poor business case for consumers to invest. However, growing government concerns regarding energy security and the consumer bill implications of the falling utilization of networks is giving rise to subsidy programmes aimed at developing distributed storage in conjunction with PV development. These schemes have assisted to deliver over 1 GWh of household battery storage capacity by the end of 2019.

A number of feasibility trials are under way to assess the role of small consumers in assisting to address emerging grid security concerns. A large proportion of the batteries installed by consumers have been offered to operate as Virtual Power Stations (VPPs), being a condition of the generous government subsidies, with the intent to assist the development of grid support services. A number of energy retailers, battery suppliers, and demand aggregators have set agreements with consumers regarding the operation and dispatch of these devices (Sioshansi, 2020).

In 2019—20, a number of system events requiring frequency support or energy support occurred. A study released under the Australian Renewable Energy Agency Advancing Renewables Programme (AEMO, 2020a,b,c) drew insights from the data on how the consumer VPP is interacting with the power system. Despite the relatively small pool of data to draw from, the data received so far indicates that VPPs can effectively respond to power system events and price signals and in so doing, can deliver a reasonable economic return to consumers participating in these offerings.

In the trial, the majority of the benefit arose from two periods with high-priced frequency control (FCAS) services following regional transmission separation, as shown in Fig. 22.6.

Commercially, it is clear that the economic returns to customers offering active energy sources into the grid stabilization market is extremely sensitive to the number of significantly priced, infrequent ancillary service extremes that may occur.

In addition to the South Australian trial demonstrating the technical capability of the VPP arrangement, it highlighted a number of very practical issues for consumers regarding their involvement in providing DER flexibility, including:

a. What arrangements can exist between the consumer and the market participant (retailer or aggregator) to ensure a fair proportion of the market payments (and risk) are passed to the owners of the battery system? Will consumers accept the risks?

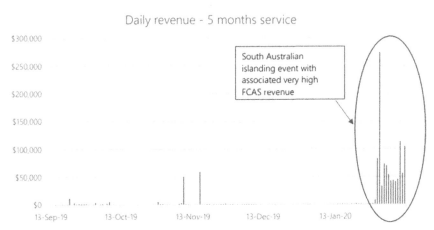

Figure 22.6 Daily revenue from the South Australian VPP trial, 2019—20. *(Source: AEMO.)*

b. Would similar results be achieved if the consumer chose to use their battery system for a range of applications, such as arbitrage or local energy security (blackout protection)? How available are batteries at times of high demand?

c. Is the local network capable of accepting undiversified, simultaneous energy feed-in on rare occasions without compromising power quality limits?

d. Are the market rules and regulations appropriate to allow customer installations (individual or aggregated) to provide ancillary services through export to the grid?

e. Are communication and cybersecurity arrangements sufficiently robust and trusted by consumers to participate in wide-area automated processes?

These are issues critical to the success of a VPP framework. VPPs, like all behind-the-meter investments, are owned and operated by consumers who have specific and often highly variable expectations around payment, surety, risk exposure, and empowerment. As the VPP platforms continue to expand and enrolment programmes for customers remain active, market operators and regulators must consider the consumer point of view.

Ongoing and effective consumer engagement, an acceptance of compromise, and the expectation of variability in the consumer response remain prominent features critical to the success of the VPP proposal (Reneweconomy, 2020).

5. Adapting old tools to a new environment

Consumer energy demand has for some time faced signals to be more flexible, or, at least, present the opportunity to curtail peak demand.

In Australia, states with a high penetration of electric storage water heating (an excellent passive "battery") such as Queensland and New South Wales embraced demand management as early as the 1970s. With energy tariffs initially pegged as low as 40% of the anytime flat usage tariff, off-peak hot water found its way into almost 70% of homes throughout the 30 years from the late 1960s. Dry cleaners, bakeries, and product manufacturing also adopted the attractive off-peak tariffs, changing working hours and processes to capitalize on the savings on offer through reduced "controlled load" tariffs.

Over time, the controlled load capability expanded to include other home appliances that can be flexible regarding the time of day it is operated, including pool filtration, clothes washing, and some air-conditioning. Consumers embraced these new forms of off-peak controlled load, accepting lower energy costs, mainly through a lower tariff for energy through a separately metered circuit, for little impact on amenity. Another attraction was the fact that controlled load was not "netted off" when calculating net feed energy for consumers with solar PV or peak demand limits in the case of the few consumers on demand-type tariffs. The widespread application of controlled load was also promulgated by trusted third-party suppliers of appliances—plumbers, pool builders, air conditioner retailers—which has the effect of reinforcing the trust of the service in consumer's minds.

The benefits of peak demand reduction through demand control have been immense across the generation, transmission, and distributor sectors through the deferral of network and generation investment to meet peak demand. In return for participating, consumers enjoyed a total bill reduction of typically 20% through off-peak tariffs.

Larger customers—commercial, industrial, and other nonresidential customers—continue to face pricing signals from both the wholesale market either directly by being exposed to the spot price or indirectly through retail back-to-back hedging arrangements. Network congestion signals exist via demand and capacity charges (being charged on a $/kVA basis) and curtailable load agreements.

In 2007, more elegant forms of demand response were envisaged. New standards for the demand response of appliances were published, with the intent that a basic level of control (off, on, reduce 50%, reduce 25%) could be implemented when coupled with an appropriate signaling equipment (Standards Australia, 2012). The "peak smart" initiative by Energex offered an attractive capital cost rebate in return for the consumer implementing the demand reduction capability. The arrangement allowed the utility to "turn down" the air conditioner in times of network stress (Energex, 2019).

5.1 Demand management (temporarily) fell out of favor

Enrolment of customer load to deal with peak-demand events fell somewhat out of favor from the late 1990s and early 2000s. Essentially, the commercial attractiveness faded, and fewer customers saw the value of offering up demand to be flexible. A number of behaviors led to the significant reduction in the take up of controlled load for about a decade, with pricing incentives and marketing taking a backseat to attractiveness of solar PV to reduce bills and the complexity of emerging market competition.

These factors are relevant today as we consider how to remove the obstacles to increasing the amount of flexible demand. The predominant factors that contributed the level of customer demand response remaining static, or in some cases falling in key market segments, were:

- The stratification of the energy market introduced barriers to "stack the value" of peak energy reduction across the market segments. The network operator, the primary beneficiary of demand response, was "hidden behind" the retailer who managed the relationship with the consumer. Marketing the benefits of "off peak energy" to consumers became fragmented and ineffective.
- The demand management priorities of retailers, networks, and generators did not necessarily align. Even across retailers, the varied risk profiles in their energy purchase contracts meant some retailers valued flexible demand more than others. The cost differential for consumers to switch retailers in order to find a retailer offering the benefits of flexible demand is seen as too great.
- The increasing trend to recover network charges through increased fixed charges reduced the ability for consumers to manage energy bills through changing demand patterns. Whilst stable bills are often seen as attractive, the downside is that it "disempowers" consumers' control of the bill.
- To keep transition costs low, the introduction of advanced metering infrastructure (AMI) included a preference for the use of single-element meters, with the intent that time-varying tariffs such as ToU would be in widespread use as the primary mechanism to encourage off-peak energy use (Faruqui, 2018).
- Time-varying prices were seen as next viable form of demand management incentives; however, their adoption was met with significant customer resistance. In any case the pricing signals were weak. AMI is to play a large part in the resurgence of demand management as demand response some years later.
- Customers chose rooftop solar PV as the predominant vehicle to deliver energy bill reductions, with clear and strong return on the investment.
- Instantaneous water heating and other technologies proved to be a lower capital cost to install than appliances more appropriate to demand response needs.

Through that period though, some important foundations were being laid that are now proving especially useful in supporting the resurgence in demand response. Advanced metering infrastructure and associated communications permitted the measurement of energy in 30-minute slices as well as creating a communications channel to every home with a smart meter. Also, the requirement that all inverters and many consumer appliances now include Demand Response Management (DRM) capability makes its implementation much more cost-effective.

6. Demand response—new needs, new opportunities, new benefits

With the significant increase in the proportion of renewable energy in Australia's networks, particularly nondispatchable "passive" generation, the demand management capability is staging a resurgence. Whilst peak demand reduction remains a major part of a demand response capability, the ability to signal price information in more granular 30-minute intervals emerges as a powerful and necessary precursor to implementing flexible demand.

Consumers are now more comfortable with smart meters, having largely accepted the initial concerns about data privacy and the introduction of health risks from the proximity to a radio frequency—emitting device. Building from the capability of Advanced Metering Infrastructure (AMI) and new control capability for some consumer appliances, demand response is taking on a new, more dynamic form, including a greater focus on shifting load to consume energy at times when it is most abundant.

In Chapter 17, Hochsettler highlights the way markets can adjust to draw out the required responses from consumers. The dynamic shifting of storage charging, including electric vehicles, and significant residential and commercial load (such as water heating, air-conditioning, and water pumping) and large industrial loads to the act as a "sink" for excess passive generation in the daytime have developed into a hierarchy of ways the consumer can choose to benefit from offering flexible demand, summarized in Fig. 22.6.

Passive discretionary consumer action is available to all energy consumers, requiring just the appropriate metering to determine the quantum of the benefit. With the adoption of interval meters as a standard offering, flexible demand is becoming available for all new customers, including those with solar PV.

Consumers are responding through shifting loads to times when network pricing reflects availability, such as daytime solar peaks. In Chapter 14, Alan Rai discusses the development of market-based responses and innovative tariffs. The "solar sponge" approach taken by South Australia Power Networks is not unlike the "Sunshine tariff" offered by Western Power Distribution to residential customers in the South West of England during 2016, and similar residential tariffs in parts of North America (Faruqui, 2018).

6.1 Hosting capacity—flexible demand unlocking behind the meter resources for all consumers

Before embedded generation was common, networks considered their capacity to deliver energy at appropriate quality; termed "demand capacity." As DER developed, hosting capacity emerged as a parameter to describe the ability for a network to accept feed-in energy, namely PV, without the risk of that energy creating adverse network conditions.

Now, the term hosting capacity is a proxy for the ability of a distribution network to accept, transport, and in some ways control the net energy flows in local networks. For consumers installing PV or accessing the higher order functions of flexible demand that involve feed-in energy, hosting capacity matters.

Limited hosting capacity has a detrimental effect on a range of options consumers have to be involved in, and benefit from, distributed energy resources. Export energy from new solar PV is curtailed, and system capacity is limited. VPPs are restricted in their coordinated actions, and the effectiveness of local trading markets is reduced.

Increasing hosting capacity can be achieved in two ways—either by the network operator investing in greater capacity or smarter control systems, or by consumers taking demand response actions to optimize power flows make the best use of the spare capacity of the existing system—the so-called "Uber of Energy." By doing so, the allowed level of PV generation increases and therefore consumes the value of greater levels of renewable energy in local networks and communities (Fig. 22.7).

The ability to manage local "excess" generation through nearby controllable demand to minimize the adverse impacts of embedded generation on local networks is now seen as contributing to a "large footprint, customer-owned, virtual power station" through a form of control of, to some extent, where and when that local generation is consumed. This creates an aggregated generation/demand resource that can be facilitated,

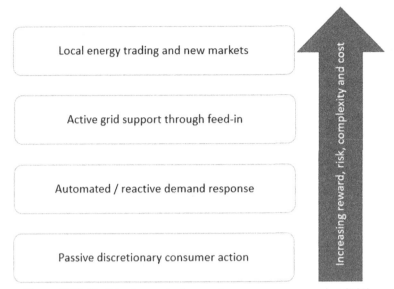

Figure 22.7 Flexible demand—forms of consumer engagement (TCA).

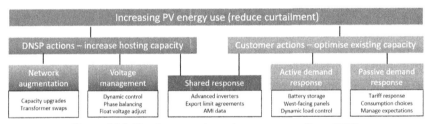

Figure 22.8 Hosting capacity enhancement options (TCA).

encouraged, and ultimately controlled and dispatched, not unlike other large renewable power sources supplying all consumers in the electricity network (Fig. 22.8).

7. Building the consumer case for benefit

Most of the change in dispatchability, supply risk, and the requisite investment in the tools and systems to manage this shift has taken place out of the gaze of the majority of energy consumers and the wider community, where the remarkable community support for renewable energy continues unabated. In the background, consumers are receiving the benefit of stable, even falling wholesale energy costs, despite average retail prices in Australia at around USD 25 cents per kilowatt-hour being relatively expensive in comparison with global averages (GlobalPetrolPrices.com, 2019).

If managed well, the advantages to all energy consumers are significant. But success, as defined by an efficient and effective adoption of demand response tools and attitudes by consumers, requires much more than a working market and good technical tools.

In reality, it is proving difficult to make good progress, but the path to flexible demand delivering significant benefits to consumers is fairly clear.

Firstly, tariff reform, critical to an effective demand response offering, whilst proven difficult to implement widely as customers tend to resist change that does not deliver the clear, unequivocal benefits seen in the solar PV phenomenon, is slowly making progress. Key to successful tariff reform is to have a clear strategy on how to address the needs of the proportion of consumers who will be worse off than currently the case.

Secondly, the many consumers predisposed to engage in energy management have to a large extent embraced rooftop solar PV as their "weapon of choice" to not only address high energy bills but also develop a sense of independence and empowerment in meeting their energy needs. Feed-in tariffs are moderate and seen a reasonable contributor to system cost payback, but as midday wholesale prices fall, consumers will be more focused on what happens to that "spare energy."

Some challenges will remain.

Technology development is at such a rate that investments can become superseded or redundant in a short time. No clear winner in the form of technological implementations or commercial arrangements exists, so the ability to create a clear direction in consumers' minds is hard. Some energy retailers offering "fixed bills" is a case in point.

Also, the relative prices of battery storage remain high, and payback of the investment in enabling flexible demand, and particularly active demand response such as Virtual Power Stations (VPPs), remains weak, especially without the significant levels of subsidy or incentives that underpin most customer investments currently.

Finally, technical and regulatory reform, including the development of so-called 2-sided markets and common interoperability standards to encourage demand response, especially for smaller consumers, remains complex and slow, especially for small consumers. Varied definitions of success across the components of the energy supply chain delay clarity and agreement on the way forward.

Clearly, engaging medium and small consumers in demand response is not a trivial exercise. That being said, the outlook for greater consumer involvement in flexible demand is positive. The reach of AMI infrastructure is expanding, which alongside time varying tariffs as the default option for all new customer connections is providing a level of infrastructure to support a regime of flexible demand. The technical standards for inverters and some appliances, in particular air conditioners, include basic demand response capability.

Most importantly, demand response is not a new concept to many consumers. Variable or demand pricing is becoming more prevalent through a range of purchases, such as public transport fares and parking. Direct load control in the form of off-peak tariffs targeted at reducing demand peaks are familiar components of energy supply to many consumers.

Key is to create an environment that recognizes that change is needed, and that the change presents net benefit to consumers and the community. There will need a level of commitment, as often any negative position in relation to the development of renewable energy draws criticism. Over time, consumers will see benefits in the form of connection agreements that allow larger embedded generators when coupled with flexible demand and the ability to reduce exposure to peak energy prices.

8. Conclusion

Whilst consumers have been able to opt in to demand response services for many years, the future is looking quite different. With the resource mix now including millions of distributed energy systems that will react to a wide range of energy use signals, new business models for consumers are emerging. Flexibility and adaptability will be features of regulatory frameworks and consumer energy use alike. Uncertainty will prevail.

In Australia, consideration of how the significant share of variable generation can technically be accommodated at a grid level is well progressed at the utility and large consumer level. Progressive jurisdictions are pursuing a nimble regulatory framework to unlock consumer's access to the more advanced returns to consumers through participation in active markets.

Without intelligent use of flexible demand, there is a high risk that less efficient solutions such as overinvesting in network capacity (with the commensurate stranded asset risk) exists, adding to the asset base that needs to be recovered through customer bills for the next 50 years. Alternatively, a poor response denies the benefits from further development of renewable energy.

There is a wide spectrum of consumers who have the ability to exercise flexible demand. Those consumers with rooftop generation or battery storage can seek new income and enhance their return on investment by participating in active demand response markets or choose to simply reap the benefits of reduced energy costs through self-consumption of their generated energy. Energy arbitrage is emerging, although the economics are not at all attractive in the vast majority of cases.

Other more passive roles for consumers include lower prices through demand management, whether it be in a passive role through adjusting appliance use under time of use pricing or simply a commitment to the value that greater use of renewable energy generally can bring. Happily, we are seeing a resurgence in more active demand management, with simple automation and control of water heating, pool pumps, and air conditioners offering lower energy retail rates.

The wide diversity in customers' understanding, expectations and appetite to become involved in demand response challenges the effective implementation of flexible demand capability to deliver the aggregate "nimbleness" of consumer energy demand required to meet the balance of variable supply in both short and long timeframes. Whilst "value stacking" remains a powerful concept, in practice the engagement of consumers is often complicated by the noncoincident nature of individual requirements along the "supply chain," where the needs of energy retailers, networks, market operators, and demand aggregators do not always align.

As the market operators and distributors establish new tools and capabilities, it is clearly time to advance the role of consumers in meeting the new electricity supply challenges. Not only is consumer engagement and action highly cost-effective in the range of responses to new energy requirements, the financial benefits to consumers have the potential to be significant. "Smart" customer generation, advanced demand response, and innovative tariffs are emerging as tools that will effectively assist the change in the generation mix and deliver benefits to consumers. In many cases, we can learn from experience through the widespread application of demand reduction capability such as off-peak load control that has been in place for many years and is well accepted by consumers.

Flexible demand is, put simply, a consumer-sourced resource to increase or decrease energy use at any point in time, and is a critical component in ensuring the reliability and security of the energy supply is efficiently maintained as the share of more variable renewable energy increases. Ultimately, flexible demand will be judged by consumers by its contribution to the energy trilemma: maintaining a reliable and secure power system, affordable energy prices for all consumers, and a sustainable reduction in carbon emissions in the energy sector.

References

ABC, December 1, 2019. The Rise of Solar Is Jeopardising the WA Energy Grid. Retrieved from Australian Broadcasting Corporation News: https://www.abc.net.au/news/2019-12-01/rise-of-rooftop-solar-power-jeopardising-wa-energy-grid/11731452.

AEMC, 2018. 2018 Residential Electricity Price Trends, Final Report. Australian Energy Market Commission, Sydney. Retrieved from: aemc.gov.au.

AEMC, 2020. Whilesale Demand Response Mechanism, Rule Determination. Australian Energy Market Commission. Retrieved from: https://www.aemc.gov.au/rule-changes/wholesale-demand-response-mechanism.

AEMO, 2019. Draft 2020 Integrated System Plan. Australian Energy Market Operator. Retrieved from: https://www.aemo.com.au.

AEMO, 2020a. Renewable Integration Study: Stage 1 Report. Australian Energy Market Operator. Retrieved from: www.aemo.com.au.

AEMO, 2020b. Virtual Power Plant Demonstration. Australian Energy Market Operator Limited. Retrieved from: aemo.com.au.

AEMO, March 2020c. AEMO Virtual Power Plant Demonstration. Australian Energy Market Operator Limited, Melbourne.

AER, 2020. State of the Energy Market. Australian Energy Regulator, ISBN 978-1-920702-53-3.

Alan Rai, R.E., 2019. The times they are a changin': current and future trends in electricity demand and supply. The Electricity Journal. https://doi.org/10.1016/j.tej.2019.05.017. Elsevier.

ARENA, 2019. United Energy Demand Response Project. Australian Renewable Energy Agency & United Energy Ltd., Melbourne, Australia. Retrieved from: https://arena.gov.au/projects/united-energy-distribution-demand-response/.

Energex, 2019. Demand Management Plan. Energex Limited. Retrieved from: https://www.energex.com.au/__data/assets/pdf_file/0006/765069/2019-20-Demand-Management-Plan.pdf.

Faruqui, A., November 9, 2018. Modernising Distribution Tariffs for Households: a Presentation to Energy Consumers Australia.

GlobalPetrolPrices.com, 2019. Retail Energy Price Data. The Global Economy Project. Retrieved from: https://www.globalpetrolprices.com/electricity_prices/.

IRENA, 2020a. Renewable Capacity Statistics 2020. International Renewable Energy Agency (IRENA), Abu Dhabi.

IRENA, 2020b. Renewable Power Generation Costs in 2019. International Renewable Energy Agency, Abu Dhabi. Retrieved from: https://www.irena.org/-/media/Files/IRENA/Agency/Publication/2020/Jun/IRENA_Power_Generation_Costs_2019.pdf.

Johnson, S., 2019. Solar Saturation: Sooner than We Thought. Energy Networks Australia. Retrieved from: https://www.energynetworks.com.au/news/energy-insider/solar-saturation-sooner-than-we-thought/.

Johnston, J., 2020. A platform for trading flexibility on the distribution network — a UK case study. In: Johnston, J., Sioshansi, F. (Eds.), Behind and Beyond the Meter. Academic Press.

Lawson, T.W., 2011. Consumer Response to Time Varying Prices for Electricity. University of Otaga School of Business, Dunedin, NZ.

Mazengarb, M., March 2020. AEMC set to approve demand response energy trading, but households left out. Renew. Econ.

Productivity Commission, 2013. The Costs and Benefits of Demand Management for Households. Australian Productivity Commission, Melbourne. Retrieved from: https://www.pc.gov.au.

Reneweconomy, 2020. Solar exports: Should households pay for the right to export more solar? https://reneweconomy.com.au/solar-exports-should-households-pay-for-right-to-export-more-solar-to-the-grid-86776/. (Accessed 16 July 2020).

Sioshansi, F., July 2020. First PVs, then EVs, now VPPs. EEnergy Informer.

SSE, 2020. Off Peak Tariffs. Retrieved from OVo energy tariffs: https://sse.co.uk/help/electric-heating/off-peak.

Standards Australia, 2012. Australian/New Zealand Standard AS4755.3.1:2012 – Demand Response Capabilities and Supporting Technologies. Standards Australia.

UKERC, 2013. Transforming the UK Energy System: Public Values, Attitudes and Acceptability. UK Energy Research Centre.

Epilogue

During the next decade or two, most of the power flowing through the power lines will come from zero carbon sources in three-quarters of the country, with California, Hawaii, and New York leading the way, with Colorado, Illinois, Michigan, and Minnesota right behind them. Of course, zero carbon resources will create new challenges for the power system. Power supply will be highly variable, switching rapidly and unpredictably among solar, hydro, natural gas, and nuclear resources.

Electricity prices will fluctuate in wholesale markets from second to second, minute to minute. To preserve reliability at a reasonable cost, it would be important to introduce load flexibility into the system. By 2040, it is likely that most homes will be meeting some of their energy needs with solar panels, and a substantial share of the homes will pair the solar panels with battery storage. It's quite possible that half of the cars on the road will be electric vehicles. Additionally, just about every major appliance in the home will be digital and Wi-Fi enabled.

Instantaneous load flexibility will have emerged as the perfect complement to renewable resource variability. Wholesale prices will flow seamlessly to the appliances. Some homeowners, who are interested in lowering their power bills, will have programmed their preferences into their smartphones. As prices rise, the least important appliance will either be turned off or toned down. As prices continued to rise, the next least important appliance will be turned off or toned down, and so on, in a predetermined loading order of end-use loads—under the emerging paradigm of a demand-side merit order. The opposite will happen when prices fall.

Contrary to the popular fears that are expressed today, homeowners will not be spending their time on their smartphones checking hourly prices. All of them will have programmed their energy lifestyle into Alexa 5.0 devices or its competing versions. Some homeowners will hedge part of their energy purchases by buying power on the forward market at a known price and buying the remainder on the spot market. Some will hedge it entirely by buying their entire usage on the forward market, essentially having a Netflix-like bill.

It is worth noting that MIT's Fred Schweppe[1] had anticipated this phenomenon back in 1978 when he coined the phrase homeostatic control. At the time, no one understood what it meant. And among those who understood the term, few believed it would ever be implemented.

Around the same time, EPRI's Clark Gellings[2] had articulated a vision of getting prices to devices through technological innovation and propounded the concept of flexible load shapes that no one seemed to understand either. Schweppe and Gellings were simply a few decades ahead of their time.

Load flexibility can properly be said to be their brainchild—a timely topic that is explored in the essays of this volume. As Luca Lo Schiavo states in the Preface, "This book is a plunge into the future." I agree with him. Those who read the book will find new ideas for coping with the challenges that lie ahead.

Ahmad Faruqui

Principal, The Brattle Group

San Francisco, CA, United States

---

[1] Spot Pricing of Electricity, 1988 at https://www.amazon.com/Pricing-Electricity-Power-Electronics-Systems/dp/0898382602.

[2] For example see, Then and now: The perspective of the man who coined DSM, in **DSM in transition: from mandates to markets, Energy Policy, F. Sioshansi (Ed.) Apr 1996** at https://doi.org/10.1016/0301-4215(95)00134-4.

Index

'Note: Page numbers followed by "f" indicate figures, "t" indicate tables and "b" indicate boxes.'

Printed in the United States
By Bookmasters